Innovation and Future Trends in Food Manufacturing and Supply Chain Technologies

Related Titles

Open Innovation in the Food and Beverage Industry
(ISBN 978-0-85709-595-4)

Robotics and Automation in the Food Industry
(ISBN 978-1-84569-801-0)

Microbial Decontamination in the Food Industry
(ISBN 978-0-85709-085-0)

Woodhead Publishing Series in Food Science, Technology and Nutrition: Number 293

Innovation and Future Trends in Food Manufacturing and Supply Chain Technologies

Edited by

C.E. Leadley

ELSEVIER

AMSTERDAM • BOSTON • CAMBRIDGE • HEIDELBERG
LONDON • NEW YORK • OXFORD • PARIS • SAN DIEGO
SAN FRANCISCO • SINGAPORE • SYDNEY • TOKYO
Woodhead Publishing is an imprint of Elsevier

WP
WOODHEAD PUBLISHING

Woodhead Publishing is an imprint of Elsevier
80 High Street, Sawston, Cambridge, CB22 3HJ, UK
225 Wyman Street, Waltham, MA 02451, USA
Langford Lane, Kidlington, OX5 1GB, UK

ISBN: 978-1-78242-447-5

British Library Cataloguing in Publication Data
A catalogue record for this book is available from the British Library

Library of Congress Cataloging-in-Publication Data
A catalog record for this book is available from the Library of Congress

Publisher: Nikki Levy
Acquisition Editor: Rob Sykes
Editorial Project Manager: Josh Bennett
Production Project Manager: Nicky Carter
Designer: Greg Harris

Typeset by TNQ Books and Journals
www.tnq.co.in

Contents

List of Contributors	**ix**
Woodhead Publishing Series in Food Science, Technology and Nutrition	**xi**
Introduction	**xxiii**

Part One Supply Chain Developments 1

1 Food Fraud and Authenticity: Emerging Issues and Future Trends **3**
Annemieke M. Pustjens, Yannick Weesepoel and Saskia M. van Ruth
1.1 Introduction 3
1.2 Background 3
1.3 Emerging Issues and Future Trends 10
1.4 Conclusion 16
References 17

2 The Role of Information Technology Developments in Food Supply Chain Integration and Monitoring **21**
Nitaigour Mahalik and Kiseon Kim
2.1 Introduction 21
2.2 IT Components in General 22
2.3 Supply Chain Management 23
2.4 Traceability 26
2.5 Food Processing and Packaging 28
2.6 Information Technology in Food Processing and Packaging 30
2.7 Nanotechnology in Food Processing and Packaging 32
2.8 Example 34
2.9 Conclusions 35
References 36
Further Reading 37

Part Two Factory Design and Operation 39

3 Emerging Trends and Methods in Food Factory Design **41**
Frank Moerman and Patrick C. Wouters
3.1 Introduction 41
3.2 Renovation of an Existing Food Factory 41

3.3 Building a New Food Factory 50
3.4 Tools to Facilitate the Renovation or Design of a New Food Factory 71
3.5 Role of Experts, Contractors, or a Department Specialized in the Hygienic Engineering and Design of Food Factories 74
3.6 Conclusions 76
References 77
Further Reading 79

4 Hygiene Concepts for Food Factory Design 81
Frank Moerman and Patrick C. Wouters
4.1 Introduction 81
4.2 Food Plant Building Concepts to Prevent Entry and Harborage of Pests 81
4.3 Interior Building Layout: Trends 100
4.4 Air Handling Systems 108
4.5 Interior Construction Elements Reducing the Likeliness of Contamination 110
4.6 Integration of Process Piping and Utilities within Zone M Process Areas 118
4.7 Relocation of Process Piping and Utilities Outside Zone H Process Areas 123
4.8 Conclusion 132
References 132

5 Factory-Friendly Approaches to Applying Multivariate Analytics for Productivity and Yield Gains 135
Sean Robinson
5.1 Introduction 135
5.2 Areas of Opportunity for the Typical Producer 137
5.3 Using Existing Data Sources or Establishing the Data Sets 139
5.4 Deriving Reliable Process Models from Historical Performance 143
5.5 Embedding Predictive Capability within Common Plant Systems 146
5.6 Developing Trends 147
5.7 Sources of Further Information and Advice 148
Reference 149

6 Emerging Methods and Principles in Food Contact Surface Decontamination/Prevention 151
Torstein Skåra and Jan T. Rosnes
6.1 Introduction 151
6.2 Methods and Principles 152
6.3 Future Trends 163
6.4 Sources of Further Information and Advice 165
References 166

Part Three Innovative Food Processing Technologies **173**

7 Emerging Refrigeration and Freezing Technologies for Food Preservation **175**
Judith Evans
7.1 Introduction 175
7.2 Refrigeration Based Technologies 175
7.3 Technologies that are Product Oriented Technologies 186
7.4 Future Trends 195
Sources of Further Information and Advice 196
References 196

8 Advanced Heating Technologies for Food Processing **203**
Norman Maloney and Michael Harrison
8.1 Introduction 203
8.2 Ohmic Heating 205
8.3 Radio Frequency and Microwave Heating 222
8.4 Safety 247
8.5 Comparison of Techniques and an Overview of Opportunities 250
References 253

9 Emerging Nonthermal Food Preservation Technologies **257**
Kumari Shikha Ojha, Brijesh K. Tiwari, Colm O'Donnell and Joseph P. Kerry
9.1 Introduction 257
9.2 Emerging Nonthermal Technologies 258
9.3 Conclusions and Future Trends 269
References 269

Index **275**

List of Contributors

Judith Evans London South Bank University, London, UK

Michael Harrison C-Tech Innovation Ltd, Capenhurst Technology Park, Chester, UK

Joseph P. Kerry Food Packaging Group, School of Food and Nutritional Sciences, University College Cork, Cork, Ireland

Kiseon Kim Communication and Sensor Network Lab, School of Information and Communication, Gwangju Institute of Science and technology, Gwangju, Republic of Korea

Nitaigour Mahalik Department of Industrial Technology, Jordan College of Agricultural Sciences and Technology, California State University, Fresno, CA, USA

Norman Maloney C-Tech Innovation Ltd, Capenhurst Technology Park, Chester, UK

Frank Moerman Catholic University of Leuven, Leuven, Belgium

Colm O'Donnell School Biosystems and Food Engineering, University College Dublin, Dublin, Ireland

Kumari Shikha Ojha Food Biosciences, Teagasc Food Research Centre, Dublin, Ireland; School Biosystems and Food Engineering, University College Dublin, Dublin, Ireland

Annemieke M. Pustjens RIKILT, Institute of Food Safety, Wageningen University and Research Centre, Wageningen, The Netherlands

Sean Robinson Kerrco Automation Ltd, Glasgow, UK

Jan T. Rosnes Nofima, Stavanger, Norway

Saskia M. van Ruth RIKILT, Institute of Food Safety, Wageningen University and Research Centre, Wageningen, The Netherlands; Food Quality and Design Group, Wageningen University, Wageningen, The Netherlands

Torstein Skåra Nofima, Stavanger, Norway

Brijesh K. Tiwari Food Biosciences, Teagasc Food Research Centre, Dublin, Ireland

Yannick Weesepoel RIKILT, Institute of Food Safety, Wageningen University and Research Centre, Wageningen, The Netherlands

Patrick C. Wouters Cargill, Food Safety, Quality and Regulatory, Schipol, The Netherlands

Woodhead Publishing Series in Food Science, Technology and Nutrition

1 **Chilled foods: A comprehensive guide**
Edited by C. Dennis and M. Stringer
2 **Yoghurt: Science and technology**
A. Y. Tamime and R. K. Robinson
3 **Food processing technology: Principles and practice**
P. J. Fellows
4 **Bender's dictionary of nutrition and food technology Sixth edition**
D. A. Bender
5 **Determination of veterinary residues in food**
Edited by N. T. Crosby
6 **Food contaminants: Sources and surveillance**
Edited by C. Creaser and R. Purchase
7 **Nitrates and nitrites in food and water**
Edited by M. J. Hill
8 **Pesticide chemistry and bioscience: The food-environment challenge**
Edited by G. T. Brooks and T. Roberts
9 **Pesticides: Developments, impacts and controls**
Edited by G. A. Best and A. D. Ruthven
10 **Dietary fibre: Chemical and biological aspects**
Edited by D. A. T. Southgate, K. W. Waldron, I. T. Johnson and G. R. Fenwick
11 **Vitamins and minerals in health and nutrition**
M. Tolonen
12 **Technology of biscuits, crackers and cookies Second edition**
D. Manley
13 **Instrumentation and sensors for the food industry**
Edited by E. Kress-Rogers
14 **Food and cancer prevention: Chemical and biological aspects**
Edited by K. W. Waldron, I. T. Johnson and G. R. Fenwick
15 **Food colloids: Proteins, lipids and polysaccharides**
Edited by E. Dickinson and B. Bergenstahl
16 **Food emulsions and foams**
Edited by E. Dickinson
17 **Maillard reactions in chemistry, food and health**
Edited by T. P. Labuza, V. Monnier, J. Baynes and J. O'Brien
18 **The Maillard reaction in foods and medicine**
Edited by J. O'Brien, H. E. Nursten, M. J. Crabbe and J. M. Ames
19 **Encapsulation and controlled release**
Edited by D. R. Karsa and R. A. Stephenson
20 **Flavours and fragrances**
Edited by A. D. Swift
21 **Feta and related cheeses**
Edited by A. Y. Tamime and R. K. Robinson

22 **Biochemistry of milk products**
Edited by A. T. Andrews and J. R. Varley
23 **Physical properties of foods and food processing systems**
M. J. Lewis
24 **Food irradiation: A reference guide**
V. M. Wilkinson and G. Gould
25 **Kent's technology of cereals: An introduction for students of food science and agriculture Fourth edition**
N. L. Kent and A. D. Evers
26 **Biosensors for food analysis**
Edited by A. O. Scott
27 **Separation processes in the food and biotechnology industries: Principles and applications**
Edited by A. S. Grandison and M. J. Lewis
28 **Handbook of indices of food quality and authenticity**
R. S. Singhal, P. K. Kulkarni and D. V. Rege
29 **Principles and practices for the safe processing of foods**
D. A. Shapton and N. F. Shapton
30 **Biscuit, cookie and cracker manufacturing manuals Volume 1: Ingredients**
D. Manley
31 **Biscuit, cookie and cracker manufacturing manuals Volume 2: Biscuit doughs**
D. Manley
32 **Biscuit, cookie and cracker manufacturing manuals Volume 3: Biscuit dough piece forming**
D. Manley
33 **Biscuit, cookie and cracker manufacturing manuals Volume 4: Baking and cooling of biscuits**
D. Manley
34 **Biscuit, cookie and cracker manufacturing manuals Volume 5: Secondary processing in biscuit manufacturing**
D. Manley
35 **Biscuit, cookie and cracker manufacturing manuals Volume 6: Biscuit packaging and storage**
D. Manley
36 **Practical dehydration Second edition**
M. Greensmith
37 **Lawrie's meat science Sixth edition**
R. A. Lawrie
38 **Yoghurt: Science and technology Second edition**
A. Y. Tamime and R. K. Robinson
39 **New ingredients in food processing: Biochemistry and agriculture**
G. Linden and D. Lorient
40 **Benders' dictionary of nutrition and food technology Seventh edition**
D. A. Bender and A. E. Bender
41 **Technology of biscuits, crackers and cookies Third edition**
D. Manley
42 **Food processing technology: Principles and practice Second edition**
P. J. Fellows
43 **Managing frozen foods**
Edited by C. J. Kennedy
44 **Handbook of hydrocolloids**
Edited by G. O. Phillips and P. A. Williams
45 **Food labelling**
Edited by J. R. Blanchfield
46 **Cereal biotechnology**
Edited by P. C. Morris and J. H. Bryce
47 **Food intolerance and the food industry**
Edited by T. Dean

48 **The stability and shelf-life of food**
Edited by D. Kilcast and P. Subramaniam
49 **Functional foods: Concept to product**
Edited by G. R. Gibson and C. M. Williams
50 **Chilled foods: A comprehensive guide Second edition**
Edited by M. Stringer and C. Dennis
51 **HACCP in the meat industry**
Edited by M. Brown
52 **Biscuit, cracker and cookie recipes for the food industry**
D. Manley
53 **Cereals processing technology**
Edited by G. Owens
54 **Baking problems solved**
S. P. Cauvain and L. S. Young
55 **Thermal technologies in food processing**
Edited by P. Richardson
56 **Frying: Improving quality**
Edited by J. B. Rossell
57 **Food chemical safety Volume 1: Contaminants**
Edited by D. Watson
58 **Making the most of HACCP: Learning from others' experience**
Edited by T. Mayes and S. Mortimore
59 **Food process modelling**
Edited by L. M. M. Tijskens, M. L. A. T. M. Hertog and B. M. Nicolaï
60 **EU food law: A practical guide**
Edited by K. Goodburn
61 **Extrusion cooking: Technologies and applications**
Edited by R. Guy
62 **Auditing in the food industry: From safety and quality to environmental and other audits**
Edited by M. Dillon and C. Griffith
63 **Handbook of herbs and spices Volume 1**
Edited by K. V. Peter
64 **Food product development: Maximising success**
M. Earle, R. Earle and A. Anderson
65 **Instrumentation and sensors for the food industry Second edition**
Edited by E. Kress-Rogers and C. J. B. Brimelow
66 **Food chemical safety Volume 2: Additives**
Edited by D. Watson
67 **Fruit and vegetable biotechnology**
Edited by V. Valpuesta
68 **Foodborne pathogens: Hazards, risk analysis and control**
Edited by C. de W. Blackburn and P. J. McClure
69 **Meat refrigeration**
S. J. James and C. James
70 **Lockhart and Wiseman's crop husbandry Eighth edition**
H. J. S. Finch, A. M. Samuel and G. P. F. Lane
71 **Safety and quality issues in fish processing**
Edited by H. A. Bremner
72 **Minimal processing technologies in the food industries**
Edited by T. Ohlsson and N. Bengtsson
73 **Fruit and vegetable processing: Improving quality**
Edited by W. Jongen
74 **The nutrition handbook for food processors**
Edited by C. J. K. Henry and C. Chapman

75 **Colour in food: Improving quality**
Edited by D. MacDougall
76 **Meat processing: Improving quality**
Edited by J. P. Kerry, J. F. Kerry and D. A. Ledward
77 **Microbiological risk assessment in food processing**
Edited by M. Brown and M. Stringer
78 **Performance functional foods**
Edited by D. Watson
79 **Functional dairy products Volume 1**
Edited by T. Mattila-Sandholm and M. Saarela
80 **Taints and off-flavours in foods**
Edited by B. Baigrie
81 **Yeasts in food**
Edited by T. Boekhout and V. Robert
82 **Phytochemical functional foods**
Edited by I. T. Johnson and G. Williamson
83 **Novel food packaging techniques**
Edited by R. Ahvenainen
84 **Detecting pathogens in food**
Edited by T. A. McMeekin
85 **Natural antimicrobials for the minimal processing of foods**
Edited by S. Roller
86 **Texture in food Volume 1: Semi-solid foods**
Edited by B. M. McKenna
87 **Dairy processing: Improving quality**
Edited by G. Smit
88 **Hygiene in food processing: Principles and practice**
Edited by H. L. M. Lelieveld, M. A. Mostert, B. White and J. Holah
89 **Rapid and on-line instrumentation for food quality assurance**
Edited by I. Tothill
90 **Sausage manufacture: Principles and practice**
E. Essien
91 **Environmentally-friendly food processing**
Edited by B. Mattsson and U. Sonesson
92 **Bread making: Improving quality**
Edited by S. P. Cauvain
93 **Food preservation techniques**
Edited by P. Zeuthen and L. Bøgh-Sørensen
94 **Food authenticity and traceability**
Edited by M. Lees
95 **Analytical methods for food additives**
R. Wood, L. Foster, A. Damant and P. Key
96 **Handbook of herbs and spices Volume 2**
Edited by K. V. Peter
97 **Texture in food Volume 2: Solid foods**
Edited by D. Kilcast
98 **Proteins in food processing**
Edited by R. Yada
99 **Detecting foreign bodies in food**
Edited by M. Edwards
100 **Understanding and measuring the shelf-life of food**
Edited by R. Steele
101 **Poultry meat processing and quality**
Edited by G. Mead

102 **Functional foods, ageing and degenerative disease**
Edited by C. Remacle and B. Reusens
103 **Mycotoxins in food: Detection and control**
Edited by N. Magan and M. Olsen
104 **Improving the thermal processing of foods**
Edited by P. Richardson
105 **Pesticide, veterinary and other residues in food**
Edited by D. Watson
106 **Starch in food: Structure, functions and applications**
Edited by A.-C. Eliasson
107 **Functional foods, cardiovascular disease and diabetes**
Edited by A. Arnoldi
108 **Brewing: Science and practice**
D. E. Briggs, P. A. Brookes, R. Stevens and C. A. Boulton
109 **Using cereal science and technology for the benefit of consumers: Proceedings of the 12PthP International ICC Cereal and Bread Congress, 24–26PthP May, 2004, Harrogate, UK**
Edited by S. P. Cauvain, L. S. Young and S. Salmon
110 **Improving the safety of fresh meat**
Edited by J. Sofos
111 **Understanding pathogen behaviour: Virulence, stress response and resistance**
Edited by M. Griffiths
112 **The microwave processing of foods**
Edited by H. Schubert and M. Regier
113 **Food safety control in the poultry industry**
Edited by G. Mead
114 **Improving the safety of fresh fruit and vegetables**
Edited by W. Jongen
115 **Food, diet and obesity**
Edited by D. Mela
116 **Handbook of hygiene control in the food industry**
Edited by H. L. M. Lelieveld, M. A. Mostert and J. Holah
117 **Detecting allergens in food**
Edited by S. Koppelman and S. Hefle
118 **Improving the fat content of foods**
Edited by C. Williams and J. Buttriss
119 **Improving traceability in food processing and distribution**
Edited by I. Smith and A. Furness
120 **Flavour in food**
Edited by A. Voilley and P. Etievant
121 **The Chorleywood bread process**
S. P. Cauvain and L. S. Young
122 **Food spoilage microorganisms**
Edited by C. de W. Blackburn
123 **Emerging foodborne pathogens**
Edited by Y. Motarjemi and M. Adams
124 **Benders' dictionary of nutrition and food technology Eighth edition**
D. A. Bender
125 **Optimising sweet taste in foods**
Edited by W. J. Spillane
126 **Brewing: New technologies**
Edited by C. Bamforth
127 **Handbook of herbs and spices Volume 3**
Edited by K. V. Peter

128 **Lawrie's meat science Seventh edition**
R. A. Lawrie in collaboration with D. A. Ledward
129 **Modifying lipids for use in food**
Edited by F. Gunstone
130 **Meat products handbook: Practical science and technology**
G. Feiner
131 **Food consumption and disease risk: Consumer–pathogen interactions**
Edited by M. Potter
132 **Acrylamide and other hazardous compounds in heat-treated foods**
Edited by K. Skog and J. Alexander
133 **Managing allergens in food**
Edited by C. Mills, H. Wichers and K. Hoffman-Sommergruber
134 **Microbiological analysis of red meat, poultry and eggs**
Edited by G. Mead
135 **Maximising the value of marine by-products**
Edited by F. Shahidi
136 **Chemical migration and food contact materials**
Edited by K. Barnes, R. Sinclair and D. Watson
137 **Understanding consumers of food products**
Edited by L. Frewer and H. van Trijp
138 **Reducing salt in foods: Practical strategies**
Edited by D. Kilcast and F. Angus
139 **Modelling microorganisms in food**
Edited by S. Brul, S. Van Gerwen and M. Zwietering
140 **Tamime and Robinson's Yoghurt: Science and technology Third edition**
A. Y. Tamime and R. K. Robinson
141 **Handbook of waste management and co-product recovery in food processing Volume 1**
Edited by K. W. Waldron
142 **Improving the flavour of cheese**
Edited by B. Weimer
143 **Novel food ingredients for weight control**
Edited by C. J. K. Henry
144 **Consumer-led food product development**
Edited by H. MacFie
145 **Functional dairy products Volume 2**
Edited by M. Saarela
146 **Modifying flavour in food**
Edited by A. J. Taylor and J. Hort
147 **Cheese problems solved**
Edited by P. L. H. McSweeney
148 **Handbook of organic food safety and quality**
Edited by J. Cooper, C. Leifert and U. Niggli
149 **Understanding and controlling the microstructure of complex foods**
Edited by D. J. McClements
150 **Novel enzyme technology for food applications**
Edited by R. Rastall
151 **Food preservation by pulsed electric fields: From research to application**
Edited by H. L. M. Lelieveld and S. W. H. de Haan
152 **Technology of functional cereal products**
Edited by B. R. Hamaker
153 **Case studies in food product development**
Edited by M. Earle and R. Earle
154 **Delivery and controlled release of bioactives in foods and nutraceuticals**
Edited by N. Garti

155 **Fruit and vegetable flavour: Recent advances and future prospects**
Edited by B. Brückner and S. G. Wyllie
156 **Food fortification and supplementation: Technological, safety and regulatory aspects**
Edited by P. Berry Ottaway
157 **Improving the health-promoting properties of fruit and vegetable products**
Edited by F. A. Tomás-Barberán and M. I. Gil
158 **Improving seafood products for the consumer**
Edited by T. Børresen
159 **In-pack processed foods: Improving quality**
Edited by P. Richardson
160 **Handbook of water and energy management in food processing**
Edited by J. Klemeš, R. Smith and J.-K. Kim
161 **Environmentally compatible food packaging**
Edited by E. Chiellini
162 **Improving farmed fish quality and safety**
Edited by Ø. Lie
163 **Carbohydrate-active enzymes**
Edited by K.-H. Park
164 **Chilled foods: A comprehensive guide Third edition**
Edited by M. Brown
165 **Food for the ageing population**
Edited by M. M. Raats, C. P. G. M. de Groot and W. A. Van Staveren
166 **Improving the sensory and nutritional quality of fresh meat**
Edited by J. P. Kerry and D. A. Ledward
167 **Shellfish safety and quality**
Edited by S. E. Shumway and G. E. Rodrick
168 **Functional and speciality beverage technology**
Edited by P. Paquin
169 **Functional foods: Principles and technology**
M. Guo
170 **Endocrine-disrupting chemicals in food**
Edited by I. Shaw
171 **Meals in science and practice: Interdisciplinary research and business applications**
Edited by H. L. Meiselman
172 **Food constituents and oral health: Current status and future prospects**
Edited by M. Wilson
173 **Handbook of hydrocolloids Second edition**
Edited by G. O. Phillips and P. A. Williams
174 **Food processing technology: Principles and practice Third edition**
P. J. Fellows
175 **Science and technology of enrobed and filled chocolate, confectionery and bakery products**
Edited by G. Talbot
176 **Foodborne pathogens: Hazards, risk analysis and control Second edition**
Edited by C. de W. Blackburn and P. J. McClure
177 **Designing functional foods: Measuring and controlling food structure breakdown and absorption**
Edited by D. J. McClements and E. A. Decker
178 **New technologies in aquaculture: Improving production efficiency, quality and environmental management**
Edited by G. Burnell and G. Allan
179 **More baking problems solved**
S. P. Cauvain and L. S. Young
180 **Soft drink and fruit juice problems solved**
P. Ashurst and R. Hargitt

181 **Biofilms in the food and beverage industries**
Edited by P. M. Fratamico, B. A. Annous and N. W. Gunther
182 **Dairy-derived ingredients: Food and neutraceutical uses**
Edited by M. Corredig
183 **Handbook of waste management and co-product recovery in food processing Volume 2**
Edited by K. W. Waldron
184 **Innovations in food labelling**
Edited by J. Albert
185 **Delivering performance in food supply chains**
Edited by C. Mena and G. Stevens
186 **Chemical deterioration and physical instability of food and beverages**
Edited by L. H. Skibsted, J. Risbo and M. L. Andersen
187 **Managing wine quality Volume 1: Viticulture and wine quality**
Edited by A. G. Reynolds
188 **Improving the safety and quality of milk Volume 1: Milk production and processing**
Edited by M. Griffiths
189 **Improving the safety and quality of milk Volume 2: Improving quality in milk products**
Edited by M. Griffiths
190 **Cereal grains: Assessing and managing quality**
Edited by C. Wrigley and I. Batey
191 **Sensory analysis for food and beverage quality control: A practical guide**
Edited by D. Kilcast
192 **Managing wine quality Volume 2: Oenology and wine quality**
Edited by A. G. Reynolds
193 **Winemaking problems solved**
Edited by C. E. Butzke
194 **Environmental assessment and management in the food industry**
Edited by U. Sonesson, J. Berlin and F. Ziegler
195 **Consumer-driven innovation in food and personal care products**
Edited by S. R. Jaeger and H. MacFie
196 **Tracing pathogens in the food chain**
Edited by S. Brul, P. M. Fratamico and T. A. McMeekin
197 **Case studies in novel food processing technologies: Innovations in processing, packaging, and predictive modelling**
Edited by C. J. Doona, K. Kustin and F. E. Feeherry
198 **Freeze-drying of pharmaceutical and food products**
T.-C. Hua, B.-L. Liu and H. Zhang
199 **Oxidation in foods and beverages and antioxidant applications Volume 1: Understanding mechanisms of oxidation and antioxidant activity**
Edited by E. A. Decker, R. J. Elias and D. J. McClements
200 **Oxidation in foods and beverages and antioxidant applications Volume 2: Management in different industry sectors**
Edited by E. A. Decker, R. J. Elias and D. J. McClements
201 **Protective cultures, antimicrobial metabolites and bacteriophages for food and beverage biopreservation**
Edited by C. Lacroix
202 **Separation, extraction and concentration processes in the food, beverage and nutraceutical industries**
Edited by S. S. H. Rizvi
203 **Determining mycotoxins and mycotoxigenic fungi in food and feed**
Edited by S. De Saeger
204 **Developing children's food products**
Edited by D. Kilcast and F. Angus
205 **Functional foods: Concept to product Second edition**
Edited by M. Saarela

206 **Postharvest biology and technology of tropical and subtropical fruits Volume 1: Fundamental issues**
Edited by E. M. Yahia
207 **Postharvest biology and technology of tropical and subtropical fruits Volume 2: Açai to citrus**
Edited by E. M. Yahia
208 **Postharvest biology and technology of tropical and subtropical fruits Volume 3: Cocona to mango**
Edited by E. M. Yahia
209 **Postharvest biology and technology of tropical and subtropical fruits Volume 4: Mangosteen to white sapote**
Edited by E. M. Yahia
210 **Food and beverage stability and shelf life**
Edited by D. Kilcast and P. Subramaniam
211 **Processed Meats: Improving safety, nutrition and quality**
Edited by J. P. Kerry and J. F. Kerry
212 **Food chain integrity: A holistic approach to food traceability, safety, quality and authenticity**
Edited by J. Hoorfar, K. Jordan, F. Butler and R. Prugger
213 **Improving the safety and quality of eggs and egg products Volume 1**
Edited by Y. Nys, M. Bain and F. Van Immerseel
214 **Improving the safety and quality of eggs and egg products Volume 2**
Edited by F. Van Immerseel, Y. Nys and M. Bain
215 **Animal feed contamination: Effects on livestock and food safety**
Edited by J. Fink-Gremmels
216 **Hygienic design of food factories**
Edited by J. Holah and H. L. M. Lelieveld
217 **Manley's technology of biscuits, crackers and cookies Fourth edition**
Edited by D. Manley
218 **Nanotechnology in the food, beverage and nutraceutical industries**
Edited by Q. Huang
219 **Rice quality: A guide to rice properties and analysis**
K. R. Bhattacharya
220 **Advances in meat, poultry and seafood packaging**
Edited by J. P. Kerry
221 **Reducing saturated fats in foods**
Edited by G. Talbot
222 **Handbook of food proteins**
Edited by G. O. Phillips and P. A. Williams
223 **Lifetime nutritional influences on cognition, behaviour and psychiatric illness**
Edited by D. Benton
224 **Food machinery for the production of cereal foods, snack foods and confectionery**
L.-M. Cheng
225 **Alcoholic beverages: Sensory evaluation and consumer research**
Edited by J. Piggott
226 **Extrusion problems solved: Food, pet food and feed**
M. N. Riaz and G. J. Rokey
227 **Handbook of herbs and spices Second edition Volume 1**
Edited by K. V. Peter
228 **Handbook of herbs and spices Second edition Volume 2**
Edited by K. V. Peter
229 **Breadmaking: Improving quality Second edition**
Edited by S. P. Cauvain
230 **Emerging food packaging technologies: Principles and practice**
Edited by K. L. Yam and D. S. Lee
231 **Infectious disease in aquaculture: Prevention and control**
Edited by B. Austin

232 **Diet, immunity and inflammation**
Edited by P. C. Calder and P. Yaqoob
233 **Natural food additives, ingredients and flavourings**
Edited by D. Baines and R. Seal
234 **Microbial decontamination in the food industry: Novel methods and applications**
Edited by A. Demirci and M.O. Ngadi
235 **Chemical contaminants and residues in foods**
Edited by D. Schrenk
236 **Robotics and automation in the food industry: Current and future technologies**
Edited by D. G. Caldwell
237 **Fibre-rich and wholegrain foods: Improving quality**
Edited by J. A. Delcour and K. Poutanen
238 **Computer vision technology in the food and beverage industries**
Edited by D.-W. Sun
239 **Encapsulation technologies and delivery systems for food ingredients and nutraceuticals**
Edited by N. Garti and D. J. McClements
240 **Case studies in food safety and authenticity**
Edited by J. Hoorfar
241 **Heat treatment for insect control: Developments and applications**
D. Hammond
242 **Advances in aquaculture hatchery technology**
Edited by G. Allan and G. Burnell
243 **Open innovation in the food and beverage industry**
Edited by M. Garcia Martinez
244 **Trends in packaging of food, beverages and other fast-moving consumer goods (FMCG)**
Edited by N. Farmer
245 **New analytical approaches for verifying the origin of food**
Edited by P. Brereton
246 **Microbial production of food ingredients, enzymes and nutraceuticals**
Edited by B. McNeil, D. Archer, I. Giavasis and L. Harvey
247 **Persistent organic pollutants and toxic metals in foods**
Edited by M. Rose and A. Fernandes
248 **Cereal grains for the food and beverage industries**
E. Arendt and E. Zannini
249 **Viruses in food and water: Risks, surveillance and control**
Edited by N. Cook
250 **Improving the safety and quality of nuts**
Edited by L. J. Harris
251 **Metabolomics in food and nutrition**
Edited by B. C. Weimer and C. Slupsky
252 **Food enrichment with omega-3 fatty acids**
Edited by C. Jacobsen, N. S. Nielsen, A. F. Horn and A.-D. M. Sørensen
253 **Instrumental assessment of food sensory quality: A practical guide**
Edited by D. Kilcast
254 **Food microstructures: Microscopy, measurement and modelling**
Edited by V. J. Morris and K. Groves
255 **Handbook of food powders: Processes and properties**
Edited by B. R. Bhandari, N. Bansal, M. Zhang and P. Schuck
256 **Functional ingredients from algae for foods and nutraceuticals**
Edited by H. Domínguez
257 **Satiation, satiety and the control of food intake: Theory and practice**
Edited by J. E. Blundell and F. Bellisle
258 **Hygiene in food processing: Principles and practice Second edition**
Edited by H. L. M. Lelieveld, J. Holah and D. Napper

259 **Advances in microbial food safety Volume 1**
Edited by J. Sofos
260 **Global safety of fresh produce: A handbook of best practice, innovative commercial solutions and case studies**
Edited by J. Hoorfar
261 **Human milk biochemistry and infant formula manufacturing technology**
Edited by M. Guo
262 **High throughput screening for food safety assessment: Biosensor technologies, hyperspectral imaging and practical applications**
Edited by A. K. Bhunia, M. S. Kim and C. R. Taitt
263 **Foods, nutrients and food ingredients with authorised EU health claims: Volume 1**
Edited by M. J. Sadler
264 **Handbook of food allergen detection and control**
Edited by S. Flanagan
265 **Advances in fermented foods and beverages: Improving quality, technologies and health benefits**
Edited by W. Holzapfel
266 **Metabolomics as a tool in nutrition research**
Edited by J.-L. Sébédio and L. Brennan
267 **Dietary supplements: Safety, efficacy and quality**
Edited by K. Berginc and S. Kreft
268 **Grapevine breeding programs for the wine industry**
Edited by A. G. Reynolds
269 **Handbook of antimicrobials for food safety and quality**
Edited by T. M. Taylor
270 **Managing and preventing obesity: Behavioural factors and dietary interventions**
Edited by T. P. Gill
271 **Electron beam pasteurization and complementary food processing technologies**
Edited by S. D. Pillai and S. Shayanfar
272 **Advances in food and beverage labelling: Information and regulations**
Edited by P. Berryman
273 **Flavour development, analysis and perception in food and beverages**
Edited by J. K. Parker, S. Elmore and L. Methven
274 **Rapid sensory profiling techniques and related methods: Applications in new product development and consumer research**
Edited by J. Delarue, J. B. Lawlor and M. Rogeaux
275 **Advances in microbial food safety: Volume 2**
Edited by J. Sofos
276 **Handbook of antioxidants for food preservation**
Edited by F. Shahidi
277 **Lockhart and Wiseman's crop husbandry including grassland: Ninth edition**
H. J. S. Finch, A. M. Samuel and G. P. F. Lane
278 **Global legislation for food contact materials**
Edited by J. S. Baughan
279 **Colour additives for food and beverages**
Edited by M. Scotter
280 **A complete course in canning and related processes 14th Edition: Volume 1**
Revised by S. Featherstone
281 **A complete course in canning and related processes 14th Edition: Volume 2**
Revised by S. Featherstone
282 **A complete course in canning and related processes 14th Edition: Volume 3**
Revised by S. Featherstone
283 **Modifying food texture: Volume 1: Novel ingredients and processing techniques**
Edited by J. Chen and A. Rosenthal

284 **Modifying food texture: Volume 2: Sensory analysis, consumer requirements and preferences**
Edited by J. Chen and A. Rosenthal
285 **Modeling food processing operations**
Edited by S. Bakalis, K. Knoerzer and P. J. Fryer
286 **Foods, nutrients and food ingredients with authorised EU health claims Volume 2**
Edited by M. J. Sadler
287 **Feed and feeding practices in aquaculture**
Edited by D. Allen Davis
288 **Foodborne parasites in the food supply web: Occurrence and control**
Edited by A. Gajadhar
289 **Brewing microbiology: design and technology applications for spoilage management, sensory quality and waste valorisation**
Edited by A. E. Hill
290 **Specialty oils and fats in food and nutrition: Properties, processing and applications**
Edited by G. Talbot
291 **Improving and tailoring enzymes for food quality and functionality**
Edited by R. Yada
292 **Emerging Technologies for Promoting Food Security: Overcoming the World Food Crisis**
Edited by C. Madramootoo
293 **Innovation and Future Trends in Food Manufacturing and Supply Chain Technologies**
Edited by C. E. Leadley
294 **Functional Dietary Lipids: Food Formulation, Consumer Issues and Innovation for Health**
Edited by T. A. B. Sanders

Introduction

The imperative for rapid and significant innovations in all aspects of food production was starkly illustrated in Professor John Beddington's report of 2011, "The future of food and farming: challenges and choices for global sustainability". This report categorized "major failings" in the current food system notably:

- Hunger remaining widespread with 925 million people experiencing hunger and a further billion being deficient in important micronutrients. Much of this hunger is not due to a global food shortage but due to poor distribution, poverty, political instability, and poor governance.
- Many systems of food production being unsustainable, impacting in areas such as climate change (due to a heavy reliance on fossil fuels and the production of pollutants), reducing biodiversity, contributing to soil degradation and impacting on water availability due to extraction for irrigation.

The food industry must (and is) working toward addressing these deficiencies but has to do this in a time of profound global shifts creating significant pressures for innovation and change. These global shifts include:

- A growing population predicted to rise from around seven billion to nine billion by 2050
- An aging population in developed countries which will impact on both nutritional requirements and the physical capabilities of the work force
- Rising energy costs making it essential to increase productivity
- Increasing competition for land, energy, and water
- The impact of climate change becoming more apparent and requiring adaptation on the part of food producers.

The Beddington report identified five challenges that are critical to address in future food production. These were:

- Balancing future demand and supply sustainably, to ensure food supplies remain affordable
- Ensuring adequate stability in food supplies, and protecting the most vulnerable from the volatility that does occur
- Achieving global access to food and ending hunger
- Managing the contribution of the food system to the mitigation of climate change
- Maintaining biodiversity and ecosystems while feeding the world.

Trying to make predictions about what food production might look like in the future is distinctly high risk but it can be said with some degree of certainty that there must be innovations in products, production methods, and technologies that address the global shifts and five challenges noted above.

This book explores existing and future perspectives around three main topics, supply chain technologies, factory design and operation, and innovations in food processing technologies, specifically preservation technologies.

Supply Chain Technologies

This section of the book reviews the current and future status of food fraud and authenticity techniques and the growing importance of information technology in food production. Regrettably, issues such as raw material substitution, dilution, and fraudulent labeling (e.g., passing off nonorganic materials as organic materials) are a significant challenge for the food industry. Rising production costs and the impact of climate change as outlined previously increase the risk of food fraud. Current and future methods for food fraud detection are outlined as well as the emergence of risk management techniques such as TACCP and VACCP (threat and vulnerability assessment critical control point). Developments in information technology for the purposes of supply chain and production monitoring are outlined. Such developments offer enormous potential for improving the traceability, quality and safety of raw materials, intermediate and finished products, as well as impacting on production efficiency. An example of how intelligent packaging might interact with information technology systems serves as an insight into what developments we might expect to see in the future.

Factory Design and Operation

This section outlines the current state of the art and potential future developments in the areas of factory design, hygienic operation, and process monitoring and control. Despite there being a wealth of knowledge and information available around good practices regarding factory and equipment design—most notably from organizations such as Campden BRI and the European Hygienic Engineering and Design Group—examples of poor implementation are not uncommon. It is hoped that a significant development in terms of future food manufacturing will be a more widespread adoption and implementation of existing knowledge on the topic. After adopting good factory and equipment design principles, effective cleaning, and sanitation remains an essential component of day-to-day food manufacturing operations. This section also explores current and emerging methods for decontamination of food contact surfaces.

Finally, this section explores the enormous potential for developments in multivariate analytics to be applied more widely in food production. We are at a point in time where we have an unprecedented capability for process measurement. Sensors are now available to measure virtually any aspect of raw materials, product, or process performance. Measurement is of limited value without the tools to monitor and interpret the data which is generated. Multivariate statistical analysis is a powerful tool to gain an understanding of the key factors that contribute to product and process variability. While such methods have been available for some time, their use has been restricted to statisticians. Software developments have now opened up these tools for more widespread implementation. Such techniques offer the potential to improve productivity and reduce waste.

Innovative Food Processing Technologies

Food waste remains a significant issue and must urgently be addressed, particularly when considered against the challenges outlined in the Beddington report. Global food waste is estimated to be something like 30–50% of all food grown. With millions suffering from hunger, and a growing population, technologies and working practices that reduce food waste and extend food shelf-life will be critical areas of future innovation.

The traditional techniques of thermal processing, chilling, and freezing will play an important role in reducing food waste. However, with consumer demands for high quality convenience products, food waste, and shelf-life extension cannot just be tackled by these conventional processes alone. Technologies that can deliver shelf-life extension while retaining nutritional, sensory, and functional quality will become increasingly important. It is in this domain that advanced thermal, nonthermal, and freezing technologies will have an impact and these technologies are the focus of this section.

This book serves as an introduction to the current state of the art and potential future developments in food production. It should be of use to research and development managers working in the food industry, as well as food processors, manufacturers, postgraduate students, and academic researchers with an interest in advances in food manufacturing methods.

Craig E. Leadley

Reference

Beddington, 2011. Foresight. The Future of Food and Farming. Executive Summary. The Government Office for Science, London.

Part One

Supply Chain Developments

Food Fraud and Authenticity: Emerging Issues and Future Trends

1

Annemieke M. Pustjens[1], *Yannick Weesepoel*[1], *Saskia M. van Ruth*[1,2]
[1]RIKILT, Institute of Food Safety, Wageningen University and Research Centre, Wageningen, The Netherlands; [2]Food Quality and Design Group, Wageningen University, Wageningen, The Netherlands

1.1 Introduction

Food fraud is a collective term that covers the "deliberate substitution, addition, tampering or misrepresentation of food, food ingredients or food packaging, or false or misleading statements made about a product for economic gain," as defined by the United States Pharmacopeial Convention (USP). The deliberate substitution or addition of a fraudulent component to a finished product is called adulteration. This is the most common type of food fraud. Although food fraud or economically motivated adulteration has financial gain as the main goal, adulteration may result in serious public health consequences when the adulterant is toxic or allergenic (Everstine et al., 2013). Food fraud is a significant and growing problem, driven by globalization, economic opportunity, and the low probability and severity of punishment (Kennedy, 2012). Therefore, in this chapter, the occurrence of different types of food fraud, vulnerability of the supply chain, analytical verification within a laboratory (as has been the focus in recent years), and emerging handheld and in-line screening equipment will be discussed. Such an overview will help to gain more insight in the vulnerable products and processes of the food chain and to evaluate whether the appropriate analytical and screening methods are available in defense of food fraud in the future.

1.2 Background

1.2.1 Types of Food Fraud

Different types of food fraud can be distinguished (Figure 1.1) and will be discussed in the paragraph below with some examples.

1.2.1.1 Composition

Most of the types of food fraud influence the composition of a product, e.g., substitution, dilution, unapproved enhancements, and concealment. Changing the composition of a food product can be either done by the addition of poor-quality own-product material, product-foreign material, or adulteration with other constituents.

Innovation and Future Trends in Food Manufacturing and Supply Chain Technologies
http://dx.doi.org/10.1016/B978-1-78242-447-5.00001-0

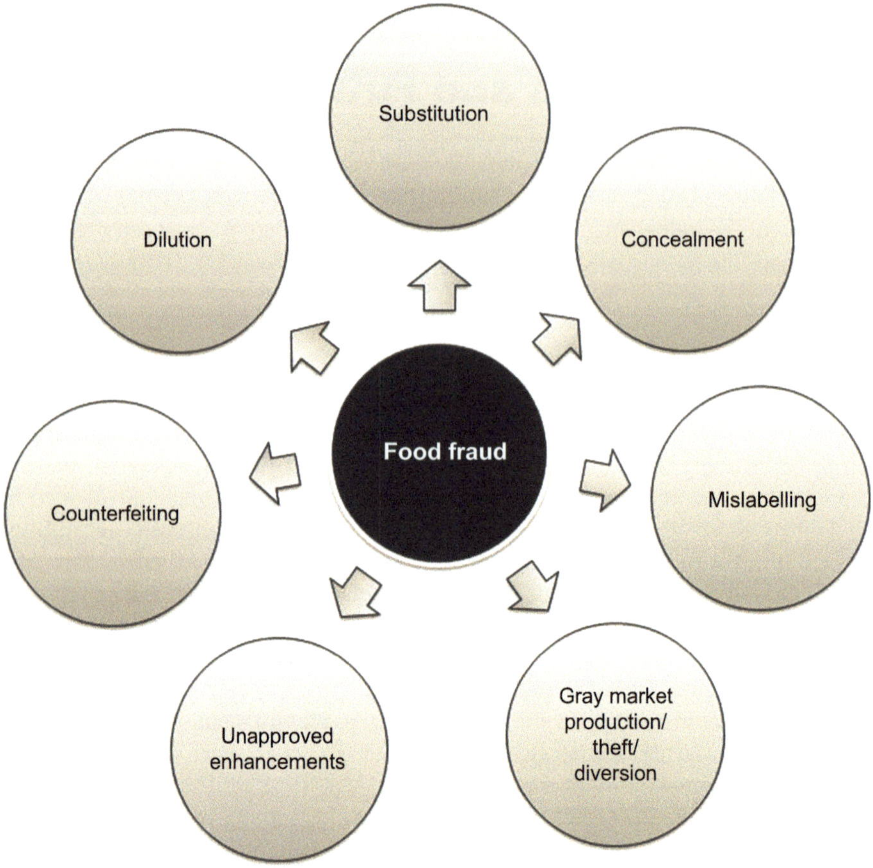

Figure 1.1 Food fraud terminology.
Adapted from Food Fraud Think Tank (2013).

Food producers try to gain more money by adding cheaper nonauthentic ingredients in more expensive or authentic products like honey (Strayer et al., 2014) or fruit juices (Ehling and Cole, 2011). Another form of adding product-foreign material is the addition of melamine to enhance the apparent protein content (Domingo et al., 2014).

Adulteration with constituents include, for example, diluting milk with water (Kasemsumran et al., 2007). Also in poultry, meat, or fish, water can be added. For poultry, the water/protein ratio is controlled, therefore fraudsters mask the addition of water by the addition of protein or phosphate salts, for example (Ballin, 2010). Hamburgers in Mexico were found to contain soy protein, while in 30% of the cases this was not mentioned on the label (Ballin, 2010).

1.2.1.2 Products of Plant Origin: Cultivar and Variety Identification

Certain food and feed products may possibly originate from different botanical origin, which influences their applicability in food and feed. For example, dried distillers grains with solubles, which are a by-product from the bioethanol production and mainly used in animal feed, can originate from different cereals, such as maize and wheat. The large variation in the composition can contribute to imbalanced feed formulations. Therefore, it is important to control its composition (Tres et al., 2014). Also, molecular biological methods can be used to differentiate types of cereals in food. For example, DNA sequencing can distinguish between the protected Farro della Garfagnana (emmer) and common bread wheat (Prins et al., 2010). Biological methods can also be used to evaluate whether plant material, like soy beans, maize, rice, and sugar beet, is genetically modified or not (Querci et al., 2010).

1.2.1.3 Products of Animal Origin: Species Identification

Authenticity investigations revealed that in certain countries for 15–39% of the meat products, like hamburgers and sausages, animal species was not declared on the label (Ballin, 2010). Also, more expensive, more authentic meat can be (partly) substituted with cheaper, less authentic meat tissues or species. For example, different cuts of beef that were interchanged can be analytically distinguished (Al-Jowder et al., 1999). The horsemeat scandal is a more recent example where beef was (partly) substituted with horsemeat for economic gain (Zając et al., 2014).

1.2.1.4 Geographical Origin

Geographical origin has become of higher importance for economic and safety reasons, but also in view of sustainability issues. Food products are sometimes mislabeled; therefore, analytical methods should be developed to verify the geographical origin of a food product. This is often done in a nontargeted approach, for example, the verification of the geographical origin of olive oil by the analysis of the volatile components by proton transfer reaction–mass spectrometry (PTR–MS) or the fatty acid profile by gas chromatography (GC) (Araghipour et al., 2008; Ruiz-Samblás et al., 2012, 2013).

Counterfeiting occurs with products with European Union protected designation of origin, like the typical cheese from the Dutch city of Leiden. In this case, the typical cheese can be distinguished from other (cheaper) imitation cheeses in a nontargeted approach by the analysis of the volatile components by PTR–MS (Galle et al., 2011). In the case of fish, DNA is extracted from the microorganisms present. The variety of microorganisms is determined by the geographical origin of the fish (El Sheikha and Montet, 2014).

1.2.1.5 Production System

Food can be produced in different ways, e.g., conventional, organic, Halal/Kosher, sustainable, considering animal welfare, and rainforest friendly. Consumers want to be assured that the product they buy is produced in the way that is stated on the package and that the extra money they pay for a specific production method is appropriate. Methods have, for example, been developed to distinguish conventional from organic dairy products (Capuano et al., 2014a,b), conventional from organic meat (Castellini et al., 2002), organic from conventional eggs (van Ruth et al., 2011), and to distinguish wild from farmed fish (Capuano et al., 2013). For the assurance of Halal food production several methods have been described to verify the origin of animals, animal products and proteins, and the presence of alcohol (van der Spiegel et al., 2012).

1.2.1.6 Processing

During production of a food product, it often undergoes some kind of processing. For example, meat or fish that has been frozen and thawed but is sold as being fresh (Uddin et al., 2005). Another example is coffee, from which the beans have undergone an extra processing step, like in the expensive Indonesian Kopi Luwak, which can be distinguished from normal espresso (Özdestan et al., 2013). Also, cheeses that are made from raw milk instead of thermally treated milk can be distinguished based on the available lysine content. A novel amperometric sensor has been developed for rapid detection of raw milk cheeses and their imitations (Ciriello et al., 2015).

1.2.2 Food Fraud Occurrence and Usage of Databases

Since the 1990s, the European food industry has encountered several crises that harmed consumer confidence in the food industry and regulatory bodies (van Ruth and Huisman, 2014). In order to register historical and ongoing food fraud events, two databases are available, which are leading nowadays: (1) the USP Food Fraud Database, and (2) the National Center for Food Protection and Defense (NCFPD) Food Fraud Incident Database (Foodfraud.org, 2014; FoodSHIELD.org, 2014). The USP Food Fraud Database is a public database that lists both scientific and media records on fraud to identify vulnerable food products and/or ingredients. The NCFPD FoodSHIELD Database mainly keeps general records of the main fraud incidents occurring worldwide. When studying both databases, the number of records per problematic food ingredient might represent its fraud susceptibility and may provide an outlook to future food fraud. Therefore, in Table 1.1, the top five of most reported product groups were listed from both databases. Per product group, both total records and records from the last 5 years were compared. Note that often data from 2013 were only partially available, and data from 2014 were not available at all.

The USP Food Fraud Database is composed of both scholarly and media reports, which has been collected from scholarly and undocumented industrial analyses and mainstream media in the English language (Foodfraud.org, 2014). From Table 1.1, it immediately becomes clear that of all reported fraud incidents in the database, 50% of the scholarly work and 40% of the media cases occurred in the period 2009–2013. Although, apart from illustrating the upcoming research, funding, and public awareness on food fraud, distilling future trends for individual product categories from this data is precarious. The records in the scholarly database do not represent actual fraud incidents but instead list reports on detection methods and analysis. Therefore, the food fraud scholarly database represents the food categories most researched. Similarly, media entries represent the attention of the mainstream media in the English language on fraud reports. Topic selection is, therefore, subjected to several undefined criteria. When assessing the diversity of individual foods present in the main categories, a clear imbalance is present among the records. For example, research has been primarily

Table 1.1 **Overview of the Top Five Reported Food Categories in the USP Food Fraud Database and the NCFPD FoodSHIELD Database**

Database[a]	**Start of 2013 Reports (#)**[b] **Subcategory (%)**[b]	**2009–2013**[a] **Reports (#)**[b] **Subcategory (%)**[b]
USP food fraud scholarly	**1444**	**664**
Oils	411	228
Olive oil	62%	62%
Sesame oil	3%	3%
Spices	200	110
Saffron	51%	35%
Turmeric	8%	8%
Black pepper, chili powder[c]	7%	11, 13%
Fruit juices, concentrates, jams, purees, and preserves	139	28
Other	135	68
(Instant) coffee	38%	32%
Tea	10%	19%
Cocoa powder	9%	4%
Sweeteners	129	52
Honey	84%	96%
Maple syrup	12%	0%
USP food fraud media	**332**	**151**
Natural flavoring complexes	74	0
Spices	69	27
Chili powder	20%	15%
Turmeric	19%	37%
Saffron, black pepper[c]	12%	19, 7%
Seafood	46	31
Oils	29	16
Olive oil, cooking oil[c]	38%	19, 38%
Other	29	26
(Instant) coffee	17%	19%
Tea, salt[c]	14%	8, 15%
NCFPD foodSHIELD	**310**	**79**
Fish and seafood products	93	16
Oils and fats	33	14
Olive oil	61%	50%
Honey and other natural sweeteners	31	4
Honey	65%	100%
Meat and meat products	28	8
Alcoholic beverages	25	10

[a]Data from October 21, 2014 (Foodfraud.org, 2014; Foodshield.org, 2014). Data from 2013 was not complete yet; 2014 data was lacking at the date of collection. For the NCFPD database, year categorization was made on the "Year began" category if available; otherwise, the entry was not considered in the 2009–2013 column.
[b]Reports represent the individual entries per food category; subcategory represents the percentage of reported cases on the top individual food subcategories.
[c]Subcategories had equal entries.

focused on oil issues since 2009, and in particular on olive oils, which represent 62% of all reports in this food category in the period 2009–2013. The scholarly reports on spices mainly focus on the expensive spice saffron (51% and 35% in the start of 2013 and 2009–2013, respectively), while in the media saffron is only mentioned in 12% and 19% of the cases. In the case of sweeteners, this is even more prominent, with honey representing 84% and 96% of all scholarly sweetener entries, while in the media sweeteners are not present in the top five. In the past, the media focused mainly on natural flavoring complexes, but focus has moved more toward spices and seafood. This under- and overrepresentation of certain food categories results in imbalance and may give the impression that these food categories are most vulnerable to food fraud. However, the number of scholarly and media reports cannot represent fraud vulnerability; they merely sketch trends in research and media instead. On top of that, there seems to be little coherence between the scholarly and media database, making fraud prediction merely impossible.

The NFCDP FoodSHIELD Database lists food fraud incidents as "a documented, isolated occurrence of food fraud in a single food product or group of associated food products occurring within a defined time frame and with a distinct group of perpetrators" and therefore provides a completely different top five of food categories than in the USP Food Fraud Database (Foodshield.org, 2014). As an example, 50 single records comprising "dairy" and "melamine" exist in the USP Food Fraud Database as a result from the melamine incident in China 2007–2008. In the FoodSHIELD Database, this incident is recorded as a single event and therefore has a different impact on record statistics. Studies have been conducted on finding similarities in food fraud cases recorded in the FoodSHIELD Database, concluding that prevention of food fraud cannot rely on traditional food safety strategies (Everstine et al., 2013). Individual trends of food fraud in food categories cannot be established, as long as effective analytical methods are not widely implemented by governments.

As an alternative system in the European Union, the Rapid Alert System for Food and Feed (RASFF) has been introduced. RASFF includes cases of food fraud if they are of health concern and are expected to reach multiple European Union member states. In 2013, 168 individual cases of adulteration or fraud have been recorded (European Commission, 2014); however, it is not clear which product groups were most involved in food fraud. For 2015, a new food fraud reporting system was announced in the European Union, which might dismiss the political hurdles for reporting fraud (RASFF, 2014).

In conclusion, in order to estimate fraud occurrence or vulnerability of commodities to food fraud, the extrapolation of data from the existing databases is not a suitable tool because these databases do not purely show occurrence but rather have other pillars of information. However, there is a great need to estimate food fraud occurrence worldwide and to understand what determines food fraud vulnerability.

1.2.3 Food Fraud Vulnerability: The Fraud Triangle

Food fraud vulnerability can be explained by the fraud triangle, indicating that three components are important to fraud: opportunity, motivation, and rationalization (Figure 1.2). Opportunities are created by weak internal controls, poor management

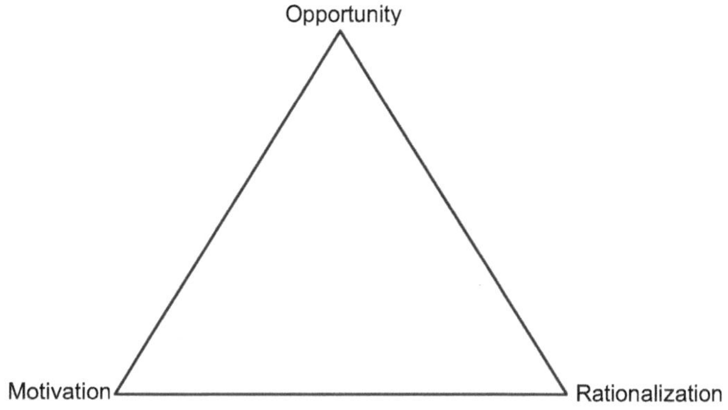

Figure 1.2 The fraud triangle.

oversight, and/or through the (ab)use of one's position and authority. In the case of food fraud, every actor in the food chain poses a potential opportunity for the occurrence of food fraud. The motivation is a pressure or a need felt by the person who commits fraud, which is economic gain in the case of food fraud. Rationalization involves a person reconciling his/her behavior with the commonly accepted notions of decency and trust (Rodgers, 2012). The vulnerability of the food supply chain to fraud will continue as long as the potential for profit exceeds the chance of getting caught, and the potential consequences do not act as a barrier (Everstine et al., 2013).

1.2.4 Quality Management Systems

Regular testing at critical points along the supply chain with targeted, specific testing methods is the first line of defense for verifying that process control systems are working (Everstine et al., 2013). Worldwide, several food quality management systems are available, for example, Hazard Analysis and Critical Control Points (HACCP), International Organization for Standardization (ISO 22000), British Retail Consortium Global Standards, and Food Safety System Certification (FSSC 22000). These systems guide manufacturers to implement and manage an effective food quality and safety program. Global food manufacturers have established the Global Food Safety Initiative (GFSI). The GFSI has set requirements for food safety schemes through a benchmarking process in order to improve cost efficiency throughout the food supply chain (Hoorfar et al., 2011).

1.2.5 Traceability

Food products are often combinations of different food ingredients that all need to be controlled among their production chain. In the European General Food Law (EC Regulation No. 178/2002), product tracing is defined as "the ability to trace and follow a food, feed, food-producing animal or substance intended to be, or expected to be incorporated into a food or feed, through all stages of production, processing and distribution." Traceability systems help to control food chain integrity as well as reducing the withdrawal costs of a product (Hoorfar et al., 2011). Extrinsic cues/markers used in traceability systems are package materials, labels, bar coding, Radio Frequency Identification (RFIDs), place of purchase, price, and country of origin (Northen, 2000). Furthermore, intrinsic product markers may be used in tracking and tracing in the food chain as well (Raspor, 2005). The latter poses an extra challenge to find analytical tools to enable verification of these intrinsic markers.

1.2.6 Analytical Verification

In general, there are two strategies for detecting food fraud: (1) testing for the presence/absence of a specific known adulterant (targeted approach); and (2) testing the identity, authenticity, or purity of a food ingredient (untargeted approach) (Moore et al., 2012). There is a growing concern that in some ways food fraud may be more risky than traditional threats to the food supply, as the adulterants used in these activities are often unconventional (Moore et al., 2012). Under those circumstances, targeted tests will fail. Therefore, fast and untargeted anomaly testing is developing rapidly.

Analytical techniques that can be used in a laboratory are classic wet chemical techniques to determine fat, protein, or moisture content, gas or liquid chromatography to determine the type of fats, proteins, or carbohydrates present, mass spectrometry, immunological tests to determine the presence of product-foreign protein, spectroscopic techniques to determine the concentration of an adulterant, isotope ratio analyses to determine the geographical origin, DNA-based techniques to determine meat species, and enzymatic reactions to determine the freshness of a product (van Ruth and Huisman, 2014).

Statistical analyses of the data are often performed using chemometrics as a multivariate data analysis tool. Chemometrics is a powerful tool used either qualitatively for grouping or classifying unknown samples with similar characteristics or quantitatively for determining the concentration of adulterant analytes in samples (Moore et al., 2012). Modeling and visualization of results can be done in different ways. Usually, one starts to check natural clustering of genuine samples versus adulterated samples using principal component analysis. Then data can be clustered (e.g., hierarchical) and classified (e.g., using discriminant analysis). To predict the concentration of a specific adulterant, regression models can be developed using principal component regression, partial least square regression, and many more available algorithms.

1.3 Emerging Issues and Future Trends

Classical authenticity assessment is based on the analysis of a specific marker compound and determining whether the analytical parameter determined is violating the established limit or is in compliance. However, from the point of view of regulatory bodies and food industry this is both tedious and expensive. There is an increasing need for the correct food safety management systems, supported by nontargeted and nondestructive analytical methods, which can be used outside the laboratory, like in-line or portable methods.

1.3.1 General Issues: Fraud Vulnerability Assessments

Food fraud prevention is fundamentally different from combating food safety or food defense incidents (US Government Accountability Office, 2011). However, prevention of food fraud should be integrated with all aspects of food safety management. Focus has changed from assessing the exposure to danger (risk) itself toward assessing the possibility of the risk (vulnerability) (Food Fraud Think Tank, 2013). Therefore, according to the GFSI's food safety management umbrella, three systems should be addressed to assess food safety (HACCP), food defense (Threat Assessment Critical Control Points, TACCP) and food fraud (Vulnerability Analysis and Critical Control Points, VACCP) (Figure 1.3). Where HACCP addresses unintentional or accidental adulteration, TACCP and VACCP address intentional adulteration, which is either ideologically motivated (in the case of TACCP) or economically motivated (in the case of VACCP). The British Standard Institution has introduced TACCP as a tool for "systematic management of risks through the evaluation of threats, identification of vulnerabilities and implementation of controls" (The British Standard Institution, 2014). VACCP is applying the HACCP-system specifically to a food fraud incident. The Food and Drug Administration introduced CARVER + Shock (Criticality, Accessibility, Recuperability, Vulnerability, Effect and Recognizability), which includes prioritization of the risks and consequences of a potential attack (Maillot, 2010). GFSI will introduce a food fraud vulnerability assessment tool based on context factors and the company's fraud measures in early 2015. The tool comprises a questionnaire of 50 questions, which will allow a self-assessment of food fraud vulnerability for individual companies. The context factors include the technological situation, the supplier and company environment, the chain and branch environment, and the economic circumstances.

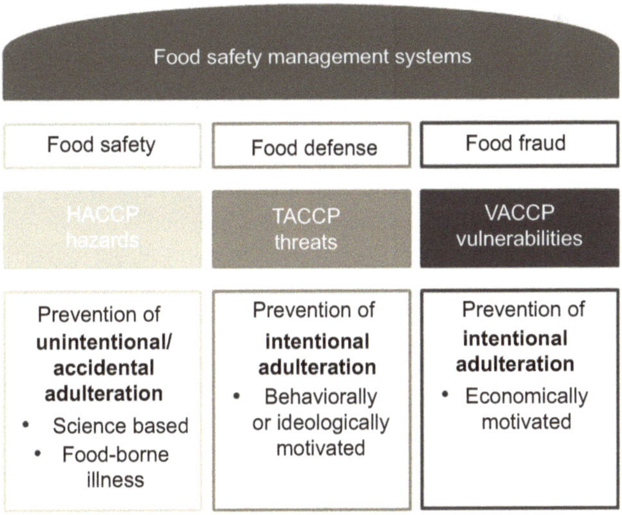

Figure 1.3 GFSI's food safety management umbrella.
Adapted from Food Fraud Think Tank (2013).

In conclusion, both governments and food manufacturers continuously try to develop suitable management tools and keep them up-to-date to prevent food fraud in this rapidly changing environment.

1.3.2 Specific Issues: Fraud Analytical Measurements

Analytical detection systems support the food fraud management tools as mentioned above in section1.3.1. However, their major bottleneck is the limited number of samples that can be investigated in specialized laboratories for analysis, due to both time and financial restrictions. These restrictions reduce the proportion of food products that can be verified and will therefore lower the threshold for food fraud. Fortunately, due to the more general development of handheld and in-line analytical equipment, more applications are being developed to carry out a rapid screening onsite with non-destructive measurement tools. In this section, an outlook is provided on future implementation of these methods in the food fraud detection. This involves, in most cases, state-of-the-art developments in the field of the respective benchtop equipment.

1.3.2.1 Spectroscopic Equipment

Spectroscopic equipment used in food fraud detection is Raman, Fourier transform (FT), and near- and mid-infrared spectrometers (NIR, MIR), and hyperspectral imaging. In the case of Raman, FTIR, and NIR, portable equipment is available in the 0.8–1.8 kg range (including computer hardware) (Sorak et al., 2011). Measurements by handheld equipment, in combination with spectral libraries and chemometrics, showed good identification and discrimination capabilities, as well as excellent quantitative determination of additives, active ingredients, and blend components in the liquid and solid state (Sorak et al., 2011). Increasing numbers of methods and spectral libraries have been developed for applications in the food-agro industry. The variety of topics presented in the literature is staggering: characterization of internal and external parameters of fruits and vegetables, conservation state and fat content of meat and fish, distinguishing among and quality evaluation of beverages and dairy products, protein content of cereals, evaluation of grape ripeness in vineyards, soil analysis, etc. (Teixeira dos Santos et al., 2013). This methodology has been primarily developed for food quality aspects, but to some extend will also allow food authentication. It seems that the major bottlenecks for the development of new spectroscopic methods for food fraud detection is only limited to either the diversity of the products to be investigated, development of the spectroscopic library, the accompanying chemometric model or availability and development of correct reference materials for calibration. The most important challenge for the coming decade in handheld spectroscopic equipment is the lack of compatible or sharing databases to exponentially expand food fraud controls. At this moment, European 7th Framework projects, e.g., food integrity, have identified these issues (FERA, 2014).

1.3.2.2 Nuclear Magnetic Resonance (NMR) and Magnetic Resonance Imaging (MRI)

One of the most widely used analytical techniques for food fingerprinting and authentication is NMR spectroscopy. Typically in food fraud research, NMR is used for both species and origin verification of numerous of plant and animal matrices in combination with the development of accompanying chemometric models (Esslinger et al., 2014; Laghi et al., 2014; Monakhova et al., 2013).

A completely new and exciting application of NMR is its use as an MRI tool for determining food structure relationships. Van As and van Duynhoven (2013) showed that by improvement of the cutting-edge technical advances and integration of MRI accessible parameters, nano- and microstructures of food matrices can be related to their processing method. Although, so far, no specific reports are available on the application of MRI for food fraud detection, a good example of the potential of this technique is shown by studies on the impact of freeze-drying on microstructure and rehydration properties of carrot and the effect of rice kernel microstructure on cooking behavior (Mohoric et al., 2009; Voda et al., 2012). By means of MRI imaging, rice preprocessing like milling, parboiling, wet-processing, and extrusion of rice kernels can be revealed (Figure 1.4). Even combinations of puffing with the individual preprocessing can be visualized. Clearly, these ideas can be swiftly extrapolated for food fraud investigation as multiple fraud parameters (e.g., fresh versus thawed meat products or fresh versus rehydrated plant matrixes) will result in differences in nano- and microstructures. The reason that this technique has not been implemented for fraud assessment is that MRI has always been the relatively more expensive technique compared with other light and mass spectroscopic methods. Interestingly, due to technical advances, MRI is moving from highly specialized equipment to mobile benchtop and downscaled versions to be used by a broad user base in the field, greenhouse, food laboratory, or factory (Van As and van Duynhoven, 2013). For example, Rahmatallah et al. (2006) published a preliminary study on one-sided portable magnetic resonance probes for noninvasive investigation of large food objects. Furthermore, benchtop low-field NMR has been used to measure water distribution, mobility, and quality aspects of muscle and processed meats (Pearce et al., 2011). Ongoing developments on improving benchtop NMR and MRI equipment will in the future lead to better onsite insight in the history of food matrices on a nano- and microstructure level and therefore contribute to defeat food fraud.

1.3.2.3 Mass Spectrometry

The development of handheld mass spectrometers is readily developing, thereby also affecting the food fraud detection segment (Nielen et al., 2011). In combination with the breakthrough of ambient ionization MS techniques, such as direct analysis in real time (DART), desorption electrospray ionization (DESI), extractive electrospray ionization, and other related techniques, the way is paved for food fraud deterrence applications. The advantage of these methods is that they allow rapid, direct measurement of compounds present on the sample surface. Onsite, samples are exposed to the ionization medium under ambient conditions and do not require extraction and separation processes, making the analysis of small molecules (pesticides, hormones) easy and efficient (Wang et al., 2013). Many applications would be possible with these types of instruments, also in the food fraud area. Lohne et al. (2012) showed the detection of quinolone antibiotic residues in catfish, shrimps, and salmon by laser diode thermal desorption atmospheric pressure chemical ionization MS. Cajka et al. were able to separate chicken bone meal (a banned feed component in the European Union)-fed

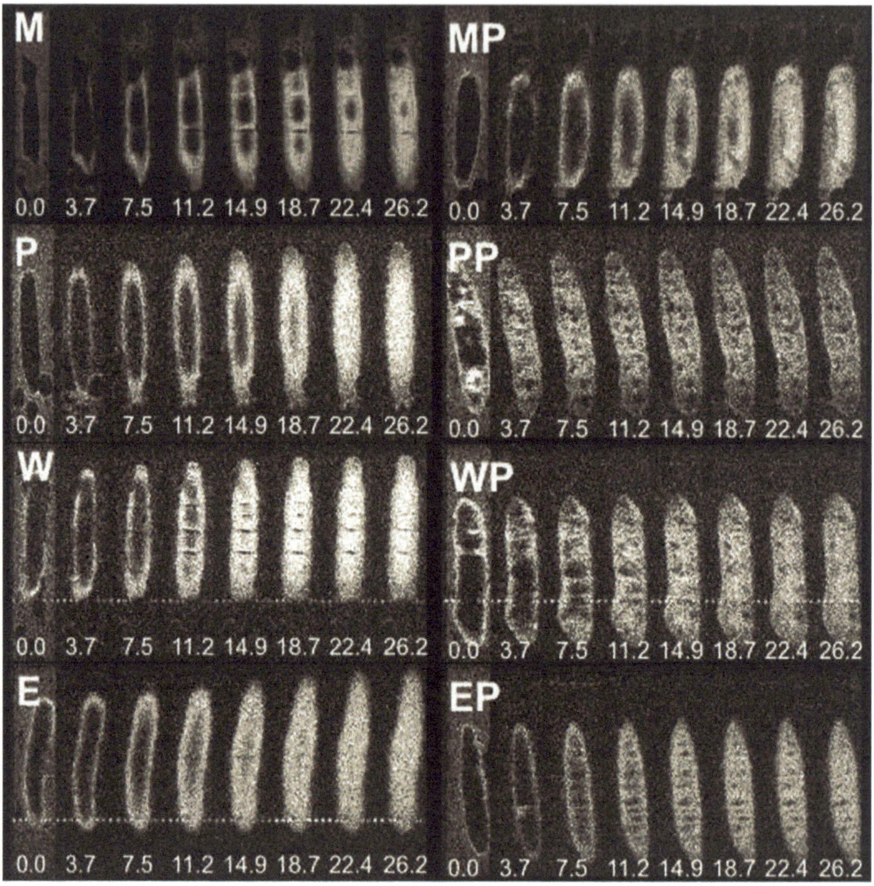

Figure 1.4 Series of MRI images through the center of the kernel representing the cooking process at different times indicated at the bottom in minutes. The left column represents from top to bottom, milled (M), parboiled (P), wet processed (W), and extruded (E) kernels, respectively. The right column represents milled and puffed (MP), parboiled and puffed (PP), wet processed and puffed (WP), and extruded and puffed (EP) kernels, respectively.
Reproduced with permission from Mohoric et al. (2009).

(5–8% (w/w)) chickens from those who were deficient in their diet of this by DART time-of-flight MS (Cajka et al., 2013). Although both applications need to be developed further, and they still require a simple extraction procedure, onsite control for microcomponents in meats and seafood is emerging.

Besides handheld equipment, in-line equipment for the food industry is of interest, for example, proton transfer reaction coupled to a mass spectrometer (PTR–MS), especially in the case of monitoring volatile compounds. With the help of relatively mild ionization of the volatiles present in the headspace of the samples, PRT–MS analysis monitors a volatile fingerprint that enables distinguishing specific characteristics of the product (Heenan et al., 2013). In the case of food fraud, PTR–MS, in combination with a multivariate statistical analysis, has proven itself to be a powerful tool to discriminate products from, for example, geographical origin or speciality products (Özdestan et al., 2013; Ruiz-Samblás et al., 2012). At the moment, PTR–MS techniques are still in the semi real-time stage, which means that individual samples can be measured within a few minutes. However, from a producer's point of view, real-time or in-line PTR–MS would be desirable in order to detect deviations while raw materials are being processed. Published work on PTR–MS techniques indicate that suitable interfaces for in-line detection are being developed, although they are again not applied in the food fraud area yet (Papurello et al., 2014). Finally, gas chromatography combustion–isotope ratio mass spectrometry (GC–C–IRMS) is being used more often in applications for traceability and authenticity in foods and beverages. GC–C–IRMS combines separation of individual volatiles with isotope ratio mass spectra of carbon and nitrogen of these volatiles and an accompanying chemometric model. This solves the problem of the data loss occurring with traditional elemental analyzer–isotope ratio mass spectrometry, where the entire sample is combusted (van Leeuwen et al., 2014).

1.3.2.4 Lab-on-a-Chip

The spectroscopic methods as mentioned above have their limitations when it comes to, for example, identification of species of (parts of) animal carcasses, (processed) bush meat, and complex mixtures of (traditional) medicine. Often prohibited species of animals and plants are present in these products and require DNA testing for proper identification, which normally would proceed in specialized laboratories. Lab-on-a-chip applications present a possible answer to onsite identification problems of these complex products. Lab-on-a-chip equipment is based on microfluidics and micro–nano fabrication techniques, and contain moving parts ranging from nanometer to millimeter scale. These are considered to be sensitive, more specific, low cost, energy-efficient, robust, fast, and can also provide real-time information (Kalia and Parshad, 2014). In the case of species identification, polymerase chain reaction–restriction fragment length polymorphism can be used in combination with microfluidic lab-on-a-chip capillary electrophoresis systems for end-point analysis and accurate sizing of DNA fragments, also in mixtures of species (Chen et al., 2014). The applications for the rapid and accurate identification and quantification of microbial pathogens and their toxins are now increasing, while the medical applications of all types of lab-on-a-chip and other nanotechnological devices are enormous (Duran and Marcato, 2013; Kalia and Parshad, 2014). The field of food fraud has been barely touched by these techniques so far. A few successful studies have been conducted on differentiation of several (mixtures of) fish species assisted by lab-on-a-chip analysis (Chen et al., 2014; Dooley et al., 2005). The development of this field is mainly dependant on the gathering and identification of unique (mitochondrial) DNA parts in order to discriminate within species and subspecies.

1.3.2.5 Cell Phone-Based Detection and Crowd Analysis

With over 6.5 billion cell phone subscribers worldwide and approximately 1.6 billion new devices being sold each year, cell phone technology is creating new opportunities in food fraud investigation. Many cell phone-based applications available now are targeting diabetes management, weight management, monitoring of blood pressure,

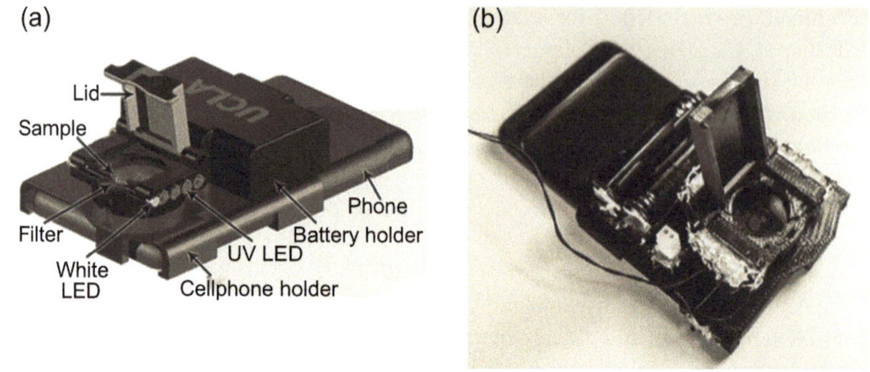

Figure 1.5 Schematic overview (a) and a picture (b) of the cell phone attachment for the detection of antirecombinant bovine somatotropin biomarkers in milk extracts.
Reproduced with permission from Ludwig et al. (2014).

and pulse rate. At present, analytical applications comprise mostly microscopy, spectroscopy, flow cytometry, lateral flow assays, colorimetric assays, immunoassays, and biochemical sensing (Vashist et al., 2014). Most applications require an additional cell phone attachment to perform the measurements. Among many applications in the food safety field, direct applications for the detection of food fraud are scarce. A microsphere fluorescent immunoassay cell phone application was developed to detect antirecombinant bovine somatotropin antibodies (illegal in the European Union) in milk (Ludwig et al., 2014). Using an optomechanical cell phone attachment (Figure 1.5), these antibodies could be detected in cleaned-up milk extracts by fluorescence and darkfield microimages evaluated by custom-developed Android application running on the same cell phone. Another promising application on fraud detection is the detection of porcine immunoglobulin G using the cell phone camera as the detection instrument (Gallegos et al., 2013). Alternatively, instead of using the cell phone camera as detection instrument, an external detection apparatus can be used, e.g., miniaturized spectroscopic MIR and NIR detectors. By means of wireless communication, a cell phone application can analyze the acquired data and communicate with an in-line spectroscopic library.

Simultaneously, with the development of cell phone food fraud detection applications, crowd analysis (alternatively, "crowdsourcing" or "citizen scientist") comes within the vicinity for usage as food fraud detection tool. This will disclose a completely new concept in the analysis of food fraud occurrence, as uploaded sample data can be linked to a cell phone's geographical position. This creates the opportunity to gather a significant amount of information beyond that collected in the existing systems.

1.4 Conclusion

Prediction of the next food fraud incident is tricky, as producers (logically) do not provide insight in malicious practices. The use of fraud incident databases often sketch a distorted image, as they are mostly filled with records that do not directly reflect the actual fraud incident. Still, by combining proper traceability management systems and cutting edge analysis techniques, an improvement in fraud deterrence can be achieved. Furthermore, as counterpart of the advanced food analysis technology, one finds highly creative, simplified solutions with potential for rapid onsite testing. It is expected that fraud deterrence will be "moving toward portable food analysis systems, eventually even on such an ordinary device as a mobile phone" (Nielen et al., 2014).

References

Al-Jowder, O., Defernez, M., Kemsley, E.K., Wilson, R.H., 1999. Mid-infrared spectroscopy and chemometrics for the authentication of meat products. Journal of Agricultural and Food Chemistry 47 (8), 3210–3218.

Araghipour, N., Colineau, J., Koot, A., Akkermans, W., Rojas, J.M.M., Beauchamp, J., Wisthaler, A., Märk, T.D., Downey, G., Guillou, C., Mannina, L., Ruth, S.V., 2008. Geographical origin classification of olive oils by PTR-MS. Food Chemistry 108 (1), 374–383.

Ballin, N.Z., 2010. Authentication of meat and meat products. Meat Science 86 (3), 577–587.

Cajka, T., Danhelova, H., Zachariasova, M., Riddellova, K., Hajslova, J., 2013. Application of direct analysis in real time ionization-mass spectrometry (DART-MS) in chicken meat metabolomics aiming at the retrospective control of feed fraud. Metabolomics 9 (3), 545–557.

Capuano, E., Heenan, S.P., de la Dura, A., van Ruth, S.M., 2013. Authentication of wild processed salmon by means of fatty acid and volatile profiles. Foods & Food Ingredients Journal of Japan 218 (1), 61–73.

Capuano, E., Rademaker, J., van den Bijgaart, H., van Ruth, S.M., 2014a. Verification of fresh grass feeding, pasture grazing and organic farming by FTIR spectroscopy analysis of bovine milk. Food Research International 60 (0), 59–65.

Capuano, E., van der Veer, G., Boerrigter-Eenling, R., Elgersma, A., Rademaker, J., Sterian, A., van Ruth, S.M., 2014b. Verification of fresh grass feeding, pasture grazing and organic farming by cows farm milk fatty acid profile. Food Chemistry 164 (0), 234–241.

Castellini, C., Mugnai, C., Dal Bosco, A., 2002. Effect of organic production system on broiler carcass and meat quality. Meat Science 60 (3), 219–225.

Chen, S.Y., Zhang, Y.X., Li, H., Wang, J.H., Chen, W.L., Zhou, Y., Zhou, S., 2014. Differentiation of fish species in Taiwan Strait by PCR-RFLP and lab-on-a-chip system. Food Control 44, 26–34.

Ciriello, R., Cataldi, T.R.I., Crispo, F., Guerrieri, A., 2015. Quantification of l-lysine in cheese by a novel amperometric biosensor. Food Chemistry 169 (0), 13–19.

Domingo, E., Tirelli, A.A., Nunes, C.A., Guerreiro, M.C., Pinto, S.M., 2014. Melamine detection in milk using vibrational spectroscopy and chemometrics analysis: a review. Food Research International 60 (0), 131–139.

Dooley, J.J., Sage, H.D., Clarke, M.-A.L., Brown, H.M., Garrett, S.D., 2005. Fish Species Identification Using PCR–RFLP Analysis and Lab-on-a-Chip Capillary Electrophoresis: Application to Detect White Fish Species in Food Products and an Interlaboratory Study. Journal of Agricultural and Food Chemistry 53 (9), 3348–3357.

Duran, N., Marcato, P.D., 2013. Nanobiotechnology perspectives. Role of nanotechnology in the food industry: a review. International Journal of Food Science and Technology 48 (6), 1127–1134.

Ehling, S., Cole, S., 2011. Analysis of organic acids in fruit juices by liquid chromatography-mass spectrometry: an enhanced tool for authenticity testing. Journal of Agricultural and Food Chemistry 59 (6), 2229–2234.

El Sheikha, A.F., Montet, D., 2014. How to determine the geographical origin of seafood? Critical Reviews in Food Science and Nutrition. http://dx.doi.org/10.1080/10408398.2012.745478.
Esslinger, S., Riedl, J., Fauhl-Hassek, C., 2014. Potential and limitations of non-targeted fingerprinting for authentication of food in official control. Food Research International 60, 189–204.
Everstine, K., Spink, J., Kennedy, S., 2013. Economically motivated adulteration (EMA) of food: common characteristics of EMA incidents. Journal of Food Protection 76 (4), 723–735.
European Commission, 2014. The Rapid Alert System for Food and Feed 2013 Annual Report. http://ec.europa.eu/food/safety/rasff/docs/rasff_annual_report_2013.pdf last visited November 14, 2014.
FERA, Food Intergrity, https://secure.fera.defra.gov.uk/foodintegrity/index.cfm, last visited November 14, 2014.
Food Fraud Database, USP, www.foodfraud.org, last visited October 28, 2014.
Food Fraud Think Tank, 2013. Food Fraud Overview, GFSI Annual Conference, Presented by Petra Wissenburg. Barcelona, Spain. March 2013. The Food Fraud Think Tank was created by the GFSI Board of Directors to provide guidance on addressing Food Fraud.
FoodSHIELD, NCFPD EMA Incidents Database, www.foodshield.org, last visited October 28, 2014.
Galle, S.A., Koot, A., Soukoulis, C., Cappellin, L., Biasioli, F., Alewijn, M., van Ruth, S.M., 2011. Typicality and geographical origin markers of protected origin cheese from The Netherlands revealed by PTR-MS. Journal of Agricultural and Food Chemistry 59 (6), 2554–2563.
Gallegos, D., Long, K.D., Yu, H., Clark, P.P., Lin, Y., George, S., Nath, P., Cunningham, B.T., 2013. Label-free biodetection using a smartphone. Lab on a Chip 13 (11), 2124–2132.
Heenan, S.P., van Ruth, S.M., Kilcast, D., 2013. 10-Emerging flavour analysis methods for food authentication. In: Instrumental Assessment of Food Sensory Quality. Woodhead Publishing, Cambridge, UK. pp. 284–312.
Hoorfar, J., Pruggerl, R., Butler, F., Jordan, K., Hoorfar, J., Jordan, K., Butler, F., Prugger, R., 2011. 18-Future trends in food chain integrity. In: Food Chain Integrity. Woodhead Publishing, Cambridge, UK. pp. 303–308.
Kalia, A., Parshad, V.R., 2014. Novel trends to revolutionize preservation and packaging of fruits/fruit products: microbiological and nanotechnological perspectives. Critical Reviews in Food Science and Nutrition 55 (2), 159–182.
Kasemsumran, S., Thanapase, W., Kiatsoonthon, A., 2007. Feasibility of near-infrared spectroscopy to detect and to quantify adulterants in cow milk. Analytical Sciences 23 (7), 907–910.
Kennedy, S., 2012. Emerging global food system risks and potential solutions. In: Improving Import Food Safety. John Wiley & Sons, Inc, Hoboken, NJ, USA. pp. 1–20.
Laghi, L., Picone, G., Capozzi, F., 2014. Nuclear magnetic resonance for foodomics beyond food analysis. Trac-Trends in Analytical Chemistry 59, 93–102.
Lohne, J.J., Andersen, W.C., Clark, S.B., Turnipseed, S.B., Madson, M.R., 2012. Laser diode thermal desorption mass spectrometry for the analysis of quinolone antibiotic residues in aquacultured seafood. Rapid Communications in Mass Spectrometry 26 (24), 2854–2864.
Ludwig, S.K.J., Zhu, H., Phillips, S., Shiledar, A., Feng, S., Tseng, D., van Ginkel, L.A., Nielen, M.W.F., Ozcan, A., 2014. Cellphone-based detection platform for rbST biomarker analysis in milk extracts using a microsphere fluorescence immunoassay. Analytical and Bioanalytical Chemistry 406 (27), 6857–6866.
Maillot, E.M., 2010. Terrorism and the food chain. In: Koukouliou, V., Ujevic, M., Premstaller, O. (Eds.), NATO Advanced Research Workshop on Threats to Food and Water Chain Infrastructure. Springer, Berlin, Germany.

Mohoric, A., Vergeldt, F., Gerkema, E., van Dalen, G., van den Doel, L.R., van Vliet, L.J., Van As, H., Van Duynhoven, J., 2009. The effect of rice kernel microstructure on cooking behaviour: a combined mu-CT and MRI study. Food Chemistry 115 (4), 1491–1499.

Monakhova, Y.B., Kuballa, T., Lachenmeier, D.W., 2013. Chemometric methods in NMR spectroscopic analysis of food products. Journal of Analytical Chemistry 68 (9), 755–766.

Moore, J.C., Spink, J., Lipp, M., 2012. Development and application of a database of food ingredient fraud and economically motivated adulteration from 1980 to 2010. Journal of Food Science 77 (4), R118–R126.

Nielen, M.W.F., Hajslova, J., Krska, R., 2014. Advanced food analysis. Analytical and Bioanalytical Chemistry 406 (27), 6765–6766.

Nielen, M.W.F., Hooijerink, H., Zomer, P., Mol, J.G.J., 2011. Desorption electrospray ionization mass spectrometry in the analysis of chemical food contaminants in food. Trac-Trends in Analytical Chemistry 30 (2), 165–180.

Northen, J.R., 2000. Quality attributes and quality cues effective communication in the UK meat supply chain. British Food Journal 102 (3), 230–245.

Özdestan, Ö., van Ruth, S.M., Alewijn, M., Koot, A., Romano, A., Cappellin, L., Biasioli, F., 2013. Differentiation of specialty coffees by proton transfer reaction-mass spectrometry. Food Research International 53 (1), 433–439.

Papurello, D., Lanzini, A., Leone, P., Santarelli, M., Silvestri, S., 2014. Biogas from the organic fraction of municipal solid waste: Dealing with contaminants for a solid oxide fuel cell energy generator. Waste Management 34 (11), 2047–2056 (New York, N.Y.).

Pearce, K.L., Rosenvold, K., Andersen, H.J., Hopkins, D.L., 2011. Water distribution and mobility in meat during the conversion of muscle to meat and ageing and the impacts on fresh meat quality attributes - a review. Meat Science 89 (2), 111–124.

Prins, T.W., van Dijk, J.P., Angeline Van Hoef, A.M., Voorhuijzen, M.M., Broeders, S., Trapmann, S., Seyfarth, R., Pardigol, A., Schoen, C.D., Aarts, H.J.M., Kok, E.J., 2010. Towards a multiplex cereal traceability tool using padlock probe ligation on genomic DNA. Food Chemistry 118 (4), 966–973.

Querci, M., Van den Bulcke, M., Žel, J., Van den Eede, G., Broll, H., 2010. New approaches in GMO detection 396 (6), 1991–2002.

Rahmatallah, S., Li, Y., Seton, H.C., Gregory, J.S., Aspden, R.M., 2006. Measurement of relaxation times in foodstuffs using a one-sided portable magnetic resonance probe. European Food Research and Technology 222 (3-4), 298–301.

Raspor, P., 2005. Bio-markers: traceability in food safety issues. Acta Biochimica Polonica 52 (3), 659–664.

Rapid Alert System for Food and Feed, http://ec.europa.eu/food/safety/rasff/index_en.htm, last visited November 14, 2014.

Rodgers, W., 2012. Fraud and internal control. In: Rodgers, W. (Ed.), Biometric and Auditing Issues Addressed in a Throughput Model. Information Age Publishing Inc., Scottsdale, AZ, USA. pp. 261–318.

Ruiz-Samblás, C., Arrebola-Pascual, C., Tres, A., van Ruth, S., Cuadros-Rodríguez, L., 2013. Authentication of geographical origin of palm oil by chromatographic fingerprinting of triacylglycerols and partial least square-discriminant analysis. Talanta 116 (0), 788–793.

Ruiz-Samblás, C., Tres, A., Koot, A., van Ruth, S.M., González-Casado, A., Cuadros-Rodríguez, L., 2012. Proton transfer reaction-mass spectrometry volatile organic compound fingerprinting for monovarietal extra virgin olive oil identification. Food Chemistry 134 (1), 589–596.

Sorak, D., Herberholz, L., Iwascek, S., Altinpinar, S., Pfeifer, F., Siesler, H.W., 2011. New developments and applications of handheld raman, mid-infrared, and near-infrared spectrometers. Applied Spectroscopy Reviews 47 (2), 83–115.

Strayer, S.E., Everstine, K., Kennedy, S., 2014. Economically motivated adulteration of honey: quality control vulnerabilities in the international Money Market. Food Protection Trends 34 (1), 8–14.
Teixeira dos Santos, C.A., Lopo, M., Pascoa, R.N.M.J., Lopes, J.A., 2013. A review on the applications of portable near-infrared spectrometers in the agro-food industry. Applied Spectroscopy 67 (11), 1215–1233.
The British Standard Institution, Guide to protecting and defending food and drink from deliberate attack, http://shop.bsigroup.com/upload/Shop/Download/PAS/PAS%2096%20 (2014).pdf, last visited November 14, 2014.
Tres, A., Heenan, S.P., van Ruth, S., 2014. Authentication of dried distilled grain with solubles (DDGS) by fatty acid and volatile profiling. LWT - Food Science and Technology 59 (1), 215–221.
Uddin, M., Okazaki, E., Turza, S., Yumiko, Y., Tanaka, M., Fukuda, Y., 2005. Non-destructive visible/NIR spectroscopy for differentiation of fresh and frozen-thawed fish. Journal of Food Science 70 (8), c506–c510.
US Government Accountability Office, 2011. Better Coordination Could Enhance Efforts to Address Economic Adulteration and Protect the Public Health. http://www.gao.gov/new.items/d1246.pdf last visited November 14, 2014.
Van As, H., van Duynhoven, J., 2013. MRI of plants and foods. Journal of Magnetic Resonance 229, 25–34.
van der Spiegel, M., van der Fels-Klerx, H.J., Sterrenburg, P., van Ruth, S.M., Scholtens-Toma, I.M.J., Kok, E.J., 2012. Halal assurance in food supply chains: verification of halal certificates using audits and laboratory analysis. Trends in Food Science & Technology 27 (2), 109–119.
van Leeuwen, K.A., Prenzler, P.D., Ryan, D., Camin, F., 2014. Gas chromatography-combustion-isotope ratio mass spectrometry for traceability and authenticity in foods and beverages. Comprehensive Reviews in Food Science and Food Safety 13 (5), 814–837.
van Ruth, S., Alewijn, M., Rogers, K., Newton-Smith, E., Tena, N., Bollen, M., Koot, A., 2011. Authentication of organic and conventional eggs by carotenoid profiling. Food Chemistry 126 (3), 1299–1305.
van Ruth, S.M., Huisman, W., Mei 2014. Kwetsbaarheid voor voedselfraude in de vleessector. In: Wetenschappelijk Onderzoek- en Documentatiecentrum. Justitiële verkenningen-Voedselcriminaliteit, Jaargang 40, Boom Lemma Uitgevers, The Hague, The Netherlands. pp. 28–53.
Vashist, S., Mudanyali, O., Schneider, E.M., Zengerle, R., Ozcan, A., 2014. Cellphone-based devices for bioanalytical sciences. Analytical and Bioanalytical Chemistry 406 (14), 3263–3277.
Voda, A., Homan, N., Witek, M., Duijster, A., van Dalen, G., van der Sman, R., Nijsse, J., van Vliet, L., Van As, H., van Duynhoven, J., 2012. The impact of freeze-drying on microstructure and rehydration properties of carrot. Food Research International 49 (2), 687–693.
Wang, X., Wang, S.J., Cai, Z.W., 2013. The latest developments and applications of mass spectrometry in food-safety and quality analysis. Trac-Trends in Analytical Chemistry 52, 170–185.
Zając, A., Hanuza, J., Dymińska, L., 2014. Raman spectroscopy in determination of horse meat content in the mixture with other meats. Food Chemistry 156 (0), 333–338.

The Role of Information Technology Developments in Food Supply Chain Integration and Monitoring

2

Nitaigour Mahalik[1], *Kiseon Kim*[2]
[1]Department of Industrial Technology, Jordan College of Agricultural Sciences and Technology, California State University, Fresno, CA, USA; [2]Communication and Sensor Network Lab, School of Information and Communication, Gwangju Institute of Science and technology, Gwangju, Republic of Korea

2.1 Introduction

Information technology (IT) has extended its importance in several other sectors such as transportation, online sales and marketing, weather forecasting, online education, agriculture, and so on although it has been in use in the military sector for a long time. The role of IT in the food supply chain is seen to be vast in developed countries. Developing countries are adopting the principles and methods at a slower pace. Soon its importance will be manifested by the suppliers and consumers. Food safety is an important criterion in the entire supply system and in this respect IT helps to a great extent. IT is simply defined as use of computing platforms and internet-based communication devices and protocols to compute, manipulate, store and retrieve, and communicate (transmit and receive) data and information mostly for business and community services. IT is a networking system. Wireless-based networking systems are seen to be popular. Many business and community service systems have IT departments for managing the computing servers, the network devices, and other technical hardware and software applications of their businesses. The changing role of IT in food and beverage logistics management, beverage network optimization using intelligent agent technology has been presented by Mangina and Vlachos (2005).

Thus, IT is not a stand-alone system as its existence in the food supply chain not only combines with multidisciplinary design but also policy interventions and production practices that help with sustainability. Sustainability is important because before a consumer chooses a food item, the producer should be ready or willing to produce. This means that the supply–demand loop has reversed to a demand–supply scenario. Food is becoming a choice for the consumers and innovations regulate the choices. IT helps to facilitate not only integrating the demand–supply loop instantly but also in providing comprehensive information to preserve food safety standards and requirement indices.

This chapter focuses on the role of IT developments in food supply chain integration and monitoring. The chapter has been divided into five sections. In the first section we have reviewed components of IT. In the second part supply chain

Innovation and Future Trends in Food Manufacturing and Supply Chain Technologies
http://dx.doi.org/10.1016/B978-1-78242-447-5.00002-2

management is presented. In the next sections, traceability followed by food processing and packaging are presented. The chapter concludes with the role of IT in processing and packaging. The chapter also extends its scope by presenting a section on application of nanotechnology in food processing and packaging. The reason for introducing nanotechnology is that, a sensor detects a change in color of the coating material in the nanolayer that is used in the inner side of packaging for monitoring the spoilage of food (e.g., meat or vegetables), and it triggers the central monitoring and expert system either wirelessly or communication network.

2.2 IT Components in General

IT is the use of computer systems to process, store, manipulate, retrieve, compute, and communicate electronic data. Any system that uses digital means to manage and output data is an IT system. IT is used mainly for commercial, industrial, and business purposes. It is today applied to a variety of fields, from agriculture to commercial data processing to cloud computing. IT systems can be either specialized for one task, such as data-mining systems, or can be multipurpose computing platforms, such as mobile devices. For example, a data-mining system processes large sets of data to discover patterns, stores those patterns in memory, and communicates the processed data to other computing platforms or to researchers. Figure 2.1 illustrates a broader view of IT.

As stated above, IT consists of computing platforms and internet-based communication devices and protocols to compute, manipulate, store and retrieve, and communicate (transmit and receive) data and information mostly for business and community services. IT is thus a networking system. The two important parts of IT are computing technology and communication technology. Traditionally, computing technology

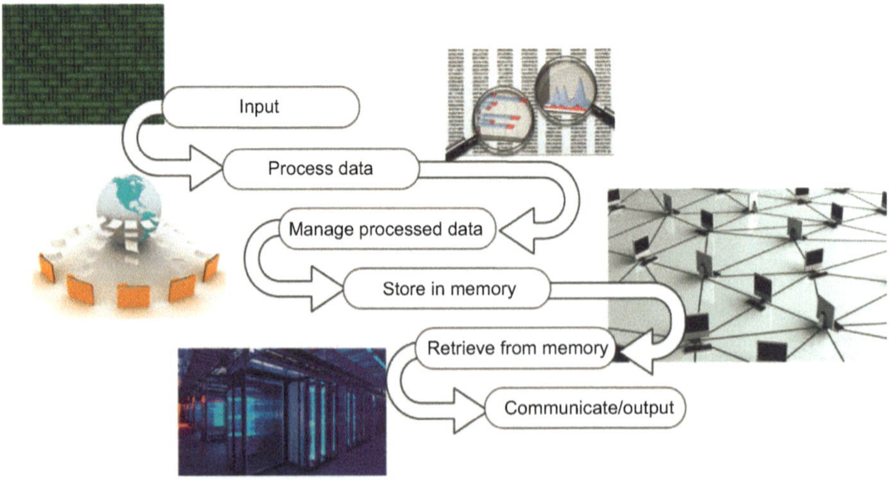

Figure 2.1 Information technology: A broader view.

consisted of computers and mainframe servers, but recently because of the advent of smaller mobile devices such as the iPad, iPhone, wearable devices, and handheld devices, computing platforms have shifted more toward a cloud-computing environment with mainframe servers as used for back-end data storage and security purposes.

The communication technology is now wireless-based. There have been significant developments in wireless technology as wireless communications systems enable people to be better connected and reachable over the space–time limitation. Because frequency resources are limited, full utilization inevitably faces interference problems. The advanced communication systems such as Long Term Evolution (LTE) Advanced, Wi-Fi, and Mobile WiMAX technologies for cellular networks sometimes suffer from intercell interference. Robust signal design, along with stringent interference aliment and mitigation, coordinate resource allocation and scheduling are the essential technologies for the cellular systems. Moreover, cooperative communication is one of the issues in communication because it provides advantages such as extended coverage as well as enhanced data rate with better Signal to Noise Ratio (SNR) beyond hardware limitations in size, battery, and costs. The cutting-edge communication technologies are now more complex than traditional cellular networks for which the devices that are developed for wireless systems need to meet the quality of services. Therefore, reliable wireless communication system design, development, and their realistic performance evaluation are important (Figure 2.2).

2.3 Supply Chain Management

A food product takes various steps from its farm to its final destination to the vendor. These intermediate steps, which may consist of inspection for pathogens, treatment for long-term storage, processing of perishables, packaging for vendor display, shipping to vendors, and finally regulatory checks, comprise the food supply chain (Figure 2.3). The food supply chain is a complex interconnection of human services and automated processes. Food supply chain management therefore is the coordination of a supply chain logistics and characteristics in order to better meet the customers' demand and improve efficiency of the product delivery while maintaining safety standards (Linton, 2015).

Proper management of the food supply chain entails the following: effective communication between different components of the supply chain; scalability of the chain to accommodate improvements in technology and changes in standards and protocols; and logistics management to improve chain efficiency and ensure convergent supply-and-demand ratios (Anderson et al., 1997).

Because the food supply chain involves human operators such as farm workers, plant managers, machine operators, delivery personnel, and safety inspectors as well as automated processes such as processing, packaging, tagging, and monitoring, it is important for proper communication between the various components to ensure smooth functioning of the supply chain as well as rapid analysis of emergency situations. Equal information sharing allows all members of the chain access to the same information. Cloud-based technology with database implementation can synchronize

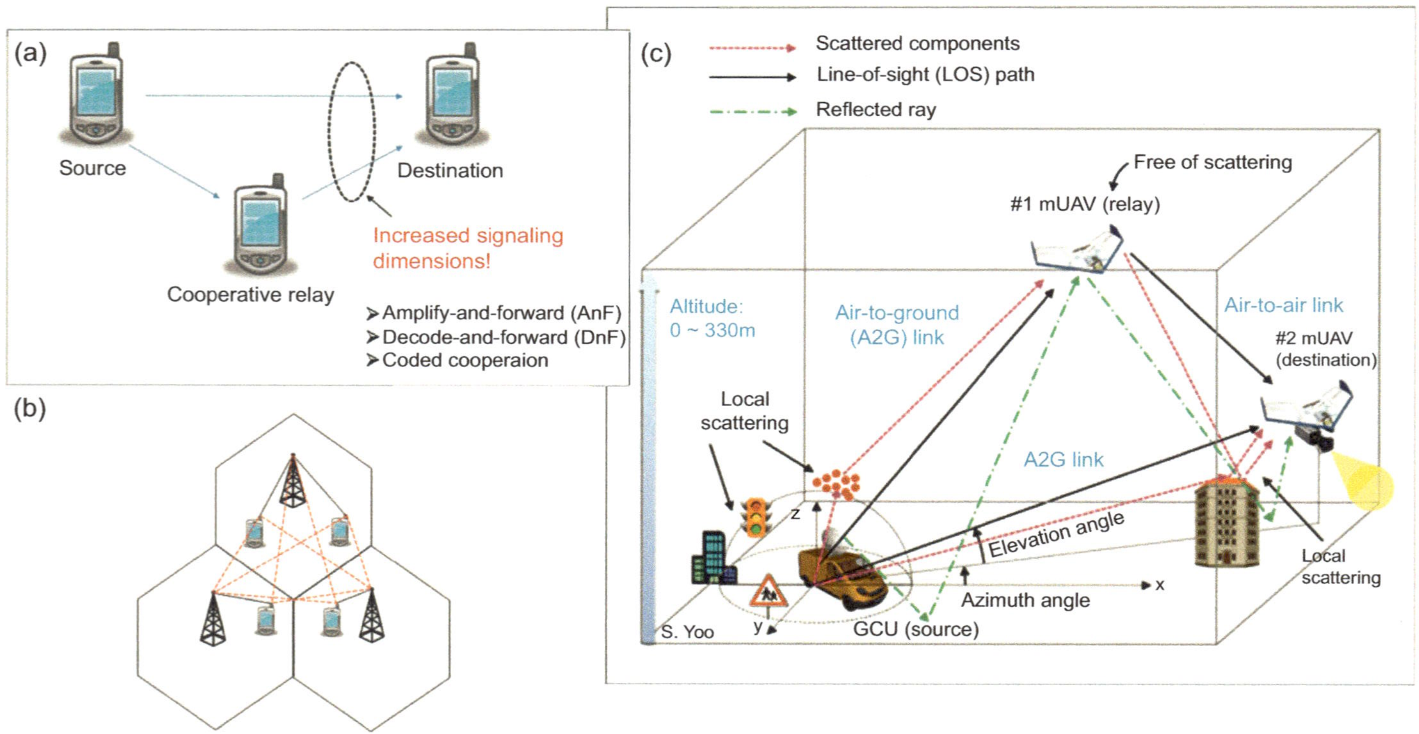

Figure 2.2 Components of IT (a) cooperative communications (relay-based communication); (b) antiinterference communications; (c) integration of channel modeling for mobile wireless communications.
Courtesy of the Communication and Sensor Network Laboratory, Gwangju Institute of Sciences and Technology, South Korea.

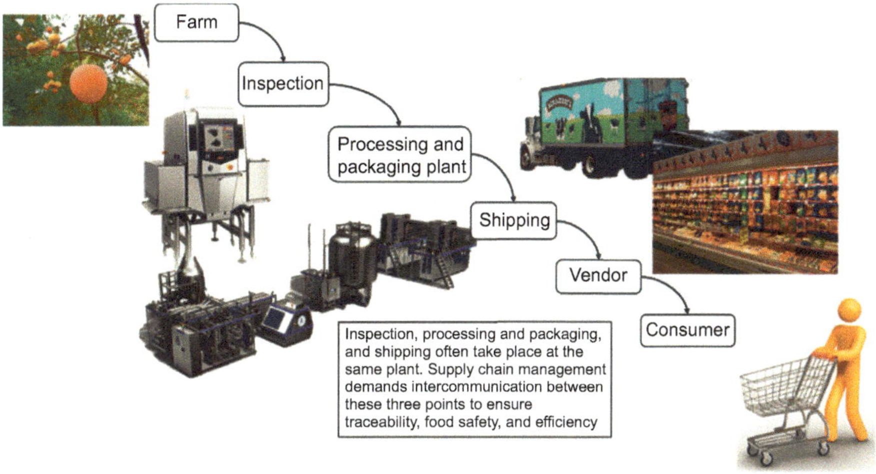

Figure 2.3 The food supply chain from farm to consumer with processing and packaging. Images courtesy of (1) Ishida Europe Ltd., (2) Tetra Pak International SA, (3) Ben & Jerry's Inc., (4) Food Lion Store LLC, and (5) Kraft Kennedy Inc.

data from all parts of the chain so that the same data is accessible to each member. Further, equal information access can pinpoint, in the case of chain emergencies (technical problems, equipment malfunctions, and software upgrades are a few examples) parts of the supply chain that need shunting or support.

Scalability is an important component of all supply chains, especially the food supply chain. With fluctuations in food supply at the farm due to changes in weather, nutritional levels, soil quality and characteristics (temperature, moisture content, acidity, etc.), and labor costs, it is important to manage a food supply change that can handle changes in supply without significant adverse effects to production and efficiency. This is particularly important for horizontally integrated producers that do not have complete control over their supply chain. External influences including market, environmental, and regulatory effects can significantly change the supply chain. Scalability can be affected by various means (CSCMP, 2015). Organization of chain transport mechanisms in order to effectively handle various lot sizes can positively impact the movement of food through the supply chain. Assembly-line analogues and chain parallelism ensures all source materials (harvested food) can progress through to treatment facilities and packaging centers without holding up any one point. It is also important to keep in mind safety standards. A scalable supply chain can also accommodate the increased volume of food products without negatively impacting safety standards.

Communication and scalability are implicitly tied to logistics management. It is necessary to use heuristics and feedback loops in a supply chain to ensure proper usage of the food supply based on consumer demand. For example, an increase in demand requires more food products through the chain, and thus more food product from the source. This can be accomplished by purchasing food products from multiple sources. Scalability comes into effect so that the increased number of raw food can be processed and packaged efficiently without backlogs in the chain.

Communication is important in ensuring that subsequent control points in the chain are aware of and have access to data regarding the increased number of food products. These three components—communication, scalability, and logistics management—are integral aspects of any supply chain. In the food supply chain, where external political, environmental, and regulatory effects have a tremendous impact on the products, supply chain management is especially important in ensuring timely delivery of food products while ensuring safety and enforcing standards.

2.4 Traceability

In the food supply chain, traceability is becoming increasingly important to ensure proper documentation of the processed food for financial, regulatory, and safety purposes—financial purposes may involve using traceability data to improve efficiency and reduce production congestion. Traceability itself is defined as the capability to evaluate a product from inception to delivery in the production process and verify its identity with the use of documentation, reports, codes, or other tracking media (Levinson, 2009). Its broad definition reflects both the facts that traceability can be achieved via various approaches and that complete traceability involving all possible information about a food product is impossible to maintain (Golan et al., 2004). Good traceability systems, however, are possible and already in effect.

In United States, the Food and Drugs Administration (FDA) reported on two tracing pilot projects that were conducted by the Institute of Food Technologists: one for processed foods and one for produce. They were required under section 204 of the FDA Food Safety Modernization Act to establish recordkeeping for high-risk foods. Note that the projects were strategized to demonstrate methods and procedure for rapid as well as effective tracking and tracing of food. This includes types of data meaningful for traceability. For example, they have defined objectives and protocols, such as raw materials, pest control tracking, facilities tracking, wage tracking, and/or delivery tracking. They link disparate aspects of the supply chain, from the harvest location to the initial processing facility to shipping center to finally the vendor (Figure 2.4). In fact, the FDA has made both trace-back and trace forward requirements in food supply chain. Trace-back refers to determining the source of a food product by using traceability documentation from the vendor or facilities. Conversely, trace forward refers to the ability to determine the final destination of a food product at the farm source based on preexisting protocols and orders. The former is necessary in the rare but serious case of contamination of food products with pathogens such as *Salmonella* or *Escherichia coli* or other industrial contaminants and carcinogens.

Traceability depth can be either multilevel or shallow. Multilevel traceability is associated with tracking food from the farm to vendor. It is common with vertically integrated companies, which own all aspects of their supply chain (Golan et al., 2004). Multilevel traceability is also available through third-party service companies that can implement the correct standards and ensure scalability of the implemented systems. While the FDA does not require complete, lot-specific traceability for all food products, shallow traceability is not recommended as it hampers tracing during "food emergencies" (Levinson, 2009).

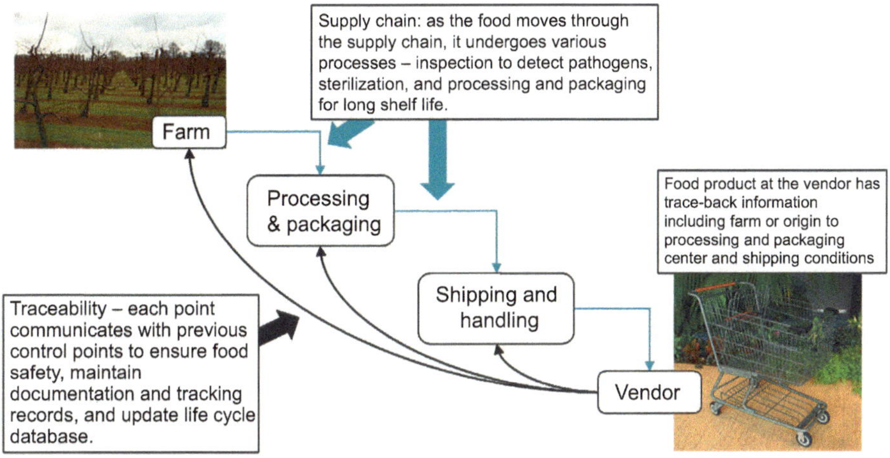

Figure 2.4 Traceability in food supply chain.
Images courtesy of (1) Loddington Farm, and (2) National Cart Co.

Traceability can be affected with tagging or codes, electronic records, or reports (Setboonsarng et al., 2009). Tagging food requires either Radio Frequency Identification (RFID) or barcodes at the farm source such that these can be accessed at any point in the supply chain. The tags/codes can be used in conjunction with shared electronic records and databases to ensure synchronicity. The breadth, depth, and precision (Golan et al., 2004) of a traceability system are especially important and can be fine-tuned with these subcomponents. The breadth is the different types of data recorded—low breadth systems record a specific statistic, such as preservative content, while high breadth systems track statistics encompassing treatment facilities, pathogenic content, delivery methods, preservative content, ingredients, and worker wages. Depth, as has been mentioned, is necessary for many food products under FDA regulations. It can extend to either the last point in the supply chain or to the beginning at the farm. Finally, precision is the amount of information recorded per statistic—either a high-level overview or an in-depth report is recorded based on the standard's requirements.

To be effective, traceability systems must be used in conjunction with other safety and quality control systems. Traceability gives the information necessary to improve efficiency in the supply chain. The data it provides can be coupled to several systems: freeze control systems for storage and transport ensure safety; real-time processing and packaging monitoring improves production efficacy; and intelligent routing algorithms ensure efficient delivery to vendors or other facilities downstream of the supply chain. Traceability protocols and standards have markedly improved food safety and production efficiency since their inception. These standards are developed by the World Health Organization and the International Standards Organization and documented by both the FDA for most food products and the United States Department of Agriculture for meat and poultry (Flynn, 2013). Where these standards are not in effect, third-party organizations ensure traceability systems are used by food production companies.

2.5 Food Processing and Packaging

The application of IT is important in food processing and packaging. In the next section details are presented. However, an introduction to food processing and packaging is needed. In the sequel, a brief introduction is given. Bearing in mind that in advanced countries about 90% of the foods including specialty crops and vegetables are processed and packaged using machineries and automation technology, software and hardware standardization, which is a part of application of IT, play major role for open system implementation (Cheruvu et al., 2008; Mahalik, 2014) not only in the packaging domain, but also for supply chain management. For example, traceability can also include the identification of the packaging machine. Further, machine failures cause decrease in productivity. In this respect, machine suppliers constantly monitor the machines and equipments used in the processing and packaging applications. Online monitoring is achieved via IT backbones (Figure 2.5).

The path of food from the farm to the vendor and thus to the consumer involves processing and packaging, arguably the foundation of the modern food industry. Food processing transforms raw materials (grain, barley, poultry products, etc.) into consumer-friendly products (oatmeal, cereal, cake mix, etc.) (Mathavi et al., 2013). Processing is a wide umbrella that not only includes preservation, cooking, liquefaction, and freezing but also various operations and transformations of food raw materials to final products along the automated lines. Food packaging provides protection for food from detrimental environmental elements. Advanced food packaging ("smart packaging") can also perform monitoring and environment management to ensure the best possible conditions for the food product.

Food processing is effected either in a minimal or maximal capacity (JHBSPH, 2015). Minimally processed food products are very similar to their unprocessed forms. Most fruit, vegetable, and meat products qualify as minimally processed food—in these cases, processing is conducted to reduce pathogenic contaminants (washing and heating) or to prolong shelf-life (freezing and adding preservatives). Maximally (or highly) processed food products differ significantly from their raw materials. Examples include portable, ready-made foods such as energy bars, cereals, and chips. Processing is composed of several machineries and mechanisms (Schmidt, 2012). Some of the more ubiquitous processing machines include

- Decontamination process: this process is instrumental in removing and/or limiting pathogenic contaminants in food. Decontamination can be effected on the food surface with basic washing. This requires analysis of the food product to determine threshold temperature to retain the characteristics of the food, while removing as much contamination as possible.
- Inspection machines: inspection machines are a second step in food processing. These machines are widely varied in their capabilities; they determine food health, grade classification, quality, spoilage, and infection with a variety of sensors including cameras, X-rays, and sorters by weight, size, and color. Image processing plays an important part in identifying and classifying food.
- Conveyor systems: food processing and packaging is conducted on an assembly line. Thus, conveyor systems are an efficient method to sequentially analyze and perform tasks

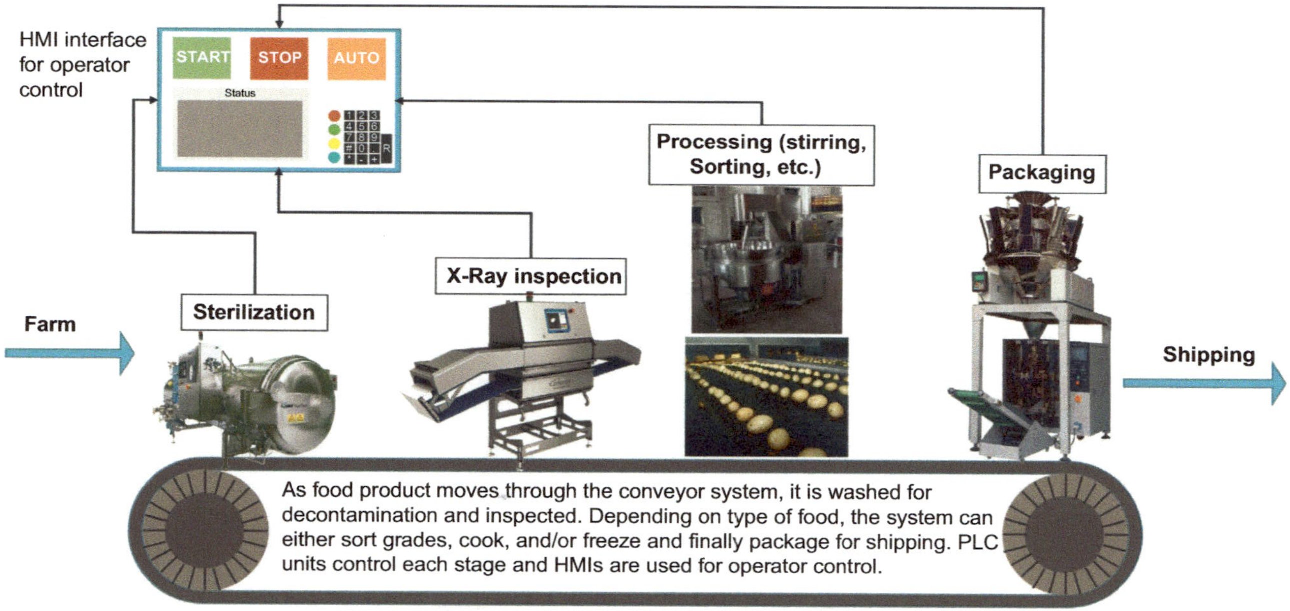

Figure 2.5 Automation of food processing and packaging on conveyor line.
Images courtesy (1) JBT FoodTech, (2) Marel Food Systems, (3) HongDe Tong Technology, (4) Compac Sorting Equipment, and (5) XunJie Packaging Machinery Co. Ltd.

on various food materials. Conveyor systems also help sort food as it is being processed. Accessory machines such as pickers, splicers, and ejectors separate the food. Separation is conducted as per the assembly line requirements, i.e., based on grade, size, color, health, or a combination of these and other characteristics.

The processing and packaging system is designed to yield products that are absent of pathogens. There are incubation tests done to check that there are no spoilage organisms that could cause premature food spoilage. Processing is extensively automated to reduce costs and improve efficiency. Processing is followed by packaging. The role of packaging in food production is to protect food products against pathogens, environmental effects, and spoilage (Marsh and Bugusu, 2007). Conventional packaging's protection against these effects can involve various technologies including controlled external and internal environments, smart monitoring, and regular inspections (Mahalik, 2009). Controlled environments remove significant amounts of guesswork involved in food longevity by ensuring the best possible temperature, air content, and pressure for the food product. Smart monitoring involves real-time monitoring and adjustment of environmental properties to ensure that the food product's surroundings match the current state of the food, for example, riper fruits may require different surroundings than less ripe fruits. Inspections, whether manual or automated, satisfy federal regulations and are a means of hardware monitoring and diagnosis. This inspection method ensures that smart monitoring and environment controllers are functioning property. Sanitation is done to all food processing equipment before the food products are passed through. Such machine sanitation is achieved at regular intervals. The sanitation of the entire packaged product is carried out once they are packaged.

2.6 Information Technology in Food Processing and Packaging

Of the various technologies used in food processing, barcoding and RFID are considered integral. Barcoding involves tagging food with barcodes. As the name represents, a barcode has black colored vertical lines with different widths separated from each and constitute a holistic code. Figure 2.6(a) shows a barcode. It is an optical machine-readable graphical representation of numbers or character relating to a product or object. An optical reader designed to read barcode uses a semiconductor-based photosensor to read the code. Basically, the reader converts the barcode into an equivalent electrical signal when project onto the code. The role of the reader is to scan and measure the relative widths of the black colored bars and the spaces between them. It than translates the patterns back into numbers or characters. Depending upon the pattern, several configurations are possible. Thus, the packets can be uniquely numbered or barcoded for its identification. The barcode reader sends the converted electrical signals to the computer which displays the code in the form of numbers on the computer screen. Under the bar coding scheme, a tracking machine requires direct line of sight in order to identify the tag. As such, RFID tags, which do not require line of sight

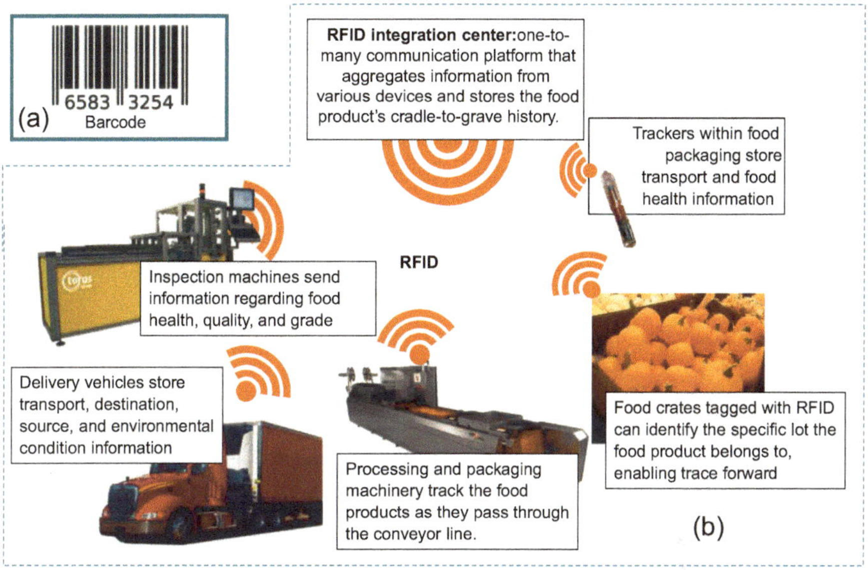

Figure 2.6 One-to-many communication scheme in RFID networks allows a single integration center to collate data from multiple ground units. (a) A sample barcode, (b) RFID integrated devices and systems.
Images courtesy of (1) Torus Group, (2) Carrier Transicold, and (3) VC999 Inc.

as they communicate through radio waves, will be the next generation tagging system. They are however becoming popular for large packaging or bulk containers tracking rather than single food unit and ubiquitous in the packaging industry. They are used for not only identification but also for sorting, tracking, traceability, and supply chain management and monitoring (Weis, 2007).

RFID-based identification is more cost-effective than barcoding or manual tagging. There are two types of RFID system. Mostly, passive RFID systems are used in packaging. The passive types draw power from radio waves broadcast by a tracker or scanner they enable one-to-many tracking. As such, a single scanner with a linked database can manage multiple food products in the same packaging line. Hence RFID can function effectively post packaging as well (Figure 2.6(b)).

RFID tag miniaturization has led to small sizes. Today, microsensors have been incorporated into these RFID tags to provide improved monitoring capability (Faudzi et al., 2013). As such, passive RFID tags can provide on-the-spot information about food environment, composition, temperature, and other relevant information. This can be especially important during long-term storage and transportation. Active RFID tags can store these data and transfer data to a linked database as and when necessary.

The presence of RFIDs from the cradle-to-grave in food allows both trace-back and trace forward (Infosys, 2015). In trace-back, an infected food product is traced back through the supply chain, using data collected from the RFID such as plant location, delivery vehicle, and lot number, to the farm source. In more robust RFID implementations, the specific batch that contained the contaminated food can also be determined. Trace-forward can then be used to determine the final locations of all food products within the same batch using the same data, i.e., lot number, delivery vehicle, processing and packaging plant location, and final vendor. As such, the traceability efficacy increases.

RFIDs have become an integral component of supply chain management. Communication within the supply chain is effected by the RFID tags themselves (Infosys, 2015). Due to the one-to-many communication scheme, it is more efficient and cost-effective to keep track of not only food products and their characteristics (temperature, content, etc.) and location but also other plant machinery such as delivery vehicles, sorting, packaging, and picker machines, conveyors, and sterilization systems. Communication among plant machinery involves diagnostics, self-repair, and alarm monitoring. Management of plant machinery is thus self-effected.

Supply chain scalability is considerably more straightforward in more robust automated and intelligent systems. In RFID systems, there can be either device-level scalability or central processing scalability (Kuhno, 2015). Device level pertains to intercommunication between plant machinery without a central processor. Advantages of this scheme are faster decisions and higher throughput. New devices are added by communication with other devices within spatial communication range. In the central processing scheme, all devices communicate with a command-and-control unit that organizes and manages each device, including device intercommunication. New devices are directly added to the central processor. The former schema is more dynamic—new devices are registered only with related devices. However, management is difficult as each spatial zone contains independent communication networks. The latter schema, central processing, may be slower; however, chain management and monitoring is easier due to the more modular approach.

The prevalence of RFID integration in the food processing and packaging industry provides a more robust and communication, management, and monitoring platform for both food products and the processing and packaging machinery. Consequently, food safety improvements are expected due to better diagnostics, traceability, and intercommunication.

2.7 Nanotechnology in Food Processing and Packaging

In this section the application of nanotechnology in food is presented. The current application of nanotechnology in food and its link to IT may not be relevant for discussion, but it has been predicted that there would be a link in future. Although IT part of the connection has not yet been interfaced, the current research has moved a step forward to establish the interface. In that respect this section presents the scenario at the rudimentary level. For example, as mentioned in the introduction section that a sensor can detect the change in color of the coating material in the form of nanolayer that are used in the inner side of packaging for monitoring the spoilage of food (e.g., meat or vegetables) and can trigger the central monitoring and expert system either wirelessly or communication network. In fact, this author has demonstrated this experiment in a lab-room project. A research team at University of California, Riverside, USA (FoodQuality.com, 2013) in their collaboration work with eCrowd Company mentions that a nanosensor can detect airborne toxins in the food supply chain is to be further refined in a bid for commercial launch (FoodQuality.com, 2013). Moreover, futuristic RFID tags can incorporate nanosensors in order to gather information such as food temperature, acidity, salinity, and limited pathogen content. Note that nanosensors have already been developed, however, their roles in food processing and packaging have not extensive, although some applications are found (http://www.iopp.org/files/public/LamDerekSanJose.pdf). Nanotechnology itself is the utilization of materials at the nanoscale (1 nm is 10^{-9} m). It has been applied in various capacities, from monitoring and pathogen detection to packaging and storage. It has already been integrated into existing technologies including RFID tags, smart packaging, and antimicrobial surfactants (Joseph and Morisson, 2006).

The nanosensors may either be located inside the food product or on the surface. Since their inception, food processing and packaging have been improved and explored in the macroscale sense. Today, the micro and nanoscale approaches have entered the picture. Nanotechnology in food processing has the potential to transform the industry. Nanosensors may be biological assays, probes, or nanostructures coated with recognition particles. Biological assays can be used for generalized detection of pathogens or other organic particles. Probes coated with palladium have been used as pH sensors—the palladium forms conductive pathways in the presence of hydrogen protons. Finally, nanostructures can be used for more specific detection of organic molecules such as proteins or deoxynucleic acid.

Smart packaging performs environment monitoring to manage and maintain the optimal environment for food products. Nanotechnology integrated (Figure 2.7) into both the passive as well as active aspects of smart packaging can be especially useful (Chaudhry, 2012). Passive aspects include packaging durability, texture, flexibility, moisture retention, and strong barrier properties. Durability and flexibility improve

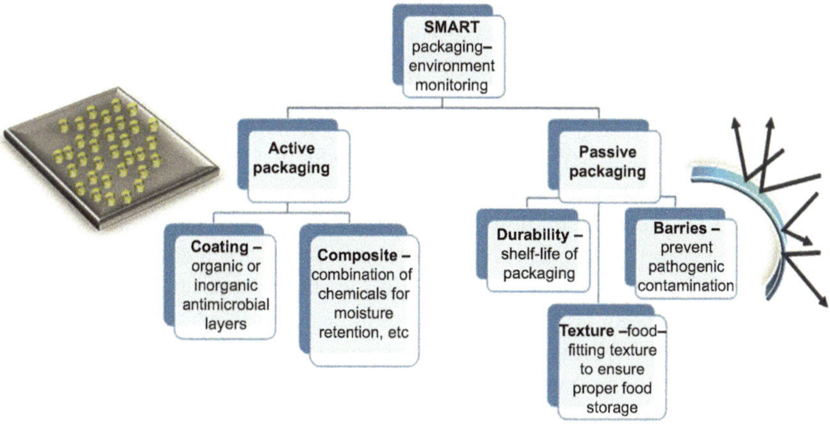

Figure 2.7 The SMART packaging hierarchy encompassing active and passive packaging.

package shelf-life. Further, they ensure the food product can survive in a variety of environments—improvements in durability have led to flameproof, waterproof, and freezeproof packaging. Texture manipulation with nanotechnology can provide a more suitable environment for food products. Moisture retention and strong barrier properties both ensure a proper temperature and environment and prevent foreign particulates, pathogens, and inorganic molecules from contaminating the packaging environment. Active aspects include antimicrobial surfactants, coatings, and composites. Antimicrobial surfactants involve nanoparticles dispersed in a surfactant medium. Silver, a potent antimicrobial, has been used with copper to prevent *E. coli* and *Staphylococcus aureus* infections and inhibit their growth. Nanocoatings such as silica with triglycerols prevent moisture due to hydrophobic properties. Other coatings include silver, as mentioned, titanium dioxide, and zinc dioxide (Bogue, 2008).

Nanotechnology has also been used to create smart labels. These printed labels indicate various analytics in real-time such as temperature, expiration date, humidity, and freshness. The labels have reactive layers with nanoparticles or nanoassays to gather environmental information within the packaging (Teixeira, 2007). RFID itself can be considered a smart label—some miniaturized RFIDs are imprinted on silicon chips. Food processing and packaging is on the cusp of a paradigm shift due to the prevalence and increasing integration of nanotechnology into the production process. It is clear that nanotechnology is becoming a maturing field for food processing and current research has yet to reveal even greater impacts of nanotechnology-based products and devices such as smart packaging, nanosensors, and RFID and IT implications.

2.8 Example

Food also includes sea products such as fish and seaweeds. The global supply of these products is about 130 million tons out of which about one-third is from aquaculture methods. On the other hand, many food fisheries struggle with poor regulation in their countries whether it is a developed or developing countries. The role of supply chain in addressing the global seafood crisis was reported by the United Nation Environment Program (UNEP) (UNEP, 2009). The report mentions that: (1) sustainability issues in fish supply chains including value network structure; and (2) traceability is a crucial and fundamental component for sustainable supply chains and related topics. The report also suggests how technology transfer and the recognition of intellectual property rights through transparent approach may be achieved and implemented. The UNEP Division of Technology, Industry and Economics helps the stakeholders including the authorities and decision-makers in business for industry to develop and implement policies and practices focusing on sustainable development. The International Environmental Technology Center implements integrated wastewater and management programs. Besides political will and available infrastructure, technology plays significant role in aquaculture business. Many vendors and advanced farmers are interested to improve their competitiveness through increased ability not only to automate (recirculating aquaculture system, for example) but also to modernize the entire supply chain with advanced and state-of-the-art distribution channels with, e.g., fresh or just-in-time fish. In fact, because of the demand and environmental pollution, ocean and land-based aquaculture methods are shifting toward tank-based methods with automation and application of IT for the operation and supply chain, respectively. How IT plays a role in the entire supply chain can be observed from an example in the context of aquaculture-based fish farming, sailing, and transportation (Mahalik and Kim, 2013) is briefly presented in this section.

Mahalik and Kim along with company partners have developed architectures called Automated Expert System Architecture (AESA) and its extended version Automated Composite Eco-balanced Aquaculture Test-beds (ACEAT) capable of facilitating the aquaculture automation (monitoring and control) at the tank-level production stage and supply chain using online service methods and procedure. AESA and ACEAT architecture is a six-layer model: interphysical layer, process layer, function block layer, portable device link and user layer, system validation layer and services, and management layer. Different fieldbus type distributed control networking systems such as PakBus, DNP3, or LonWorkss can be used. The architecture is Supervisory Control and Data Acquisition and Human Machine Interface conformant. Internet-based operation, control, and management is the backbone to the architecture. Remote operation and management is assured. Interoperability is achieved via portable devices such as iPAD, iPhone, etc. Additional interfacings are the use of RFID technology for traceability and supply chain management including just-in-time delivery system. For details the reader can refer Mahalik and Kim (Mahalik and Kim, 2013, 2014).

2.9 Conclusions

Logistics in the food sector is becoming one of the important factors in ever increasing competitiveness. Despite the availability of automation (factory floor control systems) and distribution technology (supply chain) throughout, rural producers still face many difficulties and challenges due to the reason that appropriate level of education is not available. The stakeholders do not have abundance information on how technology systems can be used to improve not only the productivity but also just-in-time food demand and societal requirements. Individual technology systems capable of solving local problems, can be integrated via networking (LAN, WAN, wireless, etc.) and online (internet, social media, etc.) technology methods into a new backbone. The farmers, producers, suppliers, distributers, and buyers who are able to acquire and utilize the new technology methods have been able to take advantage of safety, quality, health-concerns, and societal fulfillments.

The major systems in food chains today are implemented at retail, giving retailers the opportunity for chain leadership through their information advantages (Salin, 1998). High-tech information systems have been a secure source of competitive advantage to food and agricultural sectors. With supply chain strategy, the food consumption flow suits the demand for the product. The functional food products (on-demand food, person-specific food, brand-new food products, etc.) can reach to the customers easily because of the advanced supply chain strategies. Supply chain strategies can only be possible by the usage of IT. For example, innovative products that have higher value-added will be able to justify the customers about the reason for higher value via strategic IT methods. Thus, IT in food and agricultural sectors can be a source of competitive advantages to the entire supply chain.

The role of IT developments in the food supply chain integration and monitoring is seen to be high. This chapter focused on presenting subtopics such as components of IT, supply chain management, traceability, food processing and packaging, and the role of IT in processing and packaging. The chapter also presented a section on application of nanotechnology in food processing and packaging.

References

Anderson, D.L., Britt, F.F., Favre, D.J., 1997. The seven principles of supply chain management. Supply Chain Management Review 1 (1), 31–43.

Bogue, R., 2008. Nanosensors: a review of recent progress. Sensor Review 28 (1), 12–17.

Chaudhry, Q., 2012. Nanotechnology applications for food packaging. In: Workshop on the Outlook and Challenges of Nanotechnology for Food Packaging. International Life Sciences Institute.

Cheruvu, P., Kapa, S., Mahalik, N.P., 2008. Recent advances in food processing and packaging technology. International Journal of Automation and Control, Inderscience Publications 2 (4), 418–435.

CSCMP (Council of Supply Chain Management Professionals), 2015. Supply Chain Management Essentials. Available at: https://cscmp.org/online-courses/scm-essentials (retrieved on 20.06.15.).

Faudzi, N.M., Ali, M.T., Ismail, I., Ya'acob, N., Jumaat, H., Sukaimi, N.H.M., May 15–17, 2013. UHF-RFID tag antenna with miniaturization techniques. In: Electrical Engineering/Electronics, Computer, Telecommunications and Information Technology Conference 2013, pp. 1–5.

Flynn, D., May 6, 2013. The Changing World of Food Traceability. Food Safety News.

Golan, E., Krissoff, B., Kuchler, F., Calvin, L., Nelson, K., Price, G., 2004. Traceability in the U.S. food supply: economic theory and industry studies. In: Economic Research Service. Agricultural Economic Report No. 830. United States Department of Agriculture. Government Printing Office, Washington, DChttp://www.iopp.org/files/public/LamDerekSanJose.pdf (retrieved on 22.06.15.).

Linton, I., 2015. What are the Four Element of Supply Chain Management? Houston Chronicle – Small Business. Available at: http://smallbusiness.chron.com/four-elements-supply-chain-management-52355.html (visited on 20.06.15.).

Infosys, 2015. Supply Chain Traceability with RFID and SAP. Infosys White Paper. Available at: http://www.infosys.com/industries/retail/white-papers/Documents/supply-chain-traceability.pdf (retrieved on 20.06.15.).

JHBSPH, 2015. Food Processing. Teaching the Food System. Johns Hopkins Bloomberg School of Public Health. Available at: http://www.jhsph.edu/research/centers-and-institutes/teaching-the-food-system/curriculum/_pdf/Food_Processing-Background.pdf (retrieved on 20.06.15.).

Joseph, T., Morission, M., 2006. Nanotechnology in Agriculture and Food. European Nanotechnology Gateway, Institute of Nanotechnology.

Kuhno, M., 2015. Scalability is key to RFID compliance. RFID Journal. Available at: http://www.rfidjournal.com/articles/view?2862/6 (retrieved on 20.06.15.).
Levinson, D.R., 2009. Traceability in the Food Supply Chain. Office of Inspector General, US Department of Health and Human Services, Washington, DC. Government Printing Office.
Mahalik, N.P., 2009. Processing and packaging automation systems: a review. Special Issue on Advances in Food Automation Journal of Sensing and Instrumentation for Food Quality and Safety 3 (1), 12–25.
Mahalik, N.P., Kim, K.S., 2013. Retrofitting high-tech systems in land-based aquaculture to improve production efficiency: an automated expert system architecture, IETE technical review. Taylor and Francis 31 (2), 153–161.
Mahalik, N.P., 2014. Advances in packaging methods, processes, and systems. International Journal of Challenges, Multidisciplinary Digital Publication Institute (MDPI) Publishing, 5 (2), 374–389. http://dx.doi.org/10.3390/challe5020374. Switzerland.
Mahalik, N.P., Kim, K.S., 2014. Aquaculture monitoring and control systems for seaweed and fish farming. World Journal of Agricultural Research, Science and Education Publishing 2 (4), 176–182. Hong Kong.
Mangina, E., Vlachos, I.P., October 2005. The changing role of information technology in food and beverage logistics management: beverage network optimisation using intelligent agent technology. Journal of Food Engineering 70 (3), 403–420.
Marsh, K., Bugusu, B., 2007. Food packaging: roles, materials, and environmental issues. Journal of Food Science, Institute of Food Technologists 72 (3), R39–R55.
Mathavi, V., Sujatha, G., Bhavani Ramya, S., Karthika Devi, B., 2013. New trends in food processing. International Journal of Advances in Engineering and Technology 5 (2), 176–187.
Salin, V., 1998. Information technology in agri-food supply chain. The International Food and Agribusiness Management Review 1 (3), 329–334. Elsevier Sciences.
Schmidt, R., 2012. Basic Elements of Equipment Cleaning and Sanitizing in Food Processing and Handling Operations. Institute of Food and Agricultural Sciences. University of Florida.
Setboonsarng, S., Sakai, J., Vancura, L., 2009. Food Safety and ICT Traceability Systems: Lessons from Japan for Developing Countries. Asian Development Bank Instutite. Working Paper Series. No. 139.
Teixeira, V., 2007. Advanced Nanotechnology Thin Film Approaches for the Food and Medical Industry: An Overview of Current Status. Available at: nanoHUB.org (retrieved on 22.06.15.).
UNEP, 2009. The Role of Supply Chains in Addressing the Global Seafood Crisis. Available at: http://www.unep.ch/etb/publications/Fish%20Supply%20Chains/UNEP%20fish%20supply%20chains%20report.pdf (retrieved on 22.06.15.).
Weis, S., 2007. RFID (Radio Frequency Identification): Principles and Applications. MIT CSAIL Available at: http://www.eecs.harvard.edu/cs199r/readings/rfid-article.pdf (retrieved on 20.06.15.).

Further Reading

William Reed Business Media, 2013. Nanosensor Being Developed for Food Safety. FoodQualityNews.com Available at: http://www.foodqualitynews.com/R-D/Nanosensor-with-potential-food-safety-use-comes-step-closer (retrieved on 22.6.15.).

Part Two

Factory Design and Operation

Emerging Trends and Methods in Food Factory Design

3

Frank Moerman[1], *Patrick C. Wouters*[2]
[1]Catholic University of Leuven, Leuven, Belgium; [2]Cargill, Food Safety, Quality and Regulatory, Schipol, The Netherlands

3.1 Introduction

The manufacturing of safe, high quality foods against competitive costs requires that the design of a food factory is reviewed on a regular basis. With time, the well-proven "top-of-the-line" design concepts of the past may no longer suffice, due to new product developments with different food safety risks, tightening of regulations, changing in processing technologies, equipment design, materials of construction, environmental requisites, hygiene and safety requirements, and building concepts. A food manufacturer is nowadays obliged to look in the future and to develop vision of where to go, in order to meet future legislative requirements and consumer expectations, operate in a sustainable way, and remain competitive. To succeed in this review and to make the correct decisions, the food manufacturer should have knowledge of the present and emerging trends in building design requirements in relation to the identified food safety risks, taking into account the following aspects: legislation and standards, food processing, packaging and warehousing technologies, communication, control and automation systems, requirements for process aids, sustainability requirements to minimize energy, water usage, waste, and CO_2 emissions. It is the basis to successfully conceive a new or retrofit an old food factory.

The first section of this chapter deals with the drivers to renovate existing facilities and how renovation should proceed to prevent disturbances in the quality of ongoing and future food processing operations. In a second section, the rationales are described to build a new food factory and how the food factory of the future could look like. The next section describes traditional and more recent market available tools that may facilitate the renovation of an existing food factory or the design and construction of a new food factory. The last section outlines the role of experts, contractors, or a department specialized in the hygienic engineering and design of food factories.

3.2 Renovation of an Existing Food Factory

3.2.1 Drivers in Favor of Renovation (Brownfield Projects)

An established facility has many attributes that can make it attractive to continue to be the manufacturing location rather than building a greenfield new factory. The main drivers in favor of renovation could be the following:

- Increased capacity needs, or an improved and more efficient production process installation (not necessary with larger equipment).

Innovation and Future Trends in Food Manufacturing and Supply Chain Technologies
http://dx.doi.org/10.1016/B978-1-78242-447-5.00003-4

- For owners with multiple plant locations, consolidation is emerging as a trend primarily due to transportation, labor, and taxation factors. Owners can often realize a reduction in expenses and improve efficiencies by consolidating facilities. From the viewpoint of "getting more out of what you already have," they invest in the improvement, renovation, and expansion of existing facilities in lieu of developing and constructing new greenfield facilities (in contrary to a brownfield site, a greenfield site is one that has not been built on previously). In addition, there are many vacant facilities on the real estate market, and buildings are still a good bargain in today's economy.
- "Bricks-and-mortar" construction costs for the building structure, utilities, and other infrastructure are all realized and readily available for future contributions.
- Older buildings are often built better and last longer. The building structure can still be sound, the roof in good condition (without leaks), the effluent system appropriate, and the building uncontaminated. It only may require some improvements in terms of environmental performance.
- Permission to build a new factory is not generally required (unless major building works are necessary). Only permission for the change of purpose of the building will be required.
- Reuse of an existing building can reduce considerably the lead time relating to planning and construction. Renovation can be a faster way to bring new products on the market than constructing a new greenfield facility. However, if hazardous building materials (e.g., asbestos) are present, the reverse will be true.
- Capital costs for equipment have already been invested and the production operations well established.
- Long time employees acquire important and valuable experience with the facility, its systems, and existing equipment, and are adaptable to new methods and technology. The collective memory of the facility's operations staff, their knowledge of the particulars of the manufacturing process and the operations, and the spirit to succeed that comes from working with like-minded people committed to a common goal are of highest importance as the greatest operational expense of any facility is the people who work there.
- The established logistical infrastructure. However, the flow of goods must still be logical.
- The familiarity of vendors, material suppliers, and shipping/transportation services with the location of every facility and its operations, because they have integrated the specifics of a location, including established procedures, into the delivery of their services.

The mentioned features related to old factories and facilities have proven to be highly productive and profitable, and they retain the potential—with careful intervention—to remain so into the future.

3.2.2 The Need to Renovate

A complete or partial dismantlement, renovation, and refurbishment of the original food factory area are often necessary, because:

- Asbestos is present, which in the long-term may seriously affect the health of the process operators (e.g., lung and peritoneal mesothelioma) in the food factory and may bring the safety of the food produced or stored in the factory at risk (Stark and Viehl, 1995).
- Larger equipment components must be installed within the food factory. Moving large equipment components into a space can sometimes be challenging, regardless of whether

the equipment is new or used. To deal with restrictive access, new equipment can be customized to fit in the access corridor or equipment elevator, as such making the logistics far easier. Likewise, lighter materials could be selected for constructing the equipment so as to not exceed the existing structural capacities of the factory building. In contrary, when second-hand equipment is used, customizing that equipment is not possible. Facility alterations may be required to move in used equipment. So used equipment might be less desirable when space and access is tight, or the facility is of light construction (Sirabian et al., 2012).

- Major structural changes to a plant can become necessary when the structure is no longer capable of isolating the outside environment from the production environment within the plant. Facilities with an exposed steel structure are sensitive to acids, caustics, cleaning agents, oil-saturated steam, or high humidity present in the plant during operations. Also, harsh weather conditions deteriorate the integrity of structural components, which may result in moisture infiltration (e.g., leakage of roofs) and the entry of pests. A part or all the roof joists may need replacement in a particular section of the plant, due to severe rust and deterioration. The building structure also may suffer from the excessive loading of collateral equipment, requiring repair and updating of the building to protect the safety of the workers and the viability of the line at any time. Materials of construction could no longer meet current hygienic requirements, which is especially important when an existing space needs to be converted to a ready-to-eat food processing area. A walk-on ceiling or interstitial space may be required to accommodate process utilities, air handling units, and refrigeration infrastructure. For that purpose, steel roof supports will need reinforcement to have enough load capacity to support the additional roof loading. A major structural update or renovation also may be required where there is not enough lighting to effectively perform duties (Labs, 2014c).
- In older factory buildings, plumbing, utilities, and materials may be unacceptable to new codes (fire, electrical, seismic, lateral loading, hygiene, etc.) and legal standards, or they may no longer comply with the customer's hygiene requirements. Total replacement or major upgrading may be required (Stark and Viehl, 1995).
- Due to aging and long term use, corrosion and clogging of old-fashioned utilities may occur (e.g., drain traps may get blocked with food debris if drain bodies are not provided with filter baskets).
- New safety regulations may oblige food manufacturers to assure proper worker egress in situations of emergency. For the safety and health of the factory staff, a retrofit of existing safety and emergency systems may be required.
- Additional power to meet the increased electrical load of more sophisticated electronic equipment may be required. Moreover, emergency power systems and uninterruptable power supplies (UPS) are often needed to support the newer more sensitive electronic equipment. New process equipment may have higher power demands or require higher voltages. Electric and electronic upgrading of an existing facility is often demanding with regards to magnitude, complexity, labor, renovation time, and cost.
- A facility's existing telecommunication and data communication system is often inadequate to support the explosive growth in communications and process control technology. So, an upgrade and extension of existing telecommunication, data communication, and control systems is often required. Likewise, more extended power supply and distribution grids for the working of telephone, data communication, process control, and safety systems may be required (Moerman, 2011a).
- Depending on the equipment configurations, cooling and dehumidification capacities may be impacted by design conditions at the installed location. Oversized cooling and/or

dehumidification equipment may be adaptable, although such modification could reduce the operating efficiency.

- Segregation is needed to prevent that dust and contaminants from raw materials migrate to finished products. However, the addition of walls in existing plants can be challenging.
- A food manufacturer wants to make use of room pressurization as a means to combat cross-contamination, more specifically by elimination of incorrect airflows between adjacent areas. It may oblige him to review the existing equipment for capacity requirements, review the floor plan for air gaps (doorways, vents, etc.), and develop a corrective action plan. Installing new ductwork and equipment and making building corrections can improve hygienic airflow (Labs, 2014c).
- Air handling/heating, ventilation, and air-conditioning (HVAC) systems may need upgrading because HVAC systems frequently have a limited service lifespan or do not longer satisfy the clean air criteria standards of regulatory authorities. Moreover, new heat-generating equipment may place additional demand on the system. Hence, renovation of the HVAC infrastructure may be required to bring the system capability up to current hygiene and environmental requirements. Increased air quality requirements may oblige cleaning and relining of existing ductwork, and the installation of air filters with higher quality performance rates (removal of 90–95% of the supply air contaminants).
- The utilities to support the new processes could not be available, the capacity can be too small, or the flow rate, temperature, pressure, voltage, etc., of the supply can be either too little or too large. Undersized equipment is usually more problematic and requires more costly modifications. During the upgrading of old facilities, a contingency factor must be applied to the design margin to accommodate occasional peak demand. The capacity requirements of the various utilities and process equipment have to be estimated by using existing inhouse data and data from equipment vendors, so as to allow for expansion in the future (Sirabian et al., 2012).
- An existing facility will normally have main supplies already laid, but due to the nature of constantly improving legal and hygiene standards it is usually found that all distribution systems and sometimes even the main incoming control/metering systems do not meet the legal requirements.
- Floor, wall, and ceiling finishes and partition layouts may require total replacement to suit the proposed production requirements and to meet current and increasingly stringent hygiene, health, and safety legislation. Failing paint on walls and ceilings must be completely removed and replaced with a washable surface that can withstand cleaning chemicals. When the floor gets damaged to a degree that it is difficult to clean, a floor coating can be applied over the old floor so as to increase the floor's strength, cleanability, and appearance. Peeling floor surfaces must be scraped and/or shot-blasted to subsequently install a self-leveling subsurface and apply a food- and temperature-grade (urethane, epoxy, etc.) floor coating (Sutton, 2013; Labs, 2014c).
- There may be a lack of gowning and hand washing stations at all access points to production areas (Labs, 2014c).
- Hose stations to hose down equipment and build components may be inadequate: improper installation (insufficient clearances with respect to floor or walls), unsuitable water temperature, too large a distance from the equipment, and factory surfaces that must be cleaned (Labs, 2014c).
- Preliminary bad installation practices must be solved. Adding of sloped tops may be required to alleviate the problem of flat edges. Wall-mounted pipes and conduits should stand off walls at least 6cm to allow for cleaning of the entire pipe and wall surface. If this was not done during the original installation, removing, rerouting, and installing the pipes and conduits will be required. Drain lines that span from nonproduction areas to production areas are not acceptable and may require removal and replacement (Labs, 2014c).

- Doors are no longer hygienic because they are made from nonchemical resistant materials and contain water-retaining and microbial harborage points.
- Condensation problems on walls, ceilings, and pipes may require replacement of insulation and the sealing of wall/ceiling penetrations. Replacement of wall insulation and installation of a vapor barrier is time-demanding and not easy to fix. Ceiling systems like fiberglass-reinforced plastic grid and insulated metal panels often require line shutdowns (Moerman, 2011c; Labs, 2014c).
- If a building component becomes a continuous source of concern because of severe deficiencies, it is better to opt for radical renovation instead of continuous half-baked patching. Durable renovation and permanent repairs instead of temporary fixes will avoid too much downtime (Moerman et al., 2014).

3.2.3 Planning and Execution of Renovation Work

3.2.3.1 Planning of Renovations

The scope of the required renovation (modification) should be assessed. The food manufacturer should consider the requirement of an immediate complete renovation or a temporary band-aid renovation, which is usually less costly and may have a minimal negative affect on production schedules. So, first a temporary repair may be set up, followed by a planned upgrade/complete repair when schedules allow. The initial planning stage for renovation should provide a very detailed work plan and a schedule that includes tasks, duration, and responsibilities. Repairs to a building's structural systems are inherently tricky and usually involve working over, around, and under existing lines. Repairs of this type are generally time-consuming, with construction durations dependent on the extent of the work to be accomplished. As an example, replacement of floor drains will require the removal of the concrete slab, and as such will be very time demanding. Budgets for the renovation work also must be made (Labs, 2014b).

Designers and engineers should prepurchase needed materials to eliminate the possibility of construction delays. The timeline should include coordinating site activities (such as plant shutdowns, cleanup times, start-up procedures, and material deliveries) with the operational leadership. The design and construction staff must understand the scope of the work and develop a schedule that allows for tasks to be completed in the allocated time frame (Labs, 2014b).

3.2.3.2 Specific Inconveniences During the Retrofit and/or Renovation of Existing Food Factories

Retrofitting a factory building and renovation of process facilities and utilities can have a large effect on adjacent areas:

- Necessary power and utility shutdown may disrupt normal production activities and ongoing operations.
- Noise and vibrations may have an impact on the productivity of workers in neighboring manufacturing rooms.
- Dust, dirt, and welding smoke generated by the construction effort may spread to processing and/or storage areas, and could compromise the safety and quality of the manufactured food and the health of operators in the food factory. This is especially true with asbestos.

3.2.3.3 How to Cope with These Inconveniences During the Renovation Activities

The following actions can be taken to cope with the inconveniences of renovation activities:

- Segregation of construction areas from production areas.
- Applying a temporary band-aid approach with minimal inconveniences to fix a problem instead of immediate complete repair and renovation. Usually there is only a temporary shutdown of the processing line (eventually with temporarily relocation of the equipment) while the renovation work is being executed, or food-grade temporary walls can be used to encase the rework area while production continues.
- Immediate plant shutdown for complete repair or renovation, while transferring production to a sister plant, moving production to another line, or providing a redundant or duplicate line to be used during the renovation.
- Performing renovations around plant shutdowns (weekends, holidays, maintenance shutdowns, or cleanup days). As an example, a full roof replacement will require a complete plant shutdown, as dislodged dust will fall into exposed production areas below.

3.2.3.4 Outlining of Hygiene Measures Before the Onset of the Renovation Work

Besides knowledge of correct hygienic food factory engineering, design, and installation practices, it is essential that the construction management, the company and the selected construction management team (contractors) have experience and background of doing work in a controlled environment, as well as specific training on the site's access and containment requirements. To confirm that the requirements to construct adjacent to the operations are met, it is important that also operatives are involved in the project. First of all, the existing building space must be assessed with regards to the potential impact of the construction work on the adjacent areas (e.g., routes of contamination, power, and utility shutdowns in process rooms, noise, etc.). For that purpose, the construction team should make use of physical drawings of the area of construction, including access routes, utility, and HVAC diagrams showing conditions during each phase of the construction. Secondly, the construction logistics (e.g., removal of demolition debris out of the factory, transfer of construction materials into the factory, flow of construction personnel, etc.) must be considered so as to avoid that construction and operations paths cross each other. Thirdly, the expected gowning and personnel protective equipment requirements at different points must be determined, as well as the containment and isolation barriers required for personnel traffic. A next task is defining the requirements for incoming materials and the points to inspect and clean them, as well as the containment requirements for outgoing materials (e.g., sealed debris containers may be cleaned prior to leaving the construction area). Besides this, a construction pressurization plan must be developed in function of the sequence of construction activities that will take place. Adjacent production schedules should be revisited to account for changes in production that could impact construction or, in reverse, construction activities that could disturb ongoing processes (e.g., production and utility shutdown plans should be in place where applicable). Finally, a personnel training plan must be provided (Hammond and Van Wormer, 2014).

3.2.3.5 Practical Implementation of the Hygiene Measures

If food products are still manufactured in other parts of the facility, the cleanliness of those ongoing operations must be maintained. Reduction of the spread of particles must be achieved between the contaminated construction site and the surroundings. This will require the installation of temporary partitions and dust barriers (Figure 3.1) to isolate the retrofitting and renovation activities from existing operations. The integrity of those barriers during construction must be preserved. So, prior to the start of the renovation and retrofit operations, the construction site should be segregated by building false walls consisting of either double sheet plastic or a temporary wall of plywood, metal, or plastic on the renovation side. Rigid barriers may be needed to resist the potential damage of the barrier or to support the implemented pressurization schemes. An ideal arrangement is the erection of stud walls with insulation. Fiber glass-reinforced panels on the production/storage site with caulked joints provide an impervious barrier to construction debris and other contamination. Supplementary barriers may be needed to encase activities that create large amounts of particulates in the construction area, to further minimize migration of dust particles even inside the construction space. Point extraction of dust by means of vacuum is another means to reduce the spreading of dust in the construction area. All branches of utilities (e.g., HVAC) and process piping in the construction site should be isolated and disconnected from the main headers that also have branches in the adjacent areas with ongoing operations (Stark and Viehl, 1995; Hammond and Van Wormer, 2014).

Figure 3.1 Cleanliness of ongoing process operations should be maintained by construction of temporary partitions and dust barriers. The construction side should be under negative pressure relative to areas with ongoing process operations. RenovationIMG_7846.

Another option to limit cross-contamination is the creation of larger differential room pressures to provide a correct flow of air from the cleanest to the dirtiest areas. In the processing area with the highest hygiene requirements relative to the adjacent areas, a positive pressure should be considered through a ventilation system that pumps a higher volume of air into this area as compared to the volume of air that is exhausted. At the same time, by maintaining the correct balance of supply and exhaust air, a negative air pressure within the workspace could be implemented so as to prevent renovation debris from getting into the surrounding cleaner process areas. Ventilation of the construction air to the outside may reduce the number of dust particles and contaminants in this area, but care should be taken not to locate the exhaust too close to the plant's fresh air intake. If required, filtration of the exhausted air should be done (Stark and Viehl, 1995; Moerman et al., 2014). During the execution of the project, periodic or continuous monitoring of the construction area pressurization should be considered to control that the pressure stays negative compared with the surrounding production areas. Additionally, particle and/or bioburden monitoring of the air in the adjacent process areas can be done to collect evidence that the construction is not negatively impacting the air quality in the production areas (Hammond and Van Wormer, 2014).

Personnel hygiene is essential during the renovation activities. Gowning may be required before entering and working in the construction area, followed by removing and new gowning before returning through a classified production area. When materials pass through a classified area, they may need to be inspected prior to entry and/or put in sealed or enclosed containers that are cleaned. It is an advantage that construction materials are clean prior to entry in the construction site, as any debris entering the construction site must be removed afterward. To reduce the spreading of construction debris in adjacent process areas, doors should be kept closed, and swarf mats and boot washes can be used. The frequency of floor cleaning and wipe downs around the construction area may be increased. Construction tools may not be used or left in adjacent areas where processing continues. Construction debris and construction materials should not be stored in the food factory, as they are a source of contamination. The food manufacturer also must prohibit the entry of pests in the construction area and its surroundings, which means that all potential access openings must be properly protected and/or sealed. It is the responsibility of the food safety and quality function to regularly conduct audits to verify if the construction staff or contractors have adopted the correct hygiene practices during the renovation operations. Once renovation activities are finished, the construction side, adjacent areas, and branches of utility systems must be cleaned and, where required, disinfected. After the modifications and before resuming process operations, validation and testing must ensure that all systems are performing well and within acceptable limits (Moerman et al., 2014; Hammond and Van Wormer, 2014).

The environmental health and safety function must survey if existing building and utility supply and distribution systems under renovation contain asbestos. Removal of asbestos will require supplementary measures to contain all asbestos debris within the envelope of the construction site. The staff performing the renovation/retrofit work must receive additional breath and skin protection (Moerman et al., 2014).

3.2.4 Sustainable Renovation

For the purpose of this chapter, sustainability is considered to be "the wide range of measures considered necessary to help avert issues associated with climate change, the increasing world population, and the effect of carbon on the atmosphere." Other associated equally relevant topics include "eco-friendly" and "biodiversity." The explosive growth of the world population has put pressures on the global capacity to maintain growth economically and provide the resources to match the increasing requirements. However, the need for infinite growth and resource has given rise to a lot of environmental concerns.

The demolition and renovation of old manufacturing facilities requires the removal of building material. Construction waste and demolition debris are considered to be industrial waste. Some of this waste is dumped in municipal solid waste landfills along with household garbage, some is incinerated in combustion facilities, but most goes into landfills dedicated to construction waste. However, as most of the construction waste is nonhazardous, it can be diverted. It can be directed back into use with little or no modification (materials reuse) or redirected back into the manufacturing process through recycling (materials are used as raw materials to generate new products). The extra time to salvage, separate, sort, and arrange hauling of removed elements during demolition negatively impacts labor costs, but at the end they are nullified through revenue from selling recycled materials and by reduced landfill tipping fees. For example, removed concrete can be sold to local concrete companies to be crushed and used as aggregate in new concrete, or metals such as copper piping, steel, and copper wire can be sold to a metal recycling company.

Even more sustainable is the reduction of construction waste by proper renovation techniques. The more an existing building is left intact, the more waste is reduced. If a wrecking ball is not used during demolition, it is possible to leave major building elements such as the roof and structural steel intact. The simple act of reusing the building itself and maintaining these existing elements may create considerable cost savings. Many building components can be reused in other construction. Architectural items such as cabinets, light fixtures, hardware, metal stairs, and platforms, windows, doors and frames, if removed carefully, can be reinstalled. The same is true for furniture and some finishing materials, including wood trim and flooring. It is common practice to reuse expensive equipment such as air handler units, compressors, and chillers. Also, smaller HVAC, plumbing, and electrical items, such as sinks, toilets, faucets, diffusers, junction boxes, outlets, fittings and valves, are reused. With forethought and knowledge of available resources, a new project can be designed to incorporate used or salvaged building elements in the new construction. Waste reduction is an effective strategy to lower costs, while protecting the environment and its natural resources (Cochran and Maurer, 2013).

3.3 Building a New Food Factory

3.3.1 Rationales to Build a New Food Factory

3.3.1.1 Legacy Facilities

Managers of legacy food manufacturing facilities continually struggle against the ever evolving triumvirate of aging infrastructure, improvements in technology, and the evolution of good manufacturing practices. Legacy facilities are defined as those existing manufacturing facilities that have developed over an extended period of time, and in the process, have accrued incremental changes from their original configuration. Changes were made to create more capacity and space for new product lines, interior spaces, and the movement of material and personnel were repeatedly reorganized, new production equipment added, and all the while the expectations for good manufacturing practices (GMPs) continued to evolve. But at a certain moment, legacy facilities are burdened with outmoded infrastructure, equipment that has reached the end of its useful life, and production facilities that have not kept pace with current standards for good manufacturing practices. The Food and Drug Administration (FDA), however, states on his website about cGMPs—the "c" in cGMP stands for "current"—requiring companies to use technologies and systems that are up-to-date in order to comply with the regulations. Systems and equipment that may have been "top-of-the-line" to prevent contamination, mix-ups, and errors 10 or 20 years ago may be less than adequate by today's standards. Manufacturing managers often opt for the expedient solution of spot renovation to resolve immediate problems. However, this approach can exacerbate existing deficiencies. Renovations may often be very complicated, may cause a drastic change of the layout of the production area, and are usually time and cost demanding (Stark and Viehl, 1995; Bohn, 2013).

3.3.1.2 General Problems with Legacy Facilities

Legacy facilities present many challenges to maintain effective and compliant operations. Due to the incremental growth and periodic investments in new technologies and equipment, they often become a tangle of inefficiencies. Key areas of concern are the crossing of flows, the danger of cross-contamination, the aging of critical utilities, changes in equipment, and the introduction of new products/processes:

- Manufacturing facilities ideally contain clearly-defined circulation patterns that are sequential and reinforce an efficient operation. Raw materials are assembled from the warehouse, enter into the production process, and ultimately emerge as finished goods in a linear sequence that minimizes or eliminates crossing of the work-in-progress and the corresponding risk of mix-ups and contamination. Over time, and as a result of multiple additions and internal reorganizations, circulation patterns that originally followed a logical path through defined yet isolated zones can become compromised by changes that may have developed as a result of localized alterations, equipment changes, and random placement of production functions. These disjointed and circuitous material and personnel circulation patterns, as a result of designing within existing constraints, may give rise to bottlenecks or crossed movement of material and personnel, and hence product mix-ups, cross-contamination, and operational inefficiencies (increased handling and staging of material). A common symptom of this condition is when a simple and clear gowning sequence does not exist. As an example, in some infant food manufacturing facilities, corridors are accessed by personnel in both gowns and street clothes, which regulating agencies nowadays consider as a problem in need of correction. In infant food manufacturing facilities, proper segregation between personnel in gowning and street clothes is recommended (Bohn, 2013).

- Problems of cross-contamination can occur in two principal areas: air systems and product handling. Incremental additions to buildings or repurposing of existing spaces from one activity to another may result in air distribution and handling systems that lack adequate segregation between activities. This mixing of air systems, often resulting from an expedient modification of an existing system, can lead to airborne contamination of raw materials or in-process products. Problems also can arise when raw materials or finished products are mixed between process streams leading, at best, to confusion and at worst, product contamination. Modern systems of bar coding and serialization help to significantly reduce the risk of cross-contamination from handling errors, but they are not foolproof (Bohn, 2013).
- A third major category of problems associated with legacy facilities include those related to aging infrastructure, equipment, and utilities. When initially constructed, the legacy facility was likely designed and constructed with "state-of-the-art" environmental controls, electrical equipment, production machinery and their related controls, telecommunications and data services, and business operations technology. As equipment ages, breakdowns become inevitable. Over time, mechanical systems and control equipment degrade at a relatively predictable rate, and replacement can be anticipated. Incremental additions to the building structure, localized renovations for new product lines, piecemeal upgrades to product manufacturing systems or utilities, or localized system replacements can leave the legacy facility with a patchwork of mismatched and incompatible equipment. Although routine and timely maintenance can delay the inevitable, there comes a time when equipment has reached the end of its useful service life and must be replaced (Moerman, 2010; Bohn, 2013).
- Over time, new technologies can rapidly lead to obsolescence of existing systems and equipment and hinder the facility's efficiency, effectiveness, and ability to maintain compliance with cGMPs. As more efficient production, control, and handling equipment becomes available, the challenge becomes one of maintaining competitiveness and profitability within a legacy facility. Similarly, existing building utility systems and infrastructure become less efficient over time, and newer models often incorporate advances in energy efficiency, flexible operating controls, and more efficient use of resources. Thus, energy consumption and cost of goods remains hostage to outdated technology with little opportunity to effect enhancements and cost reductions that accrue from bringing online more efficient equipment (Moerman, 2010).
- Product development and marketing teams are continuously either improving existing products or introducing new products to an existing facility. Sometimes these changes can be accommodated with the existing equipment and within the infrastructure. But eventually the problem becomes one of how to fit new equipment and processing systems into the current footprint without compromising the food safety and quality. If not carefully considered in a manner that reinforces or even improves the integrity of existing flows and infrastructure, serious compromises can result.

3.3.1.3 Specific Examples of Problems with Legacy Facilities

Specific problems and drivers that may encourage a food manufacturer to build a new factory:

- Insufficient space for expansion at the existing factory site. Notice that together with the expansion of the process infrastructure, usually additional amenities (office rooms, changing rooms, toilets, welfare facilities, delivery and dispatch areas, parking for good vehicles and staff, etc.) are required. Another example is the need to install a wastewater treatment plant (Sutton, 2013).
- Quite often old factories are located in the middle of a city, surrounded by new residential areas. Additional size/work force/noise/effluents/transport, etc. may have an unacceptable effect on the surrounding community. Due to the tightening of environmental legislation and regulations, the company may be forced to abandon further operations in the existing factory and to build a new factory at a location for which local authorities have given permission to continue operations.
- New activities in the surroundings may compromise further production of safe food: neighboring companies exhausting pollutants, a newly built waste water plant, etc.
- Building materials (e.g., asbestos) in the old factory may be unacceptable to codes. The removal or treatment of these materials can take several months, which, from a time and cost point of view, could make renovation of the existing factory unacceptable. This is especially true if the food manufacturer wants to resume the production of an existing line of products quickly, or produce new products to bring them on the market in the shortest time possible.
- The factory is completely at its end of life, with building materials and elements losing their integrity. Broken cables and brittle isolation may cause short circuits (fire safety risk) or power losses.
- During the many years of production, the food factory became heavily contaminated with dust, dirt, vermin, and bioburden. Cases are known where pathogens were found in the insulation and voids behind walls and wall cladding, under flooring material, etc., due to penetrations and the lack of consistent sealant. As an example, fiber glass-reinforced plastic panels mechanically attached or glued to existing surfaces give rise to hygiene problems, because sealants eventually fail and microbes are able to grow in the recesses. Roofing material may leak, allowing water contaminated with microorganisms (bird droppings on roofs contain many pathogens) to penetrate the inside of the factory along that route. Renovation usually does not solve the problem of pathogens encapsulated within building materials and components.
- New or used equipment can be heavier, so that the existing facility could be of too light construction. Existing overhead structures/roof may not be strong enough to support newly required plant, services, utilities, and ceiling systems (Stark and Viehl, 1995; Sutton, 2013).
- Building may not be of an ideal shape or size for new process technologies and equipment. The requested process layout may not fit within the confines of the existing building without compromising to plant efficiency (CCFRA, 2003a; Sutton, 2013).
- Due to international migration patterns, there is an increased local demand for food that should not get contaminated with specific raw materials and ingredients on religious grounds. It could be that the existing factory does not allow for effective segregation between halal and nonhalal food production, kosher and nonkosher food production. Moreover, there is an increasing demand for allergen-free food production. To ensure complete allergen-free production for particular production processes, a complete segregation may be needed between products containing incompatible allergens.

- The height of the building may not be sufficient to allow clear, unobstructed processing. Sufficient floor-to-floor height is especially important when a lot of utilities, especially HVAC infrastructures, must be installed above the process area.
- There also could be no sufficient heat/steam raising, compressed air, water storage, and treatment, cooling and air-conditioning capacity in the facility to satisfy the process requirements. Hence, additional measures may be required to rectify utilities deficiencies, resulting in increased piping and additional generating capacity, transformers for different voltages, heat exchangers, etc. Those will have impact on the space, which may or may not be available in the existing facility. Additional space translates into additional costs and project time.
- The cost of renovation (removal and new installation) could be higher than that of building a new food factory. Dismantling and then restoring equipment in its original condition is time-consuming, making continual, phased expenditure likely to cost more in the long-term (CCFRA, 2003a).
- Renovation is sometimes difficult, or even impossible. Renovation of the drainage infrastructure in an existing factory building is much more complicated than when a new factory is built. In the latter case, drains are laid before the construction of the building envelope is done.
- Preference could be given to a new building because renovation may imply a complete shutdown of the existing food factory plant. Food manufacturers cannot stop production for months, and contract manufacturing by third parties can be expensive. Complete shutdown may have as a consequence that business to existing customers may be lost during a period of renovation (CFFRA, 2003a).
- Tightening of food legislation and hygiene requirements can make that renovation does no longer make sense because the existing plant is completely old-fashioned technology (CCFRA, 2003a).
- Authorities may support a new factory building project financially; the project may be self-financing from the sale of the old site (CCFRA, 2003a).
- Proximity to raw materials required for production, and suitable location for distribution, logistics, and market.
- The area in which labor, utilities, effluent, and waste handling are available at low cost.
- Lack of available turnaround space, which is room to temporarily house building occupants displaced by the renovation, is often an issue. One should have the capacity to clear at least 20–25% of the total space to be renovated, but even this would require four or five discrete construction phases, which means a greater expense (Stark and Viehl, 1995).

3.3.2 "Food Factory of the Future" Programs

With respect to food manufacturing, it is essential that new developments are tracked to identify potential issues, opportunities, and areas for improvement. The highest priority should be given to those areas of operations or infrastructure that represent the greatest risks to ongoing and future operational changes. The last decade, several organizations and even governments in many countries (United Kingdom, Netherlands, France, Sweden, Belgium, etc.) launched programs to design new or retrofit old food factories in function of future requirements, and not just in function of contemporary requirements. The initiatives taken in the frame of these "Food Factory of the Future" programs force food manufacturers no longer to think on a day-to-day basis, but to develop a vision where they need to go to be able to serve and support the future needs of their customers. They mobilize companies to analyze their current position and the intended future state, and what they have to do to reach that objective. Conceiving a new or retrofit an old food factory at least requires knowledge of the present and emerging trends in food factory design.

3.3.3 Design and Build to Reduce Present and Future Costs

An important driver to replace an old food factory by a new food factory is the reduction in costs that can be realized. Cost can be reduced by means of (Bohn, 2013):

- Modernization of equipment and facility infrastructure to eliminate bottlenecks and to increase the efficiency of manufacturing operations.
- Improving the efficiency of product and materials flows throughout the entire process, from handling of raw materials to finished goods.
- Eliminating any potential cross-contamination between materials and products due to crossed flows, deterioration in air quality as the consequence of air flowing from dirt to clean areas, etc. Less cross-contamination decreases the number of product recalls.
- Enhancing the staff efficiency and effectiveness, with a focus on saving time for performing tasks and functions.
- Creating capacity for higher production and lower cost of goods.
- Reducing the square footage of facility devoted to manufacturing.
- Decreasing the carbon footprint and emissions of polluting and noxious chemical components, allowing a reduction in environmental taxes that must be paid.

3.3.4 Design and Build for Future Regulatory and Standard Expectations

3.3.4.1 Food Safety Regulations

Annex II, Chapter I of Regulation 852/2004 describes general requirements with respect to layout, design, construction, siting, and size of food premises, while Annex II, Chapter II of this regulation describes the properties that individual rooms inside those premises must meet. Moreover, it provides rules for elements of the construction such as floors, walls, ceilings, windows, and doors. Regulation 853/2004 provides the rules for the construction, layout, and equipment of slaughterhouses, which are specific in relation to the general requirements for all food. However, these regulations do not define how to keep up with these requirements.

3.3.4.2 Food Safety Management Systems of the Past, Present, and Future

For many years, food producers largely use food safety management systems, with cGMPs (passed by the FDA in 1969), HACCP (Hazard Analysis Critical Control Point, dating from early 1960s, also found in the Codex Alimentarius Code), and prerequisite programs (PRPs) as building blocks. PRPs control basic food safety issues and prevent hazards from becoming reasonably likely to occur, therefore reducing the number of CCPs required in HACCP plans. Prerequisite programs include, for example, GMPs, building design, cleaning, equipment suitability, maintenance, training, calibrations, supplier approval, utilities, potable water, waste disposal, storage, and transport. On request of their customers, almost a decade, many food manufacturers also certify their food operations against the SQF (1st edition of code of Safe Quality Food Institute dating from 1990s), BRC (British Retail Consortium, 1st edition dating from 1998), or FSSC 22000 (Food Safety System Certification 22000, dating from 2009) standards.

Now, operating a food factory in a manner that Global Food Safety Initiative (GFSI, founded in 2000) and Food Safety Modernization Act (FSMA) guidelines are met is becoming commonplace. Both the Global Food Safety Initiative and Food Safety Modernization Act (the FDA passed FSMA in 2011) still consider cGMPs, HACCP, and prerequisite programs as the basis of any food safety management system, and food manufacturers must comply with them to obtain certification. However, with the evolution from local to national and international supply chains, cGMPS have evolved to match today's needs, and both FSMA and GFSI are still changing the cGMPs. Allergen management will become a cornerstone and make the difference between the new cGMPs and the old cGMPs. The GFSI and FSMA also require changes in the PRPs, and the FSMA wants a shift from responding to contamination to preventing it. More and more there is a shift visible from HACCP to Hazard Analysis Risk Preventive Controls, which will further impact cGMPs. Meanwhile, the existing GFSI certification schemes BRC, SQF, FSSC 22000, etc., which all may result in GFSI certification that is valid around the world, are in continuous evolution, with new editions that regularly replace the former editions. Clearly, regulations and standards are undergoing a continuous evolution, which also will reflect in food factory design and building. It is a challenge for food manufacturers to build food factories in such a way that they can cope and comply with these continuously evolving standards.

3.3.4.3 Impact of Old and New Standards on Food Factory Design and Construction

In the past, the more severe food safety standards have been reserved only for plants with high hygiene requirements, such as meat, dairy and ready-to-eat (RTE) processing. These standards that govern the design and construction of process spaces, as well as cleaning/disinfection requirements, are changing and, for the most part, are becoming stricter for all types of facilities. Moreover, under pressure from the GFSI and FSMA, also food facility finish standards that were acceptable 3 to 5 years ago are being reassessed to assure plants are capable of meeting current and future food safety requirements. With the GFSI and its numerous schemes, people have gone from just thinking about daily operations to also thinking about how they can incorporate GFSI principles into projects. When new plants are developed, BRC, SQFI, ISO/FSSC 22000, FDA, and United States Department of Agriculture regulations and standards should be routinely designed in as part of the plant planning process. Older facilities, where it is cost prohibitive to meet the new standards, present great challenges to engineering teams, especially when the construction of new facilities is not financially feasible, or the facilities are landlocked. Often, the most reasonable solution is the relocation of processes to other plant areas (e.g., warehousing space). The existing structure of a warehouse can be used as a shell structure, and by using multiple, easy-to-assemble component modules (modular design and construction), a simple factory can be built inside. As a conclusion, with the FSMA, the FDA takes food safety to a new level throughout the lifecycle of an entire renovation/building project, from design, construction, operation, and maintenance. The FSMA requires incorporation of more preventive measures into the building of new and the renovation of existing facilities by looking from floor to roof at everything that can affect food safety. Attention is paid to hygienic design and environments, which range from room finishes to improved air flow and quality within existing and new plants. Notice that today's pharmaceutical standards may be required in food production within a decade (Labs, 2014b).

3.3.4.4 New Environmental, Health, Safety, and Building Regulations

New and increasingly stringent safety and environmental regulations for both the in-plant and ambient environment are being enacted by governments worldwide, and relevant international and/or national building codes, such as the fire, electrical, mechanical, plumbing, maintenance, energy conservation, and health and safety emergency codes are regularly updated (Holah et al., 2014).

3.3.5 Design and Build for Future Introductions of New Technologies

3.3.5.1 New Technologies

The food factory design should meet current as well as likely future technology enablers (Bowen, 2013):

- Simpler process improvements.
- Improved equipment production techniques and design practices providing more economic interfaces and safer accessibility.
- New food preservation technologies.
- Smaller, modular, and flexible equipment set-ups and rigs.
- Use of isolator technology, reducing the need for further room containment.
- Laser systems increasing food safety (e.g., laser cutting) and enhancing operational efficiency (e.g., laser sealing).
- More advanced process monitoring technologies (e.g., in-line quality control by means of sensors, preferably wireless) and smarter and more visible control systems (e.g., bar-code scanning equipment, scanning-vision systems using high-speed cameras).
- Integration of advanced technologies within automation and robotics.
- Intelligent management, ensuring traceability.
- Improved hygienic engineering and design concepts.

3.3.5.2 Increasing Trend to Automation and Robotics

The automation of nonvalue-added tasks and those having an impact on quality and productivity is still in the lead. Also, robots receive more and more acceptance in the food industry due to increased speed, flexibility, and compatibility with automation. With both automation and robotics, it is possible to create a repeatable process that reduces the risk of human error, provides traceability, reduces contamination risk, and delivers consistent output. By automation and applying robots, it is possible to reduce human contact with food in the food processing area, especially as robots have low microbial counts. On the packaging side, more facilities are going toward robotics, particularly on new packing lines. These projects tend to pay for themselves in reduced labor costs fast enough to justify their additional capital cost. Declining costs and simplified programming will further accelerate the consignation of repetitive and injury-prone tasks to machines. Further, there is a trend toward tracking, creating the ability to isolate and qualify the smallest batch possible to identify inconsistencies and other issues before they reach or affect consumers (Higgins, 2013; Labs, 2014b).

3.3.5.3 Visual Communication and Remote Diagnostics

Personal communication tools that allow for visual talking will gradually do their entry in food factories. Touch technology, data collection and analysis, and handheld devices to monitor temperatures, cycle times, traceable data, and more can help with meeting operational and regulatory requirements. With real-time production information into their hands, operators and supervisors could take corrective actions when food processing starts to drift from optimal outcomes. Visual communication through sequenced imagery also can deliver the need-to-know information, with instructional images or videos explaining things without text. A mobile device, smartphone, or tablet in the palm of an operator's hand may show him how to operate a piece of equipment on the floor, and by providing remote diagnostics, they also can leverage technicians in troubleshooting, start-up, and maintenance. Visual communication tools providing an endless library of images succeed to overcome any language barriers much more easily than text messages or written instructions. Notice, however, that for security reasons, the information should still reside on a local server, not accessible from beyond the factory's walls. It is true that the today's endless possibilities of wireless communication have as major weakness that food factories may become more vulnerable to sabotage by terrorists, competitors, disgruntled employees, etc. Protection of in-company information, communication and control systems against hacking must be part of a well-thought food defense strategy (Higgins, 2013).

Closed-circuit television (CCTV) cameras—commonly used as means of "food defense" in critical process areas, reducing the risk of intentional food contamination by disgruntled employees or intruders—can serve as both a training and quality tool. Monitoring of process areas and operations by means of CCTV systems may provide ample evidence that operators have taken the right steps and added the proper ingredients to the batch. However, CCTV systems are often questioned by labor unions for reasons that employees feel continuously watched and controlled on working properly without break times.

3.3.5.4 In, On, and At-Line Sensors

Many traditional batch processes have now been replaced by automated production methods, helped by the introduction of advanced process control systems in the 1980s. The signal processing and actuating capacity of process control systems is now adequate. The full potential of these systems, however, can be realized only if they are supplied with full and up-to-date information on the process to allow feedback or feedforward control. This has prompted the development of a wider range of sensors and instruments suitable for in, on, and at-line measurements in automated continuous food processing and packaging operations running at high line speeds. In-line measurements are performed directly on the process line, while on-line measurements may be performed in a bypass loop from the main process line, which may then return the material or product to the main process line after measurement. Correct labeling of allergens is critical. Wrong labeling could lead to severe food safety issues such as anaphylactic shock due to allergens. Barcode scanning/vision systems will help to prevent mixing up of packaging material, which may lead to such incorrect labeling. Finally, at-line measurements involve removing the product from the production line and measuring it with suitable instrumentation in the production area.

New innovative technologies are under development that support both accurate and fast monitoring of food safety, product quality, hygiene, process, and environmental parameters, thanks to electronics that offer unprecedented opportunities to integrate food quality testing, process parameter monitoring, etc., in processing lines, food packaging, and storage facilities. Measurement and sensors are used for a wide range of applications:

- Monitoring of temperature and time, usually considered as critical control point (HACCP) in food maufacturing operations.
- Hygiene, with as examples pH and conductivity measurements during cleaning-in-place (CIP) operations, cleaning validation techniques, etc.
- Food safety, with as examples the detection of foreign bodies, chemical contaminants (toxins, pesticide residues, drug residues, etc.), and microbial contaminants.
- Protection of sensitive consumers, such as monitoring for allergens.
- Food quality, with as examples texture, color, flavor, etc., measurements.
- Food composition, with as examples monitoring of micronutrients, moisture content, fat content, etc.

3.3.6 Design and Build for Future Expansion

To meet a possible increase of processing activities within the food plant in the future, the building and its food processing support systems should be designed that they can either be expanded or another building and utilities could be added. Oversizing of utility systems is a common practice (Moerman, 2010; Wallin, 2013). Using rapid design-build methods to scale-out processes, such as modular design/build methods (Figure 3.2), provides for very rapid expansion if facility capacity becomes a problem. The concept of modular design and construction with onsite plug-in installation of standardized technological modules—generic platforms with well-defined sequences of unit procedures/operations—can dramatically improve facility deployment schedules. Modular construction design allows for off-site fabrication and testing of the generic platforms, while ensuring a higher degree of cleanliness during their construction as compared with onsite constructions. Onsite, the modules are then set on their respective foundations, after which the module/field connections can be made: process

Figure 3.2 The concept of modular factory construction and building by using multiple, easy-to assemble component sections is first applied by multinational pharmaceutical companies. The purpose-built factory sections are brought, ready-to-use, directly to the site and connected to each other according to requirements. In reverse, modular building construction also allows easy dismantling of a factory, as such allowing for fast transformation, renovation and refurbishment of old factories.
Courtesy of Steve Townsend, Commercial Property Photography. HMP-Oakwood – Balfour Beatty Construction Services UK.

piping, duct work, structural slab, flooring, wall partitions and ceilings, etc. As off-site construction of modular sections provides a higher degree of cleanliness, less validation work for cleaning will be required once the on-site integration activities are completed. This modular design and construction concept allows to build a food factory against aggressive time schedules (half the time as usual) for about 50–60% of the cost of a more traditional one. As the modules are standardized, precise-cost determination of projects is possible. Modular building design concepts offer a highly flexible, simple, and cost-effective solution for creating production sites in the developing world, especially because investing in Africa and some parts of Asia can be high-risk due to the potential political instability and the lack of infrastructure, reliable energy sources, and building expertise. However, modular building and construction also allow fast displacement of a factory to countries where consumer demand increases or where costs are lower (Witcher at al., 2013b).

3.3.7 Design and Build for Present and Future Segregation Requirements

In process areas designed according to the "ballroom" concept—defined as a large manufacturing area that has no fixed equipment and minimal segregation due to the use of functionally closed systems—there is an increased risk for cross-contamination, and interference may occur between different unit operations or products. Therefore, it is often required that the factory incorporates and supports optimal segregation strategies for separating processes and products manufactured. Highly segregated spaces may prevent cross-contamination between raw and processed foods, between products with and without a specific allergen, between dry and wet products, between halal and non-halal food, etc. The segregation concept also increases the flexibility to rapidly add and remove processes from the facility (Witcher at al., 2013b).

In factories originally designed according to the ballroom concept, manufacturing lines can be easily and hermetically separated from each other by means of one or —even more preferred—two series of panel walls, both of which may form a clean central corridor, eliminating any potential risk of cross-contamination (e.g., allergens) between lines. For the space to be segregated, the walls between the suite and the other rooms must extend from structure to structure with all penetrations sealed to be leak tight, and an independent HVAC system and means of backflow prevention on any utilities that serve the dedicated and segregated suite from the multipurpose facility must be in place (Wilkins, 2013).

3.3.8 Design and Build for Adaptability

Adaptability is the ability to modify the production area for other manufacturing purposes. For maximum adaptability, the process should be no longer an integral part of the facility, which means that decoupling the operating process from the facility is the new trend. Because of the decoupling of the design and construction of the building envelope and shell from the interior process components, it is possible to change the configuration of process lines without having to make facility changes. With poor adaptability just the reverse is true; it will require significant capital to reconfigure the process line and make facility changes, so as to support new manufacturing processes. Low cost unit operations changeovers, either between product campaigns or when a previous line of products is replaced by a completely new line of products, ideally can be done by making use of portable skid mounted unit operations or wheeled process equipment (Figure 3.3) instead of hard pipe interconnected stationary (read: fixed) stainless steel production systems. As such, processes can be moved in and out in function of customer demand. By using standardized unit operation skids and even

Figure 3.3 Decoupling the operating process from the facility is common sense in the biopharmaceutical industry, where wheeled and skid mounted process units are moved in and out. With this way of working, the production area can be easily modified for other manufacturing purposes.
Courtesy of Egemin Automation. egemin_life_sciences_dsc_2065_lr.

skid mounted utilities (e.g., process gases), the process also can be readily cloned to increase capacity, either by multiplying the number of skids within an existing facility, either by duplication and moving them to a new facility quickly and efficiently. However, one must notice that the production capacity of portable production systems is usually lower, and as such they are rather specific for low production volume applications. Where large production capacities are required, the use of stationary production systems is still common practice. In this approach, the facility elements (building, support functions, utilities, HVAC, etc.) are hard piped to this fixed process equipment and as such operated as part of the process, with as a drawback, increased interdependence between process and facility (Witcher and Odum, 2012; Witcher at al., 2013a,b).

Mobile process equipment also has other disadvantages. It is common practice to use tubing when mobile process equipment is used, while fixed stainless steel production systems are commonly hard pipe connected with each other and the utilities. Hoses increase the potential for tubing on the floor/slip hazards, and of more importance, they have proven to increase the risk for contamination. As process lines are continuously disassembled and assembled (requiring more operator handling in batch processing), the reliability and repeatability of the process may be compromised. Further, there is an increased risk to mix-up process equipment (components), especially if process equipment items look quite similar (e.g., erroneous use of uncleaned equipment components instead of cleaned identical equipment items including hoses, confusion between identical equipment items used for different purposes, mix-up of identical equipment parts used for respectively allergen free/allergen containing ingredients, halal versus non-halal, kosher versus non-kosher). Moving of process equipment in and out process areas also requires large doorways, wider walkways, and large storage areas for totes used to transfer non-wheeled equipment/utility skids. Moving large skids weighing hundreds of kilos with a tote also can be quite challenging, increasing the risk that operators get injured and doors, door frames, crash bars, and walls damaged (Hettenbach, 2014; Wolton and Rayner, 2014).

3.3.9 Design and Build for Versatility

Versatility indicates the ability to do different things within the same room. Utilizing closed processing allows flexible open plan layouts with the possibility for multiple products to be running in parallel within a large operating space (ballroom concept). These large operating spaces potentially reduce operating workload and create opportunities for the exchange of ideas and collaboration. When using closed process apparatuses, the product also will be contained within the borders of the process equipment, allowing to reclassify the surrounding area to a lower hygiene level, e.g., a high risk zone (zone H) may be reclassified as a medium risk zone (zone M). Requirements with regards to the quality of the air will be less stringent, and as such have a significant impact upon HVAC annual running costs. The higher the air classification of a given area, the greater the energy usage.

If the factory is designed multifunctional and versatile, which means capable to quickly and efficiently handle different multiproduct manufacturing requirements, it allows for improved facility utilization. High utilization rates almost always translate into cost-effective facilities. Building manufacturing facilities that are sized properly and have the characteristics and capabilities that allow them to run at high production rate provide high value to the enterprise (Witcher et al., 2013a).

3.3.10 Design and Build for Increased Worker Safety and Welfare

Food factories must be designed to meet current and future operational health and safety requirements, as imposed by legislation (e.g., OSHA in the United States) and requirements set by insurance companies. Ergonomic programs and communications protocols are essential components in driving employee engagement and safety. A food manufacturer can avoid a lot of injuries and absenteeism because of health problems if ergonomics are introduced into the design and planning stage of all new capital investments. By doing so, he can make the employee's work easier, safer, and more productive, often at no additional building cost. Ergonomic design can be implemented through a combination of checklists and preexisting experience in both initial plant design and all planned changes (O'Halloran, 2014). Activities that may increase worker safety and welfare are

- Implementation of the most appropriate process layout.
- Improving the light intensity while reducing glare. Eventually, allow the entry of natural light into work spaces or break areas, as—from a psychological point of view—daylight has proven to increase the welfare of the staff.
- Staff engagement and safety can be increased by interworker communication. A lack of interworker communication also may result in downtime, loss of efficiency, and other undesirable outcomes. Long distances between work stations and poorly designed process workflows can impair that interworker communication. To avoid that problem, it is recommended to involve the floor personnel in the design and engineering process to provide insights on the amount of communication required between different workstations and personnel (O'Halloran, 2014).
- Storage of all chemicals coming into a plant should be done in designated areas, secured and only accessible to authorized personnel. The area must be provided with signalization: "Authorized Personnel Only." Chemicals used for cleaning and disinfection should be stored in a well-ventilated area, preferably on pallets that will contain spills. Incompatible chemicals should be installed away from each other. Acids (e.g., phosphoric acid, acetic acid, peracetic acid, etc.) must not be kept near chlorine solutions (e.g., hypochlorite) because inadvertently mixing the two could result in the release of chlorine gas, which is extremely dangerous. Processors also often install special drainage or collection systems below chemical storage areas to ensure these materials do not get into public waste streams. Chemicals that are explosive or flammable must be stored in explosion-proof cabinets.
- Introduce features to reduce noise levels of loud equipment. Take care that they are hygienic.
- Limit the frequency and the amount of weight that operators must handle. Adjustable floor stands with all production lines provide enough adjustability for them to be set to the correct working height for each employee (O'Halloran, 2014).
- Elimination of continual bending by providing appropriate clearances between equipment and floors or walls.
- Control of the hazards associated with the use of energy (e.g., electric energy).

- ANSI Z244.1, which was established in 2003, outlines standards for achieving functional safety without disconnecting energy input, allowing machines to run in slow-speed mode or safe-off, while jams are cleared and minor maintenance is performed. Running machines at safe torque and safe velocity is about increasing safety, not short-circuiting it. Increased machine safety, especially for high-speed, large, or complex installations that require a large number of people or where there is a high risk of injury, is mainly attributed to the progress that is made in the field of automation. The dedicated safety PLC is being rendered obsolete and replaced by a single controller that—based on artificial intelligence—manages motion, logic, and safety functions. A controls platform with safety on board simplifies troubleshooting and servicing and more easily achieves lock out/tag out. Also, robots that are human aware when safe-distance protocols are breached have made the work environment a lot less hazardous (Higgins, 2013).

The food factory also must be designed from a life safety perspective by:

- Determination of the most appropriate and safe emergency routes and providing hazard and emergency signalization.
- Means for hazard communication.
- Ensuring that the site can be adequately served by fire workers, police, and rescue/first aid services.

3.3.11 Design and Build for Effective and Efficient Cleaning and Maintenance

A well-laid-out food plant can prevent or reduce contamination and allow for maximum cleanability, as well as for proper maintenance activities. Improved hygienic design of structural elements (beams, roofs, columns, walls, floors, etc.) and hygienic integration of piping and utilities within the food factory can reduce the build-up of liquids, dust, dirt, and other contaminants at the interior of the food factory building. Providing sufficient clearances around these structural elements, utilities, and process piping, as well as equipment, is a minimum prerequisite for maximal access during cleaning and maintenance. Chapter 4 gives an overview of the currently applied and emerging hygiene concepts in food factory design, at present considered as the best building practices to effectively and efficiently clean and maintain process areas in food factories.

3.3.12 Design and Build for Sustainability

The need for infinite growth and resources has given rise to a lot of environmental concerns, more specifically the increasing effect of human activity on the earth's atmosphere. The use of ozone-depleting refrigerants has led to a significant reduction of the protective ozone layer, and greenhouse gases (e.g., CO_2) produced through the excessive combustion of carbon-rich fuels bring about the threat of global warming. The reduction in the emissions of carbon and other greenhouse gases has become a legislative obligation in many nations worldwide. Also, China has started to take the necessary measures to reduce the exhaust of excessive amounts of pollutants in the atmosphere, because a lot of large cities in China are suffering from the continuous presence of smog.

Nowadays, companies make use of several methodologies to perform an environmental and lifecycle assessment of their new building projects at the design stage and subsequent to construction. The ISO 14000 series provide guidance on environmental management and the quantification and reporting of greenhouse gases. Some notable facility environmental assessment and rating methods are (Bowen, 2013):

- Building Research Establishment Environmental Assessment Method (BREEAM), an initiative taken by the state-owned Building Research Estabishment (BRE) in the UK in 1988, but which is now applied by the privatized BREGlobal (BREG).
- Leadership in Energy and Environmental Design (LEED) method, set up by the US Green Building Council (USGBC) in 1998, and launched in 2000 as the primary US assessment and rating system.
- Comprehensive Assessment System for Built Environment Efficiency (CASBEE) method, which is Japan's environmental method for building assessment, developed by the Japan Sustainable Building Consortium (JSBC). It is a self or third-party assessment tool, taking into consideration issues and problems peculiar to Japan and Asia.

Lifecycle Assessment (LCA) is a tool used in assessing the environmental impacts of a product, process, or service from design to disposal. LCA software is available from several sources, including those noted above, as either add-on services or standalone packages providing assessments for site and facilities as well as for individual product assessment.

The advantages of implementing such a facility environmental and lifecycle assessment at design and execution state are the potential savings on energy—annex operating costs—that can be realized, a more pleasant working environment, the appreciation of consumers having an increasing preference for environmental and ethical investments, etc. "Sustainability" has become a component of the marketing strategy. However, if a state makes the applicable legislation too onerous, it is less likely that global corporations are willing to invest in the country. To circumvent that potential problem, the European Union has imposed legislation to all of its member states. The EU applied what is known as the 20-20-20 target through three key directives: Energy Performance in Buildings Directive (2002/87/EC) in 2002, the Emissions Trading Scheme (2003/87/EC) empowered since October 2003, and the Renewable Energy Directive (2009/28/EC) in 2009. The energy performance directive requires the member states to achieve a 20% reduction in energy usage on 1990 levels and a similar reduction in greenhouse gases by apply minimum energy performance standards to new and large refurbished buildings using an energy performance calculation methodology and expecting displayed energy certification. In terms of renewable energy, the final 20-20-20, the target for the EU as a whole is 20% by 2020 (Bowen, 2013).

In 1990, after a period of regulatory restraint, the US government reformed the Clean Air Act to strengthen and improve existing regulations. The Clean Air Act addresses five main areas: air quality standards, motor vehicle emissions and alternative fuels, toxic air pollutants, acid rain, and stratospheric ozone depletion. The 1990 Clean Air Act Amendments require that states establish certain emission criteria in terms of the localities being "attainment" or "non-attainment" areas. Emission-generating equipment above 10^{10} J/hr generally requires permission. In many states, it is required to have emission-reducing equipment if more than 25 tons/yr of hazardous air pollutants are discharged from the equipment (referred to as a "major source"). The carbon footprint of a facility is defined as the "direct and indirect effect that individual and corporate actions have on the environment in terms of Carbon Dioxide (CO_2) emissions." Methods for determining carbon footprint status have been developed that can assist individuals and companies in determining these values. Legislation is developing to include Greenhouse Gases (GHG) into the Clean Air Act and require emissions reporting. Greenhouse gases include carbon dioxide (CO_2), nitrous oxide (N_2O), methane (CH_4), hydrofluorocarbons (HFCs), perfluorocarbons (PFCs), and sulfur hexafluoride (SF_6).

The "Food Factory of the Future" is a plant with features such as minimal energy consumption and waste production and maximal energy efficiency, while improving the overall conditions for the occupants. To reach that objective, a four-dimensional sustainable strategy should be applied: reduce demand, harvest free energy, increase efficiency, and recover waste. Bringing this strategy in practice, a sustainable building concept may consist of the following elements:

3.3.12.1 Application of Energy Best Practices

Based on the breakdown of the food manufacturing plant energy consumption by end use, energy savings should start with the large energy consumers. Facility design and layout play a dramatic part in the overall utility consumption of a facility, as the majority of the energy consumption in food manufacturing buildings lies within the process itself and the systems that support it. The consumption of energy goes beyond just powering the production equipment itself; it also encompasses increased base building utility system requirements from additional or more stringent HVAC requirements to greater lighting intensity requirements to additional utilities such as compressed air, vacuum systems, etc. One way to assess the overall performance of a factory is to compare it to other factories similar in form, function, and location. However, the manufacturing energy consumption is highly variable between factories, especially due to different requirements with respect to room temperature and humidity, room pressurization, cleanliness, containment, and other contributing factors. One of the largest energy consumers is just the air handling system required to maintain critical environments for production: air cooling, preheat and reheat loads, and fan energy consumption as determined by airflow quantity and ventilation requirements, humidification/dehumidification loads, energy to counteract the resistance of the filter during air filtration, etc. The mode of operation of similar HVAC systems can be quite different: recirculation systems versus 100% once through systems (Capparella, 2013). It is also sustainable practice to monitor energy consumption of individual users by means of submeters monitoring the consumption of electricity, gas, etc.

3.3.12.2 Energy Savings on Artificial Lighting

Energy savings on artificial lighting can be realized by the following means and methods:

- Atriums inside the food factory building, around which process rooms provided with glass walls can be built can maximize indirect entrance of daylight in the food production area. Manufacturing areas also can be placed adjacent to offices or corridors that have glazed exterior walls, hence allowing indirect entry of daylight in the food production and packaging areas by reflection and diffusion (Figure 3.4). Studies have shown that workers within the food production/packaging area psychologically can profit from optimized views to the exterior, improving their productivity and encouraging their fitting. The facility must be orientated so as to optimize daylight impact on the building (Moerman, 2011b; Holah et al., 2014).
- Food manufacturers can make use of prismatic skylights, or they can choose for a clerestory or sawtooth construction with the windows orientated to the north side (northern hemisphere) or south side (southern hemisphere) in order to avoid direct entrance of sunlight, which can warm up the food preparation, processing, and packaging area. However, windows and skylights should be kept out of processing areas (Moerman, 2011b).
- "Smart" system lighting that gradually adjusts to external conditions (e.g., daylight sensors).
- Intelligent warehousing, with occupancy sensors allowing user-controlled lighting.
- Use of more efficient lighting consuming less electricity, such as LEDs and T8 and T5 (HO) lamps.

Figure 3.4 Manufacturing areas can be placed adjacent to corridors that have glazed exterior walls, allowing indirect entry of daylight in the food production and packaging areas by reflection and diffusion. Louvered sunshades are used to maximize daylight exposure, while minimizing interior heat gains. The corridor also houses some piping, transporting heat recovered from the process toward the offices.
Courtesy of Cono Kaasmakers, the Netherlands. Clean corridor – Cono Kaasmakers.

3.3.12.3 Energy Savings by Reducing Demands on the HVAC System

The following practices may reduce loads on the air handling system, hence allowing for energy savings on air handling:

- Use of highly reflective building materials instead of black/dark building materials, the latter of which absorb almost all or high amounts of the solar radiation falling on their surface, as such gaining more heat from the sun. As an example, a roof coated externally with light-colored roofing instead of asphaltic roofing gains less heat from the sun, hence reducing the heat load on the air handling system in a plant.
- Louvered sunshades maximize daylight exposure, while minimizing interior heat gains (Figure 3.4). Prefabricated metal panels, advanced insulation, and architectural elements also may control solar gain. Panels are available with energy-efficient spray foam insulation that minimizes air infiltration, while maximizing insulation value.
- Solar shading in the Mediterranean and in subtropical and tropical regions, as such reducing the demands put on the HVAC system. Loading docks and refrigerated warehouses preferably should be orientated away from the quarter with maximum sunshine: the south in the northern hemisphere and the north in the southern hemisphere. Heat transmission and radiance also can be reduced to a minimum by providing several trees around, but not too close to the facility (Moerman and Fikiin, 2015).
- Small insulated windows with solar control glass (glass with low overall shading coefficients, usually low emissivity coated, tinted, translucent, opaque, and/or reflective colored) reduce glare and radiate and reflect away a large degree of the sun's heat.
- Insulation of hot water and steam piping to economize on energy, but also to prevent excessive heating of the food production environment above, for food safety reasons, an acceptable temperature (Moerman, 2011c).
- Equipment producing high heat loads (e.g., compressors) may certainly not be installed in process areas, as higher temperatures may compromise the safety of the food products produced. Concentration of the most energy consuming and heat producing equipment in dedicated areas allow transfer of the heat to technical areas where it can be eliminated at a reduced cost, sometimes by natural ventilation (Moerman, 2011c).
- Highly energy-efficient building envelope with ventilated façade.
- Insulating more efficiently (e.g., by using novel nano-based insulation materials).
- Use of isolator technology, reducing the need for further room containment, with as advantage that air volume exchange rates and hence the load on the air handling system can be decreased.
- Appropriately sizing of the HVAC system and provide with motor fans controlled by a Variable Frequency Driver that may reduce energy demand and use.

3.3.12.4 Energy Savings on Refrigeration Systems

Proper building design may allow for savings on refrigeration:

- Techniques to prevent infiltration of warm, humid air into and exfiltration of cold air from refrigerated areas (e.g., air curtains, fast-acting doors, vestibules, etc.), as such reducing the heat load the refrigeration system has to cope with. Loading docks should be located at that side of the food factory so prevailing winds do not blow into these docks. In case of high

heat loads, to keep up the refrigeration capacity at reasonable levels, both the compressor and fans of the refrigeration system need more power to work harder and for longer periods, decreasing the efficiency of the refrigeration system and increasing the costs of electricity (Moerman and Fikiin, 2015).
- Proper insulation of refrigerated warehouses.
- Application of industrial ammonia heat pumps may capture the heat that is extracted by the refrigeration system. So, the heat is recovered instead of wasted through condensers to the atmosphere.

3.3.12.5 Energy Recovery

The energy consumption within a food factory can be reduced by using:

- Traditional and advanced smart energy recovery techniques.
- Waste recovery systems (e.g., recovery of energy from exhaust air to hot water or air, but taking care that cross-contamination is excluded).

3.3.12.6 On-site Generation of Electrical Power and Heat

Companies can produce the necessary electric power or heat by their own means such as:

- Renewable energy production resources: solar arrays, wind turbines, etc. Photovoltaics are used for electric power generation and service water heating. Food and beverage facilities typically have big roofs that can be used for solar arrays without any impact on the usage of property, now and in the future. Although owners may not always opt to go solar at the outset, they must add 25–50 kg/m^2 to their roof loads to enable retrofitting their facility with solar panels at a later date, when the economics are more favorable.
- Cogeneration, defined as the sequential generation of heat and power from a single fuel source, has proven to be a sustainable means of heat and power production. Typical cogeneration systems are: combustion gas turbine generators, reciprocating engines, steam turbine generators, and heat recovery boilers. The implementation of cogeneration projects is considered primarily for economic reasons. During periods when electric power costs are high and fossil fuel costs are low, cogeneration tends to be an attractive option, though there are always occasional circumstances where an owner wants to own its power generating system because the electric supply is unreliable. Moreover, nowadays, environmental legislation has become a very significant element in the decision making process. The employment of cogeneration/central utility plant has the added benefit of almost universally decreasing the "carbon footprint" of the area in which the system is located (Masiello, 2013).
- Anaerobic digestion of organic waste allows for production of biogas as a source of renewable energy, which can be used for on-site power generation or heating.

3.3.12.7 Savings on Water

Cost savings can be realized by:

- Using low-flow plumbing fixtures.
- Recycling of process water rather than once through use, hence reducing waste water treatment costs and additional waste water subcharges.

- Use of gray water for heating and cooling purposes. Gray water is untreated, slightly polluted domestic wastewater originating from showers, bathtubs, hand basins, clothes washers, and dishwashers, but does not include toilet water (containing feces) and highly contaminated kitchen wastewater. Because it only contains traces of dirt, food, grease, hair, and certain household cleaning products, it is easier to treat and recycle than waste water from toilets.
- Treating its own wastewater instead of in wastewater treatment systems of the municipality. The wastewater treatment system should be designed to fit future needs (e.g., changes in the amount and composition of the waste water).

3.3.12.8 Reduction of Food and Nonfood Waste

Since the mid-1990s, sorting and recycling of nonfood waste has become common practice. Also, more attention is paid to the valorization of food processing wastes, allowing the generation of additional revenues:

- Valorization of sidestreams of food waste, allowing recovery of valuable food ingredients, phytochemicals, etc. Alternatively, another company may be willing to buy it for reprocessing or for animal feed.
- By anaerobic digestion of organic waste, the food manufacturer may reduce disposal costs, not least because digestate sludge is also a perfect fertilizer.

3.3.12.9 Reducing the Emission of Polluting Substances

Energy Performance in Buildings Directive (2002/87/EC) in the European Union and the Clean Air Act in the United States aim to reduce the emissions of pollutants, carbon dioxide (CO_2), and greenhouse gases by:

- Using environmental friendly building materials (plastics, insulation, etc.)
- Use of "clean" processes (which is better than have to resort to the use of an abatement system)
- Reducing the emission of odors, smoke, polluting, and noxious chemical components, such as nitric oxide compounds, sulfur oxide compounds, carbon monoxide (CO), and volatile organic compounds by emissions control devices. Food manufacturers often receive complaints from neighbors if unpleasant odors (e.g., smell of garlic, onions, vinegar, chlorine, cooked meat, poultry, fish, cheese, beer, poultry, curry, etc.), pollutants and noxious vapors are discharged with the exhaust air. These complaints may increase due to broken, missing, or undersized filters; due to a lack of regular cleaning and maintenance of the exhaust system; or due to inappropriate location of the exhaust outlet (e.g., too close to surrounding residences, large buildings that hinder the dispersion of odors) (Moerman, 2011d).

With respect to the latter, Singh et al. (2003) suggested the installation of fume incinerators and air pollution control devices in the path of travel of exhaust products, such as:

- System for the condensation of steam, water, and greasy vapor.
- Mist filtration of liquids, solids, and aerosols containing odorous compounds.
- Thermal incineration of the odor compounds with fuel and air at 750–850 °C.
- Catalytic combustion at low temperatures (e.g., 200 °C).

- Catalytic converters to break down chemical components. As an example, catalytic converters can break down vaporous hydrogen peroxide—used during decontamination cycles—into nonharmful constituents (Capparella, 2013).
- Biofilters with microorganisms oxidizing the volatile organic carbons and inorganic compounds.
- Adsorption on activated carbon or on activated alumina impregnated with potassium permanganate as oxidizing agent.
- Wet scrubbing (absorption) of odorous gases by means of a suitable solvent or chemical solution.
- Chemical scrubbing, where controlled quantities of acid, alkaline, and/or oxidizing agents are injected into the gas stream to neutralize odorous organic compounds.
- Photooxidation with ozone produced by means of short ultraviolet waves.
- Electrostatic precipitation to remove smoke out of a gas stream.

Many of these pollution control equipment and systems are well-developed, with the performance of many systems achieving an overall efficiency of 99.99% or better. For most systems, the major effort after installation and start-up involves maintenance.

3.3.12.10 Reducing the Geographical Footprint

In densely populated areas, there is often little space for expanding an old or building a new food factory, and due to local constraints of an environmental, conservation, industrial, residential or development nature (e.g., protected trees and sight lines, vulnerable watercourses, maximum noise emission levels, night working, etc.), it is often difficult to find a greenfield site on which to construct a factory. When the geographical footprint is limited, building in the third dimension (height) to add the necessary volume allows making the best use of space. As an example, in an effort to combat property constraints (e.g., cost of real estate, location availability, and available areas or existing operational sites), processors operate warehouses of the high-bay type to stack pallets up to 12-high, and they use more and more high-rise automatic storage and retrieval systems (AS/RS) instead of standard rack systems. With AS/RS, they can reduce operational costs affected by labor and maintenance (Wallin, 2013; Labs, 2014b).

3.3.13 Design and Build to Anticipate Climate Change

While there may be some discussion regarding the cause of climate change, there is no doubt that some areas have been affected by the phenomenon, e.g., the drying of California's Central Valley, the flooding caused by Hurricanes Sandy and Katrina, etc. Food manufacturers must strive not to build a plant that requires a heavy water supply in a dry area or locate a frozen food facility in a hot climate. The food factory of the future must anticipate for climate change and be weather-safe, as usually regulated by the building codes of the local jurisdiction (Labs, 2014d):

- In areas with heavy rainfalls, food factories should be located away from areas prone to flooding. The building must have adequate freeboard over site water levels assumed in flood risk assessment design, and there must be provided in terrain elevation for natural drainage. Where needed, there must be external drainage graded to storm sewers, which should not pass under food processing areas (storm sewers).

- Extreme rainfall also requires protected areas for the receiving and shipping of materials, covered bulk unloading stations and additional attention to roof system design, roof drainage and site drainage, plus the design of building envelope openings.
- Extreme snowfall requires preplanning for snow removal and storage (on- or off-site), so operations are not impeded. The design must allow for ease of access for shipping and receiving, grated exterior stair treads to mitigate snow build-up and the elevation of rooftop equipment. Roofs must have sufficient load bearing capacity to sustain the weight of a thick layer of snow.
- Designing for extreme winds requires careful planning, orientation, and the possible screening of building entry points to account for the prevailing wind direction. In dry areas, the winds may carry sand and dust particles, which will create the need for additional filtration of air intakes.
- Extreme temperature requires special attention to the design and construction of building envelopes and HVAC/refrigeration systems, to ensure that proper temperature and humidity requirements can be maintained to protect product from spoilage and contamination.

3.3.14 Design and Build to Protect Against Acts of Food Terrorism

Since 2001, many attacks occurred all over the globe (New York, London, Madrid, and Paris) by terroristic organizations. National security agencies are more and more aware that our food production and distribution chain is quite vulnerable to terrorist attacks with disease/death causing agents (e.g., toxic chemicals, botulinum toxin, *Escherichia coli*, and *Salmonella*). In contrary to food safety, which focuses on the reduction of unintentional contamination, food defense focuses on protecting the food supply from intentional contamination with a variety of chemicals, biological agents, or other harmful substances by people who want to do us harm. The food factory and the plant site must, respectively, be designed and developed to reduce the likeliness of intentional contamination of the food products that a food manufacturer produces. Food defense planning will continue to evolve an "all hazard approach," requiring as many experts as is necessary to determine the hazards unique to the particular situation. It is rare that a single expert will have knowledge of every possible threat (WHO, 2008).

3.4 Tools to Facilitate the Renovation or Design of a New Food Factory

3.4.1 Performing Simulations

During the conceptual stage in the design of a food factory with process equipment and utilities, simulation of processes, operations, and procedures can help to determine which equipment and utilities (e.g., required capacity and quality) are needed to manufacture a food product that is safe to consume. As an example, to implement appropriate manufacturing area conditions (e.g., low temperature and humidity) in the food plant, calculation of the heat expected to be generated by the process equipment and utility systems selected makes it possible to define the HVAC utility capacity required.

3.4.2 Three-Dimensional Visualization Programs

Three-dimensional (3D) design and visualization software (e.g., building information modeling, BIM) (Figure 3.5) and graphical computer aided design software may produce the 3D drawings and views that allow designers to give plant owners, managers, and employees a virtual walk-through the facility in 3D mode (Moerman, 2010). BIM has become the new standard in facility design, either to build new facilities or to retrofit existing ones, with many food and beverage processing plants embracing the technology's many advantages, such as:

- The 3D visualization tool allows the observation in three dimensions of all construction elements inside the facility—such as walls, windows, slabs, and roofs—from different angles and locations. The 3D model helps designers and owners to understand how the design of building components, including walls, floors, drains, columns, and lighting, function in relation to the process and cleaning/disinfection equipment, and how these elements will work together. All hygienic design and positioning problems that can negatively affect the cleanability and maintenance of process and service equipment, as well as all elements of the factory building (e.g., floors, doors, etc.) can be solved during the early stages of the project, at a moment that the ability to reduce project costs is the highest (Moerman, 2010; Eastman et al., 2011; Duff et al., 2013; Wallin, 2013).
- The simulation and visualization of employee movement and the flow of materials and products throughout the facility at floor level, including projected man/machine interactions, help owners to determine the best workflow for processes, in order to improve productivity. This again can help identify potential problems in work areas where a hygienic workflow is required. The final objective is to determine the most efficient building arrangement and layout to support the operations, to ensure that critical spaces are separated from each other, and that people and materials can travel from building entry points to the work areas without

Figure 3.5 Virtual walk-through the facility using the 3D visualization tool of building information modeling. 3D visualization of intricate process and support piping, instrumentation layouts, complex duct routings, and multilayered system routings prior to field installation is a powerful tool in the hygienic construction of food factories and the installation of the process and utility infrastructure.
Courtesy of the Stellar Group. BIM-3D visual.

the risk of cross-contamination, particularly between raw and RTE production areas (Duff et al., 2013; Wallin, 2013; Labs, 2014d).

- The possibility to integrate facility management software with a plant's BIM data allows facility managers to perform more efficient plant maintenance. A 3D, virtual view of the entire facility provides facility managers information on specific pieces of equipment and their locations within the facility. With the 3D BIM model maintenance supervisors can envision their crew navigating through the attic/interstitial space, utility rooms, and maintenance room spaces, which is more challenging on two-dimensional plans. As examples, they can verify if clearances between structural steel members, ductwork, and pipe runs in the interstitial space are wide enough, or if access to equipment in the ceiling is impeded by the supporting beams or columns. As a last example, if air units need servicing, facility managers can use BIM data to determine how to best access the equipment. Therefore, prior to the installation of the equipment and utilities, hazardous conditions in the food plant can also be defined (Duff et al., 2013).
- BIM supports increased involvement of and communication between interdisciplinary design teams. As it centralizes the design and data into a single source, it has the potential to facilitate collaboration on the different design aspects of the facility. External parties involved in different aspects of the design and construction can efficiently transfer and share information with each other through the BIM tools. Because the 3D BIM model can overlay all disciplines into one model, it also gives owners a better perspective on interdisciplinary conflict coordination. Finally, it may serve as a communication tool to stakeholders (Eastman et al., 2011; Duff et al., 2013).
- The energy modeling tool allows engineers to perform analyses and studies, so as to recommend and design the most appropriate systems for the facility from an energy use and efficiency standpoint. It can actually simulate how the building and equipment will perform once they are in use, so that the best energy-efficient solutions can be found, and the right decisions can be made. As a first example, conducting a natural lighting study on areas of the facility will help determine artificial lighting requirements for the space. Optimizing energy efficiency also can be done by selecting these processes and utility systems consuming the least amount of energy for a given process objective. Strategic placement of walls and windows also can improve the energy efficiency (Eastman et al., 2011; Duff et al., 2013).
- BIM has the potential to make budget calculations by comparing the costs of different design layouts, process, and service systems, etc. Once the appropriate design layout and equipment are selected, BIM allows for effective budgeting (Eastman et al., 2011; Duff et al., 2013).

3.4.3 Computational Fluid Dynamics

With the aid of computational fluid dynamics (CFD) software, visualization of the expected unidirectional and nonunidirectional flows in process rooms is possible. As an example, CFD may facilitate the design and construction of high hygiene areas (zone H), as it allows to visualize the need to relocate building elements, equipment, and utilities if the likelihood of poor airflow conditions (e.g., turbulences that may create unhygienic conditions) within a zone H process area is proven. With CFD, the designer may also survey the type, position, and capacity of the exhaust and HVAC system (Moerman, 2010).

3.5 Role of Experts, Contractors, or a Department Specialized in the Hygienic Engineering and Design of Food Factories

3.5.1 Need for Expertise

For a long time, many multinational food manufacturers have set up hygienic engineering and design functions within their company, but usually such dedicated functions do not exist in smaller and medium enterprises (SMEs), or only a small number of staff members has some knowledge of "hygienic engineering and design" concepts for both equipment and food factories. In quite a lot of SMEs manufacturing food products, knowledge of "hygienic engineering and design" concepts is limited, just as with most of the architects and contractors involved in their new construction projects. As a result, routes of contaminants due to poor hygienic design, building, and construction practices are often overlooked, and as a consequence foods processed in these facilities will be at risk once production starts. Too late detection of conceptual design and construction faults in a new food factory building, process line, or utility infrastructure may afterward require the permanent implementation of expensive corrective operating procedures or additional investment in infrastructure to eliminate or relief the problems caused by poor hygienic design and construction practices.

3.5.2 Role of the Hygienic Engineering and Design Experts in Project Execution

In the conceptual design phase, experts or contractors recruited from outside the company or the in-company department specialized in hygienic engineering and design must give support in the development of a building layout that complies with existing guidance for "hygienic design and construction of food factories." As an example, the working group "building design" of the European Hygienic Engineering and Design Group (EHEDG) has produced a guideline, "Hygienic Design Principles for Food Factories" (EHEDG Doc. 44), projected toward the design of new plants, though this guideline also can be adapted to existing factories and facilities (Holah et al., 2014). Guidance is also given in the handbook "Hygienic Design of Food Factories," published by Woodhead Publishing (Lelieveld and Holah, 2011). Moreover, several organizations have published their own guidelines to design and build food factories that meet the hygienic requirements dictated by national and international legislation. As an example, Campden and Chorleywood Food Research Association Group has published three guidelines (No. 39, 40, and 41) on how to design food processing factories, with emphasis on building layout, zoning, hygienic design, and construction of walls, ceilings, and floors (CCFRA, 2002, 2003a,b).

The hygienic engineering and design or food safety and quality function should be involved in the early detection of potential routes of contamination prior to the start-up of the factory construction work and the field installation of the equipment and utilities by:

- Studying the impact of the proposed building layout and zoning concept on all future operations (process, cleaning, maintenance, warehousing, loading, and unloading of goods, etc.), with emphasis on potential hygiene related problems. Full access to all equipment, utilities, and areas within the factory is required for sound cleaning, disinfection, repair, and maintenance.
- Visual estimation of the impact that a certain layout and arrangement of equipment, piping, and the utility infrastructure in the food factory may have.
- Simulating the flow of people, materials, products, and waste.

They must also check the building design concept for compliance with relevant international and/or national building codes, such as the fire, electrical, mechanical, plumbing, maintenance, energy conservation, and health and safety emergency codes.

Further, the hygienic engineering and design or food safety and quality function must be involved in vendor, contractor, and installer prequalifications, inspections, and factory acceptance tests. It is important to qualify all vendors, contractors, and installers for experience and knowledge prior to selecting a supplier, manufacturer, or installer. They must have knowledge and experience in the hygienic engineering and design of food factories. The guiding and monitoring of vendors, contractors, and installers must assure that functional and quality requirements will be met. Notice that it is unconvenient to use one general contractor for the complete job. Major subcontractors for electrical, mechanical, HVAC work, etc., must be recruited. The advantage of external experts is that they have learned from previous mistakes and see opportunities for improvements.

During construction phase, the hygienic engineering and design or food safety and quality function must continuously monitor the installation work quality. Regular walk downs by designated expert teams, as well as joint walk downs with the construction manager allow the discovery of quality issues quickly, allowing timely corrective action. The construction work must be qualified at regular time intervals, to ensure that contractors and installers meet set quality targets. Further, during the design, construction, and qualification phases, close contact with the regulatory authorities can be important.

Once the construction work is finished, all construction and service elements of the newly built food factory must be qualified. Also, the hygienic engineering and design or food safety and quality function must be involved in this installation/operational qualification, which means a check against the project documents to verify completeness and adequacy of contractor work. Tests, inspections, examinations, and reviews also must be performed to ensure that applicable codes, standards, and specifications have been met. Service systems must operate through a test cycle: airflow volumes, pressure differentials, and particulate levels (viable and nonviable), validation of control and automated systems, sufficient supply and acceptable quality of water, steam, etc.

3.5.3 Hygiene Measures During the Construction of a New Factory

The project implementation and survey department must establish stringent housekeeping and cleanliness requirements as early as possible during the construction

Figure 3.6 Keeping protective caps on open ends of the HVAC ductwork will help prevent dust and dirt generated on the jobsite from accumulation in this ductwork.
Courtesy of Fusion Tech.

effort. If the appropriate measures to maintain the cleanliness in the project environment are not taken, the ultimate quality of the project will be hampered. Some examples of a long list of hygiene measures that could be taken are:

- Cleaning of construction materials prior to entry in the construction site, as any debris entering the construction site must be removed afterwards.
- Prohibition or restriction of smoking, eating, or drinking within the project area, in order to avoid "built-in" debris.
- As the construction process proceeds, cleaning of spaces that will become concealed or inaccessible must be accomplished.
- Precleaning of utility ductwork (e.g., HVAC ductwork) and piping and keeping protective caps on open ends (Figure 3.6) may prevent dust and dirt generated on the jobsite from accumulation in this ductwork. Sometimes, large equipment has to be built and set in place, and walls, floors, ceilings, and mechanical systems are constructed around the equipment in a next stage. In order to keep this equipment as clean as possible, and prevent potential damage from the surrounding activities, temporary protective covers should be erected immediately after the equipment is installed. Maintaining the cleanliness of the equipment and materials being installed into the various assemblies will greatly help to reduce the final cleanup and commissioning effort.

3.6 Conclusions

As the food industry continues to work toward producing foods with lower or no preservatives and increases production of ready-to-eat products, including the increasing use of aseptic processing, there is not only a need for hygienically designed process equipment, but also factories and utilities that must create the correct environmental conditions to produce on a consistent basis high quality and safe foods for demanding and sensitive consumers. Food manufacturers are not only faced with the increasing expectations of consumers and food safety regulators; they also have to give answers on numerous other challenges, such as new technologies, more global competition, increasing worker safety and welfare expectations, limits on available resources, the increasing hazards of pollution and the dangers of climate change, and the threats of intentional contamination. Nowadays, food manufacturers must be flexible and develop the capability to look in the future, make predictions, and develop a vision where they need to go to answer the social, economic, technological, and environmental evolutions and developments in the years ahead. To support them in their task, they must make an appeal to a team of experts having experience in a large spectrum of disciplines.

References

Bohn, E., 2013. Master planning the legacy: meeting good manufacturing practices while using existing pharmaceutical manufacturing facilities. Pharmaceutical Engineering 33 (6), 32–38.

Bowen, R., 2013. Sustainability and its relevance to the pharmaceutical industry. Pharmaceutical Engineering 33 (5), 24–32.

Capparella, J., 2013. Energy benchmarking in the pharmaceutical industry. Pharmaceutical Engineering 33 (5), 34–40.

CCFRA, 2002. Guidelines for the design and construction of floors for food production areas. In: Timperley, A. (Ed.), Guideline N° 40 of Campden & Chorleywood Food Research Association, second ed. Chipping Campden, UK, 67 p.

CCFRA, 2003a. Guidelines for the hygienic design, construction and lay-out of food processing factories. In: Holah, J. (Ed.), Guideline N° 39 of Campden & Chorleywood Food Research Association, second ed. Chipping Campden, UK, 94 p.

CCFRA, 2003b. Guidelines for the design and construction of walls, ceilings and services for food production areas. In: Timperley, A.W. (Ed.), Guideline N° 41 of Campden & Chorleywood Food Research Association, second ed. Chipping Campden, UK, 93 p.

Cochran, J., Maurer, A.P., 2013. Construction waste reduction. Pharmaceutical Engineering 33 (5), 42–44.

Duff, J., Ross, R., Parrish, W., Phillips, J., Lane, M., 2013. Food Processing Plant Design. White Papers, Jacksonville, Florida, USA, Stellar, 7 p.

Eastman, C., Teicholz, P., Sacks, R., Liston, K., 2011. 'BIM Handbook', Guide to Building Information Modeling for Owners, Managers, Designers, Engineers and Contractors, second ed. John Wiley & Sons, Hoboken, New Jersey, USA, 648 p.

Hammond, C., Van Wormer, S.W., 2014. Construction practices: contamination risk reduction within ongoing operations. Pharmaceutical Engineering 34 (2), 52–60.

Hettenbach, J.R., 2014. Transfer station design for large scale API manufacturing. Pharmaceutical Engineering 34 (4), 24–30.

Higgins, K.T., 2013. Food Plant 2023. Food Engineering, January magazine, 89–100.

Holah, et al., 2014. Hygienic design principles for food factories. In: EHEDG Guideline N° 44, EHEDG Subgroup 'Building Design. EHEDG, Frankfurt, Germany, pp. 1–115.

Labs, W., 2014b. 'Hygienic design for safe food', 37th Annual Plant Construction Survey. Food Engineering, June magazine, 49–56.

Labs, W., 2014c. Renovate for the right reasons. Food Engineering, November magazine, 67–76.

Labs, W., 2014d. A clean slate: use it wisely. Food Engineering, December magazine, 49–56.

Lelieveld, H.L.M., Holah, J., 2011. Hygienic Design of Food Factories. N° 216. Woodhead Publishing, Cambridge, UK, 784 p.

Masiello, J.F., 2013. The application of cogeneration for pharmaceutical facilities. Pharmaceutical Engineering 33 (6), 80–92.

Moerman, F., 2010. Hygienic design of food processing facilities. Food Safety Magazine 16 (5), 50–54.

Moerman, F., 2011a. Hygienic supply of electricity in food factories (Chapter 19). In: Lelieveld, H.L.M., Holah, J. (Eds.), Hygienic Design of Food Factories, N° 216. Woodhead Publishing, Cambridge, UK, pp. 369–411.

Moerman, F., 2011b. Hygienic and innovative energy-saving lighting systems for the food industry (Chapter 20). In: Lelieveld, H.L.M., Holah, J. (Eds.), Hygienic Design of Food Factories, N° 216. Woodhead Publishing, Cambridge, UK, pp. 412–470.

Moerman, F., 2011c. Hygienic integration of process, process support and utility systems in the food factory (Chapter 21. In: Lelieveld, H.L.M., Holah, J. (Eds.), Hygienic Design of Food Factories, N° 216. Woodhead Publishing, Cambridge, UK, pp. 471–493.

Moerman, F., 2011d. Hygienic design of exhaust and dust control systems (Chapter 22). In: Lelieveld, H.L.M., Holah, J. (Eds.), Hygienic Design of Food Factories, N° 216. Woodhead Publishing, Cambridge, UK, pp. 494–556.

Moerman, F., Holah, J., Steenaard, P., 2014. Hygiene practice during maintenance operations in the food industry (Chapter 11). In: Lelieveld, H.L.M., Holah, J., Napper, D. (Eds.), Hygiene in Food Processing: Principles and Practice, N° 258, second ed. Woodhead Publishing, Cambridge, UK, pp. 384–407.

Moerman, F., Fikiin, K., 2015. Effect of hygienic design and operational parameters on frosting and defrosting of evaporators in refrigerated food processing and storage facilities. In: Gaspar, P.D., da Silva, P.D. (Eds.), Handbook of Research on Advances and Applications in Refrigeration Systems and Technologies. IGI Global, Hershey, Pennsylvania, USA, pp. 646–707.

O'Halloran, S., 2014. Employee engagement: creating a culture of operational excellence. Food Engineering August magazine, pp. 30–37.

Singh, A., Sundaravadivel, M., Kappor, M., 2003. Odour Control Technologies (Chapter 6). In: Engupta, B. (Ed.), Odour Pollution and Control, Parivesh, January 2003. Central Pollution Control Board, Delhi, India, pp. 9–14.

Sirabian, S., Matje, B., Biskup, J., Lehmann, W., 2012. Application of pre-owned equipment in pharmaceutical manufacturing operations. Pharmaceutical Engineering 32 (2), 30–44.

Stark, S., Viehl, S., 1995. Laboratory facility renovations: 15 considerations that can't be ignored. Pharmaceutical Engineering 15 (1), 28–32.

Sutton, K.P., 2013. Site considerations (Chapter 11). In: Baker, C.G.J. (Ed.), Handbook of Food Factory Design. Springer Science and business Media, New York, USA, pp. 283–296.

Wallin, P.J., 2013. Design principles (Chapter 12). In: Baker, C.G.J. (Ed.), Handbook of Food Factory Design. Springer Science and business Media, New York, USA, pp. 297–324.

WHO, 2008. Terrorist Threats to Food: Guidance for Establishing and Strengthening Prevention and Response Systems. Review 2002 edition. Food Safety Department, Zoonoses and Foodborne Disease Cluster on Health Security and Environment, World Health Organization, Geneva, Switzerland, 45 p.

Wilkins, S., 2013. Are you controlling your boundary? Pharmaceutical Engineering 33 (5), 86–89.

Witcher, M.F., Odum, J., 2012. Biopharmaceutical manufacturing in the twenty-first century – the next generation manufacturing facility. Pharmaceutical Engineering 32 (2), 10–22.

Witcher, M.F., Carbonell, R., Odum, J., Bigelow, P., Lewis, P., Zivitz, M., 2013a. Facility of the future: next generation biomanufacturing forum, part 1: why we cannot stay here – the challenges, risks and business drivers for changing the paradigm. Pharmaceutical Engineering 33 (1), 22–27.

Witcher, M.F., Odum, J., Zivitz, M., 2013b. Facility of the future – part 3: identifying facility requirements based on specific business drivers and uncertainties using the enabling technologies. Pharmaceutical Engineering 33 (3), 36–37, 40–45.

Wolton, D.A., Rayner, A., 2014. Lessons learned in the ballroom. Pharmaceutical Engineering 34 (4), 32–36.

Further Reading

Burton, F., Stewart, S., April 22, 2008. Placing the Terrorist Threat to the Food Supply in Perspective. Stratfor Global Intelligence. Security Weekly, Austin, Texas, USA, 5 p.

Centers for Disease Control and Prevention, 2003. Nicotine poisoning after ingestion of contaminated ground beef – Michigan. Morbidity and Mortality Weekly Report 52 (18), 413–416.

Chalk, P., 2004. Hitting America's Soft Underbelly – The Potential Threat of Deliberate Biological Attacks Against the U.S. Agricultural and Food Industry. National Defense Research Institute, Santa Monica, California, USA. Rand Corporation, 47 p.

Food and Drug Administration (FADA), 2007. An Overview of the CARVER + Shock Method Fro Food Sector Vulnerability Assessments. 14 p.

Federal Bureau of Investigation (FBI), 2007. Assassinations Using Poisons and Cold Steel. Lesson 16. Manchester terrorist training manual, Behavioural Analysis Program, Operational Training Unit, Counterintelligence Division, FBI Headquarters, USA. UK/BM-153 – UK/BM-160.

Hoffman, J.T., Kennedy, S., 2007. International cooperation to defend the food supply chain: nations are talking; next step – action. Vanderbilt Journal of Transnational Law 40 (4), 1171–1187.

Huff, A.G., Kircher, A., Hoffman, J., Kennedy, S.P., 2013. The development and use of the Food and Agriculture Systems Criticality Assessment Tool (FASCAT). Food Protection Trends 33 (4), 218–223.

Labs, W., 2014a. Semper Paratus: food defense depends on it. Food Engineering, May magazine, 46–55 .

McDonald, B.L., 2011. Terrorism and food security: understanding the changing landscape of threats and vulnerabilities in the food system. In: Paper, Conference "The future of Terrorism II",, 2011, October 19–21, Pennsylvania State University, Pennsylvania, USA, 15 p.

Mitenius, N., Kennedy, S.P., Busta, F.F., 2014. Food defense (Chapter 35). In: Motarjemi, Y., Lelieveld, H.L.M. (Eds.), Food Safety Management: A Practical Guide for the Industry. Academic Press, San Diego, USA, pp. 937–958.

Olson, D.M.A., 2012. Agroterrorism: threat to America's economy and food supply. FBI Law Enforcement Bulletin 81 (2), 1–9 U.S. Department of Justice, Federal Bureau of Investigation, Washington, DC, USA.

Sekheta, A.M.F., Sahtout, A.H., Sekheta, F.N., Pantovic, N., Al Omari, A.T., 2006. Terrorist threats to food and water supplies and the role of HACCP implementation as one of the major effective and preventive measures. Internet Journal of Food Safety 8, 3–34.

Taleski, V., 2006. Foodborne diseases-bioterrorist threat to public health. In: Lecture on HACCP conference: 'How to make HACCP more efficient in practice?', 2006, November 3–4, Morevska Toplice, Slovenia. 5 p.

USDA FSIS, 2003. Keep America's Food Safe. leaflet, Food Safety Inspection Service, United States Agriculture Department, Washington, DC, USA, 10 p.

USDA FSIS, 2007. Guide to Developing a Food Defense Plan for Meat and Poultry Slaughter and Processing Plants. Food Safety Inspection Service, United States Agriculture Department, Washington, DC, USA, 17 p.

USDA FSIS, 2008. Guide to Developing a Food Defense Plan for Food Processing Plants. Food Safety Inspection Service, United States Agriculture Department, Washington, DC, USA, 15 p.

Hygiene Concepts for Food Factory Design

4

Frank Moerman[1], Patrick C. Wouters[2]
[1]Catholic University of Leuven, Leuven, Belgium; [2]Cargill, Food Safety, Quality and Regulatory, Schipol, The Netherlands

4.1 Introduction

Hygiene at factory scale is as critical for combatting product contamination as at equipment level. A well-designed and built food plant will prevent contamination due to rodents, insects, birds, and microorganisms. A strategy based upon various barriers can be applied, which essentially supports the prevention of product contamination. The following elements should be taken into consideration; namely, the selection of an appropriate site, procedures to keep the exterior clean and tidy, and the design of the factory building. Attention must be given to ensure the correct foundation of the building, the supporting structure, walls, and roof in order to guarantee the structural integrity of the building. To maintain a maximum level of hygiene in all areas inside the factory building, there are several critical aspects. Firstly, proper access control at doors, windows, hatches, and air intake should support good hygiene. Secondly, the design of the structural elements such as beams, columns, and floors in the interior of the building should minimize build-up of dust and contaminants and maximize cleanability. Finally, a logical and correct flow of materials, personnel, and waste including drains, process aids, and controlled air management are essential to avoid cross-contamination. The right level of hygiene can be achieved by the implementation of a risk-based zoning concept to ensure the correct hygienic design measures are taken based on the hazards identified.

This chapter describes hygienic design concepts for the design of food factories. The overall objective is to reduce the likeliness of contamination, which will allow for the production of vulnerable food products including infant foods and ready-to-eat (RTE) products.

4.2 Food Plant Building Concepts to Prevent Entry and Harborage of Pests

Food safety regulations prohibit the entry of animals or birds, other than those essential as raw material, into food premises. They oblige food manufacturers to take effective measures against pests (rodents, birds, insects, and other vermin), so as to protect the food from contamination. The use of insecticides and rodenticides is permitted under precautions and clearly defined restrictions to prevent the contamination of

Innovation and Future Trends in Food Manufacturing and Supply Chain Technologies
http://dx.doi.org/10.1016/B978-1-78242-447-5.00004-6

food or packaging materials with illegal residues. The Food Safety Modernization Act (FSMA) even requires manufacturers to minimize the use of chemical pesticides. As an alternative to chemical pest control, pest-proof building design concepts should be applied to prevent entrance and harboring of pests and birds.

4.2.1 Surroundings of the Food Plant

4.2.1.1 Site Selection and Orientation of the Factory

Site selection is critical to food safety. The factory preferably should not be built in the proximity of sanitary landfills, junkyards, biological and chemical processing plants, municipal sewage plants, and industries that produce smoke, dust, odors, and microbiological contaminants. A lot of these activities are related to the presence of pests. Where present, a plant layout and design should be chosen that will keep these contaminants out. The plant buildings should be positioned so that prevailing winds hit the visitor parking lot and administrative side of the building. Loading docks and refrigerated warehouses should be orientated on the opposite side, away from the sunshine, wind, and/or rain (Figure 4.1). For a processor that must locate a slaughter operation as part of or near preparation and decontamination operations, the holding pens and slaughter operation should always be on the opposite side of the plant from the wind direction. This ensures that the wind will not be pushing air through the slaughter area into the processing area and also aids with odor control (Labs, 2014).

Significant pest and infestation problems also can arise in the proximity of woods, rivers, and canals (e.g., excellent spots for the breeding of insects and mosquitoes). There shall be no ponds or large standing water bodies on the plant grounds, as they attract insects, birds, and rodents. To exclude flooding and the entry of rodents, factories should be built at a higher level to the ground outside, which must be sloped for proper drainage away from structures. A site that is too flat may be difficult to drain

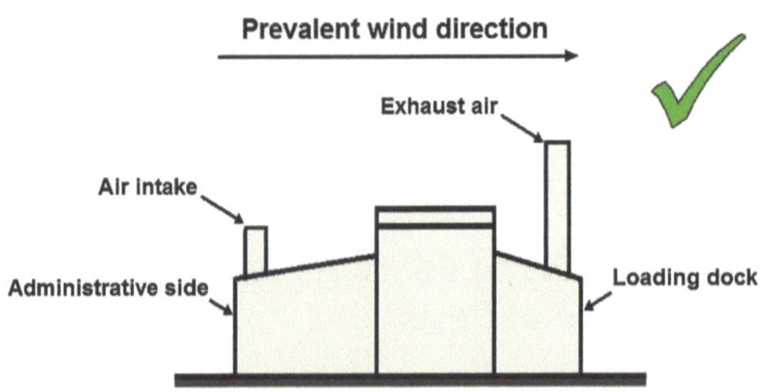

Figure 4.1 The plant building should be positioned so that prevailing winds hit the visitor parking lot and administrative side of the building. Loading docks and refrigerated warehouses should be orientated on the opposite side, away from the quarter with maximum sunshine, wind, and/or rain. The air/HVAC intake may not be positioned downwind from exhausts. Exhausted air must be immediately carried away and diluted without entering the factory's air intake system (Holah et al., 2014).

properly and lead to standing water conditions (either on the ground or in underground drainage pipes), which can become a breeding ground for mosquitoes and attract water fowl and other animals. Sloped roadways of a dense, hard, compacted, and dust-sealed material (e.g., concrete, asphalt, paving) suitable for wheeled traffic must indicate the correct routes for vehicular flow. Areas around the factory where bulk materials may be spilled upon receipt must be cleanable, as any food material may attract rodents, insects, and birds (Holah et al., 2014; Labs, 2014). More about selecting a site to build a new food factory is published by Lelieveld (2011).

4.2.1.2 Vegetation

There should be minimal vegetation and foliage around the food factory. Flowers and plants may attract insects (e.g., bees) toward the food factory. There should be weed control to reduce harborage for insects and rodents but also to prevent airborne seeds from getting into the plant. With respect to shrubs, proper earth to wood clearance should be provided. When present, shrubs and plants should be located at least 3 m and trees located at least 9 m from buildings (Figure 4.2). An area of minimum 60–90 cm around the building should be free of vegetation; Monterey and Bishop pines should not be planted near pedestrian areas (risk of blackbird nesting). The site should have a path or pebbled gravel strip around the factory (Figure 4.3) that is at least 90 cm wide and 10 cm deep. Plastic sheeting may be used below the gravel for weed control. During landscape renovation, do not raise the soil level against the building (Moerman, 2010; Holah et al., 2014).

4.2.2 Building Foundations

Foundations should be designed to prevent the access of pests to the building. They shall be designed to guarantee stability and to prohibit any movement in the building structure which may cause wall and floor finishes to crack, as such forming microbial and other hazard (e.g., pests) harborage sites. Foundations should be concrete based; wood, brick, and block foundations should be avoided. Pile foundations are

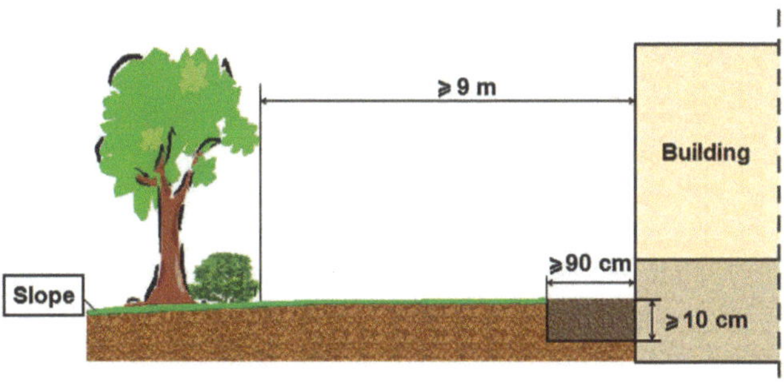

Figure 4.2 The site should have a path or pebbled gravel strip around the factory that is at least 90 cm wide and 10 cm deep, and trees should be located at least 9 m from buildings. To exclude flooding and the entry of rodents, factories should be built at a higher level to the ground outside, and there must be provided in proper drainage away from the building structures (Holah et al., 2014).

Figure 4.3 Areas in a perimeter of 3 m of the factory must be kept free of vegetation to avoid pest breeding and harborage sites. There should be a gravel strip and path around the factory building of at least 90 cm. No shrubs or plants should be located in the close vicinity of the factory building (Moerman, 2010).

commonly used if no or insufficient load bearing capacity of the subsoil is available, or if these factors are combined with a high level of groundwater. Foundations should be at least 60 cm below grade and should be coated with asphalt waterproofing. Adequate planning for storm water control should be undertaken, including the installation of drain tiles around the foundation perimeter.

For a new building, to prohibit rodents burrowing under the floor slab from entering, a concrete L-shaped foundation (Figure 4.4) should be poured, 610–915 mm below grade or ground level and with a horizontal slip extending 300 mm out from the base. For existing buildings, a 6-mm galvanized mesh screen extending 0.3 m into the ground and a right-angled, 15 cm wide, outward extending shelf should be used. Porches and ramps shall be sealed to the foundation, or shall be used to form a barrier to digging pests (rats and skunks) (Holah et al., 2014).

4.2.3 External Walls

External walls, commonly constructed out of concrete, brickwork, steel plating, or sandwich panels, should be weather-, water-, insect-, and rodent-proof. The façade surface has to be smooth and uniform, and special surface coating systems may be required to avoid dust and the penetration of moisture in the external wall elements. Building materials must be maintainable and cleanable. Walls may not contain protruding ledges or architectural elements that may attract birds. So, wall exteriors should not have horizontal surfaces; all surfaces should have gradients ≥45° (Figure 4.5). Decorative lattices (bird roosts) shall not be placed over entrances to food factories. Where needed, and if possible, bird spikes can be used. Also, overhangs must be eliminated to discourage birds from nesting (e.g., swallows) (Holah et al., 2014).

All construction from foundation to roof should be sealed to prevent access by pests such as bats, birds, and rats. Although rats and mice can gain entrance through,

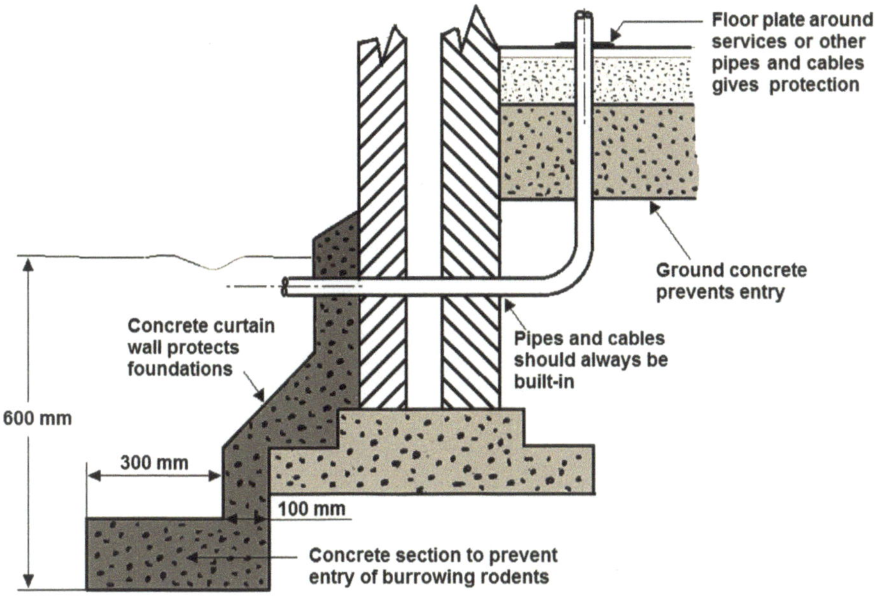

Figure 4.4 Pest-proofing measures to prevent entry through foundations. A curtain wall of at least 600 mm below ground level with a bottom member turned outwards for a distance of 300 mm prevents or reduces the ingress of burrowing rodents (CCFRA, 2003b; Holah et al., 2014).

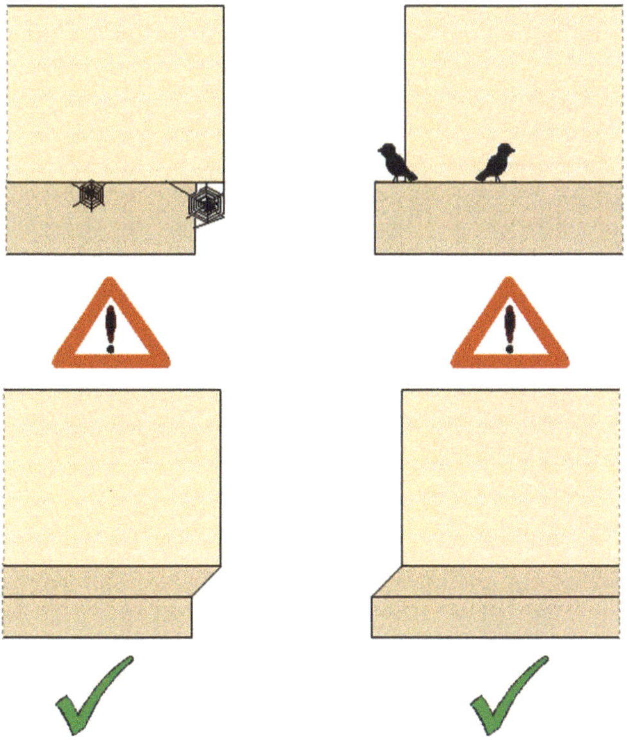

Figure 4.5 To be insect and bird-proof, wall exteriors should not have horizontal surfaces (gradients ≥45°) (Holah et al., 2014).

Figure 4.6 A cover prevents access of insects and rodents in the weep hole.
Courtesy of Tamlyn – Connected to the Industry®.

respectively, a 12mm hole and 6mm hole, façade masonry should include weep holes. Weep holes on masonry wall construction are necessary to drain and ventilate moisture absorbed by the masonry wall, but at the same time they are an open door to a building for roaches, rodents, snakes, spiders, scorpions, lizards, etc. A weep hole can be fitted with a cover (Figure 4.6) to prohibit the entry of these pests. Further, cavity walls for insulation purposes should be filled with rock-wool insulation and be capped at the base and top. All other cavities should be avoided, and if bricks or blockwork themselves have internal cavities, these should be filled with mortar, at least for the first few courses. At the exposed parts, such as verges, eaves and ridges, the connections between the inner shell of the roof and the wall elements shall be installed airtight. All openings in external walls, e.g., for pipe traverse, should be effectively sealed water and airtight.

Joints between sandwich panels should be sealed with food-grade joint sealant, e.g., silicone with added fungicide (durable for approximately 5 years) or food-grade polyurethane (durable for approximately 15 years). All fittings used to join and assemble the panels should be made of stainless steel. All compartments and openings in the sandwich panels should be properly sealed and if necessary, finished with sheeting of the same coating and colors of the panels (Holah et al., 2014).

4.2.4 Roofs

Roofs may become a major source of microbial pathogens, primarily derived from birds that roost on the roof or feed on food debris discharged through air extracts. All routes of contamination from the roof into the building, either via personnel access, air intakes, and inadvertent leakage, should therefore be controlled (Holah et al., 2014):

- The roofs should be hermetically sealed under all weather conditions, especially at expansion joints or building connections, so as to effectively prevent infiltration of water. As pooling water on a roof may be considered by birds or rodents as an optimal place to drink, roofs

Figure 4.7 The roofs should be hermetically sealed under all weather conditions, so as to effectively prevent infiltration of water. To ensure positive drainage, roof structures with a transverse joint and/or roof openings should have a slope of 9% (5°). Roof hatches should be properly protected against ingress of water (rain), debris, and pests.

should be sloped to a minimum of 1–2%, so as to ensure positive drainage. Minimum slopes apply to roofs with and without specific features (transfer joints, roof openings, etc.): 5% (3°) for roofs with transverse joint and without roof openings; 9% (5°) for roof structures with a transverse joint and/or roof openings (Figure 4.7). All downspouts should run externally of the building envelope. If this is not possible, they should run outside hygienic areas. For flat roofs, water can be transported in channels (made of roof material or stainless steel) on the roof over the hygienic areas, and further downwards by gravity through downspouts placed outside these hygiene areas and equipped with bullet-nosed roof grates that project upwards to prevent blocking. Downpipes must always be secured through traps that help prevent rodents entering the building through the downpipes (Figure 4.8).

- All roof openings (e.g., vents, air intakes, exhausts, exhaust fans) should be kept to a minimum and must be properly protected against ingress of water (rain), debris, roof rats, insects, and birds by proper screening.
- Provide uncluttered installation of utilities externally to the building envelope (Figure 4.9), with sufficient clearances between all utility piping and the roof, hence allowing proper access for cleaning and maintenance. Roofs containing stone ballast should be avoided in the food industry as they are not cleanable. Similarly, "green" or "environmentally friendly" roofs that sustain the growth of grasses and other plants have hygiene implications. These roofs must be fully sealed from the building and have external drain pipes, and all access to the roof should be from outside the building.
- Eave roof tiles eventually may have bird stops to exclude bats, bees, and umbrella wasps.
- Maintenance exits to the roofs must be well managed with the right gowning procedures in place to prevent cross-contamination from shoes.

4.2.5 Windows

At least 1.2 m above floor level, windows can be installed in close fitting frames, fitted flush, and continually sealed to the walls. Window frames should not be made of wood.

Figure 4.8 Access of rodents to the roof can be avoided by inserting a "rat stop" into the roof drainage downpipe (Holah et al., 2014).

Figure 4.9 Nonweather and water-proof asphalt roof. Liquids may leak to the inside through the many crevices in the asphalt barrier. The roof does not allow adequate drainage and cleaning. Roof openings are unscreened, permitting ingress of rain, debris, and pests. Many horizontal ledges may provide an optimal platform for nesting of birds, and the cluttered installation of utilities makes it easy for rodents to hide (Holah et al., 2014).

Windows shall be designed to prevent harborage, which means without sills or horizontal ledges. However, if unavoidable, exterior and interior window sills should be sloped to avoid accumulation of debris: outside window ledges should have a minimum 60° slope to prevent bird nesting, and inside window ledges a slope of 20°–45°. Clear passageways for pests to have access should be avoided (e.g., weep holes shall be designed in a way to prevent access of pests; window openings must be screened to exclude beetles and moths; ventilation louvers shall be screened with wire mesh to exclude birds, rodents, cats, etc.). However, windows in process areas preferably should be unopenable to prevent entry to birds and insects. It is even better to have no windows at all in process areas. If windows are present, then they should be double glazed to prevent condensation. Transparent and opaque polycarbonate glazing material, such as Makrolon®, should be used, which darkens as the sunlight becomes stronger.

4.2.6 Doors

4.2.6.1 Need for Hygienic Design

Doors play an integral role in building design by helping to segregate production areas, while presenting a barrier to contamination such as dirt, insects, and other pests and vermin. However, as food products have to move through doors, there is a possibility that contamination can drain, drip, diffuse, or be drawn into the product from door surfaces. Therefore, the food factory should have the least possible number of personnel entries while acknowledging security and fire escape requirements and the doors must be hygienically designed. Doors should not be made with hollow profiles and have hollow spaces, or at least the hollow should be filled with polyurethane foam. Also, leveling devices may not provide hollow bodies. Doors should be rust- and dust-proof, not made of absorbent materials, easy to inspect and clean, and be able to withstand state of the technology of cleaning. The use of windows in doors is not recommended, and if they are present, they should be polycarbonate or reinforced glass. Doors also must be effectively sealed to prevent microbial hazards caused by condensation. The junctions at jambs, walls, and floors must be sealed to eliminate all cracks and crevices for debris and dirt to collect (Sutton, 2013; Holah et al., 2014; Moerman and Fikiin, 2015).

4.2.6.2 Exterior Doors

Exterior doors must be primary barriers to infestation and ingress of dirt, dust, and warm humid air, and as such should not open directly into food production areas. External doors to the production bays, packaging and general stores should be of robust design, steel construction (wood is vulnerable to rodent attack) and power operated. Door closing systems (self-closing type) are important in relation to hygiene in entries for traffic, but personnel entries do not necessarily need such systems. Doors also must be defended against damage/knock-open by wind, and the best available technology for that purpose is to use a door closer. The chance that exterior doors knock-open by wind also can be reduced if they are conceived to open outwards instead of inwards. Exterior doors opening outwards instead of inwards also prevent that any insects on the door panel are swept into the building by an individual's passage. External door installations must prevent bird, rodent, and insect access between and underneath the doors when closed. Therefore, doors should have tight-fitting frames and the interfacing of door frames with walls and floors must be smooth, without cracks and holes where dirt and insects can be harbored. Gaps at exterior doors shall not exceed 6 mm, as a full grown rat can enter through a 10 mm crack and a mouse through one of 6 mm. Because plastic strip doors are not rodent-proof and only effective in keeping out birds and flying insects, they only can be used as a backup for external doors (Sutton, 2013; Holah et al., 2014).

4.2.6.3 Interior Doors

Interior doors are used to separate areas with different hygiene requirements. Interior doors also often play an important role in maintaining positive air pressure in areas requiring higher hygiene levels, requiring a door with good sealing that fits tightly in the frame. Interior doors also must prevent pests (beetles, mice, and moths) from passing through. They must not be faced with timber or glass, but a polycarbonate vision panel is recommended for doors in corridors. The closing device must not collect residues, and therefore floor level devices which have hollow bodies should be avoided. All door operating systems, especially those containing lubricants, should be effectively sealed.

4.2.6.4 Door Types

Several door types are available on the market, each with its own functional and hygiene characteristics (Holah et al., 2014; Moerman and Fikiin, 2015):

- A standard swinging door only for personnel, 0.9-m wide by 2-m high and designed for the temperature and humidity involved, is adequate and causes few hygiene problems. Horizontal double-swinging impact doors are biparting and consist of inflexible panel. However, they do not act as fast as sliding doors and are very sensitive to damage, especially in the case of forklift traffic. Once damaged, they will cause hygiene and maintenance issues.
- Vertical opening roll-up doors are single flexible panel doors that move only in the vertical plane and when open, coil tightly onto a drum assembly. Vertical opening roll-up doors act very fast at a speed of about 1 m/s and also may reduce the time that the door is open, which is especially important where frequent openings are required. Rapid roll doors are generally designed to survive impact with a forklift truck, and they also cause less damage to the truck and driver. But from a hygiene point of view, vertical opening roller shutter doors are not acceptable because debris from the bottom of the door seal (potentially containing pathogens) can drip into product/packaging, etc., passing underneath. If used, the seal should be frequently cleaned and well-maintained. It is also difficult to make roll-up doors rodent-proof, as the sides and top often provide more than adequate clearance for rats and mice to enter. Providing the door with brushes on the sides and top will prevent most rodents from entering; however, care must be taken with foreign matter cross-contamination risks as a result of loose brush hairs. Therefore, maintenance is important.
- Horizontal sliding doors (Figure 4.10) are standard when electric operation is provided. Horizontal sliding doors (either single-slide or biparting) are especially used for pallet material handling. The two-leaf biparting doors keep opening and closing time to a minimum, and the door is out of the way and quite protected from damage when open. Potential for major

Figure 4.10 Horizontal sliding doors allow for hygienic operations, at the condition that all gaps are closed between the door and the frame with gaskets of rubber or bush stripes. Sliding doors should not have channel floor tracks. Protection of the door, door frame, and wall is required. Courtesy of The Stellar Group.

damage may be reduced by proper location of pull-cord switches, which are used to activate door opening. Horizontal sliding doors allow for hygienic operations, at the condition that all gaps are closed between the door and the frame with gaskets of rubber or bush stripes. No brush strips should be used in high hygiene areas. With horizontal sliding doors, no material will be lifted up from the ground, so that the likeliness that contamination drips or is drawn into the product from door surfaces is excluded. Sliding doors should not have channel floor tracks and an inverted U channel at the top because both can act as a dust and dirt trap.

- Vertical sliding doors are useful in areas with dimensions that would not permit the use of a roll-up or horizontal sliding door. From a hygiene point of view, they suffer from the same problems as vertical opening roll-up doors in that debris from the bottom of the door seal can drip into product/packaging passing underneath. Therefore, they are more appropriate for truck receiving and shipping docks.
- Overhead garage-type doors (manual or automatic), that are hinged, multipaneled doors moving from the vertical to the horizontal plane should not be used in exposed product areas or areas subject to wet clean-up as they have spaces between the panels that could allow meat and fat, as well as other contaminants to collect.

4.2.6.5 Door Protection

With regards to door protection, if the surface of the door (or wall) becomes damaged, pests and microorganisms may be harbored, and if the surface opens onto a void or absorbent materials, fluids may penetrate, which could give rise to microbial growth. Doors should be high and wide enough to allow movement of vehicles and products without coming into contact with the door or jamb. In order to keep doors (and walls) in a hygienic condition, door (and wall) protection should be used to protect them against impacts from tray racks, pallet trucks, forklifts, containers, bins, electrical trolleys, etc. Doors needing protection from physical damage should be fitted with stainless steel or aluminum panels and edges. Exterior doors should be reinforced with a 0.6mm metal plate up to 1m from the ground, while interior doors should be equipped with kick plates and push plates, and doors in forklift or pallet areas must be of heavy duty construction, reinforced with frames and plates. They should be made from a metal plate with a minimum of 0.6mm thickness up to 1m from the ground or polyethylene crash protection layer with 15mm thickness. Galvanized or stainless steel posts are used to protect doors and door frame.

4.2.7 Loading and Unloading Docks

The number of loading and unloading docks should be minimal and 1–1.2m above ground level. To protect raw materials and food products from adverse weather conditions during loading and unloading, docks should be shielded from the elements by roofs or canopies. The canopies above the door opening reduce the infiltration of rain, wind, and warm, humid air from outside into the food factory building. However, these structures can become a serious sanitation problem due to the roosting or nesting of birds. Therefore, canopies should be completely enclosed on the underside so birds cannot gain access for roosting or nesting, or bird spikes or nets should be used. As a better alternative, dock seals (cushion-closure seals) around the truck doorways (Figure 4.11) should be used to fully seal the rear of the delivery vehicle (Moerman and Fikiin, 2015).

There should be provided in an overhanging lip with smooth and uncluttered surfaces that are sloped slightly away from the building to encourage run-off of water.

Figure 4.11 Dock shelters, such as canopies above, and dock-seal systems, such as cushion-closure seals around the truck doorways, will reduce the infiltration of warm, humid air and rain into refrigerated loading docks. These measures also prevent the entry of insects. Courtesy of Hörmann KG.

Areas beneath docks should not provide harborages for pests, should be paved and drained adequately. Gaps must be avoided, particularly beneath the leveling plate (Figure 4.12) between the truck and the dock. Ideally, dock levelers should have no pits under the plates, as this pit is open to the outside environment, and the dock leveling devices provide places for pests to harbor. Rodents that are able to access the pit may finally enter the factory by squeezing through the space between the leveler plate and the factory floor. However, they also can be made rodent-proof with brushes (Holah et al., 2014; Moerman and Fikiin, 2015).

Typical dock doors are vertical lift, garage-type "up-and-over" overhead or roller shutter type doors, which must open and close rapidly to minimize air and pest movement into the factory. When dock doors are regularly open, air curtains or plastic strip curtains are useful to prevent infiltration of warm, humid air and prohibit the entry of insects. Strip curtains are suited to packaged food, but not for open food products (e.g., cross-contamination between meat carcasses passing the strip curtain may occur). At least, the strips must be overlapping and frequent adjustment, maintenance, and replacement are required to maintain optimal effectiveness. If damaged, insects, heat, and moisture will migrate though the gaps. Air curtains are more hygienic but should extend all the way across the door and positioned on the outside to direct the air down and out. A very badly functioning air curtain can actually be worse than an open door. Moreover, the plane turbulent air jet can stir up contaminants, such as dirt and dust, if the area around the door is not kept clean. Finally, both plastic strip and air curtains lose their functionality at high wind speeds. External lighting to illuminate the truck docks should be placed in locations away from the factory building (Moerman and Fikiin, 2015).

Figure 4.12 Dock doors with leveler plates, which can be raised or lowered to match the height of vehicle loading beds. Factory floor level is a few centimeters above dock floor level. The trailer doors swing open inside the building to control infiltration of bugs as well as the loss of conditioned air.
Courtesy of Systems, Inc.- Poweramp | DLM.

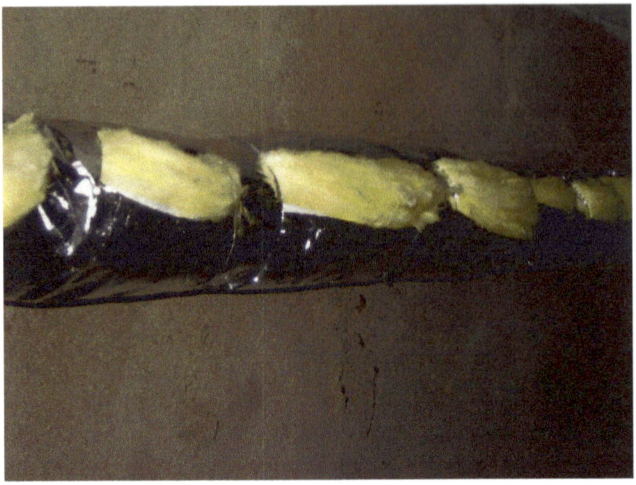

Figure 4.13 Fiberglass batting has already proven to be an excellent harborage of dust, insects, and rodents.

4.2.8 Insulated Piping

All pipe insulation must have smooth surfaces with no gaps. Use styrofoam, foam glass, or another rigid foam rather than fibrous materials (Figure 4.13) that have already proven to be an excellent harborage of dust, insects, and rodents. Of course, asbestos may never be used. Afterward, the insulation should be covered with cladding of aluminum or stainless steel sheets that have appropriate thickness and that resist tear and abrasion. The exterior of this insulation protection should be smooth, properly sealed to avoid ingress of dust and liquor, and with joints facing downwards (Moerman et al., 2014).

4.2.9 External Lighting

External lighting that illuminates the factory entrances should be placed in locations away from the building to avoid attraction of insects to the building. On the outside, car parking and building lights should be angled downward, or toward the building, never out and away from the building. Lights showing outward attract insects to the building at night. External lighting should be shielded so they are not visible from above. When external lighting is used, preference should be given to high-pressure sodium lights or an equivalent that emits low ultraviolet (UV) rays. Mercury vapor-type lights that emit high UV rays should be avoided but, if used, should be located at least 10 m from doors. Exterior light fixtures also shall be designed so that birds cannot roost or nest on or in them (Moerman, 2011b; Holah et al., 2014).

4.2.10 Floors and Drains

Use a 7.5 cm layer of sifted sand underneath slab construction to provide a permanent sand barrier to termites (both western subterranean and Formosan termites). Where such a sand barrier is not present, termites may pass through penetrating cracks in the slab construction. The use of a sand barrier is a practical proven method that is required by building codes in Hawaii and Sydney (Australia) and considered as a good alternative to termiticide chemicals. Termiticides in soil make the soil a hazardous waste and are only effective for about 10 years. Floor drains must be screened to avoid swimming rats from entering the food plant via sewers, and also the water trap must be of such a design that pests cannot enter via that route. P-traps are not suitable for that purpose.

4.2.11 Waste Handling Facilities

Most food factories produce some solid waste: unsuitable residual packaging material, rejected food products, debris captured in the sediment baskets of drains, etc. Solid wastes should not be transported through open production areas but should leave high risk areas (preferably bagged) in such a way that they minimize any potential cross-contamination with processed products. Preferably solid wastes should be routed in the reverse direction to the product, via short and direct routes. The garbage and inedible waste facilities to store the waste prior to collection, which means the dumpsters and waste treatment/pretreatment facilities, must be strategically positioned on the site but away from food areas. Invariably, waste materials should be kept at the back of a factory, so as not create an eyesore or a source for pests, insects, vermin, and microbial growth. At least waste disposal units should be sited such that prevailing winds do not blow microbial and dust aerosols into manufacturing areas (Moerman, 2010).

For packaging waste, it is usual to use a compactor to compress it and minimize its volume, hence reducing subsequent transport, recycle, landfill, or incineration costs. The compactor should be easily accessible but hidden from sight. For waste products, it may be possible to recover food that is safe to use as an ingredient in an alternative lower cost product. Alternatively, another company may be willing to buy it for reprocessing or for animal feed. If the waste is to go to landfill, it is important to select a good location for waste food containers or skips, because food waste left outside the factory building for an extended period of time will attract insects and rodents. Hence, any container for food waste should be designed to prevent the entry of vermin, and the refuse area should have a concrete floor and fast flowing water, such as a hosepipe, in order to keep that area clean (Wallin, 2013).

4.2.12 Emerging Bird Control Techniques

Pest birds can pose a major threat in and around facilities where food is being prepared, processed, or stored, and are of an as much concern to food plant operators as rats, mice, and bat or insect infestations. As an example, many food manufacturing plants have problems with birds entering through large open doors such as loading docks, after which they may start to roost or nest in rafters above food production or storage sites. When bird droppings then fall from the rafters above, they can contaminate food produced or stored below. Bird droppings can spread disease, as they harbor over 40 types of parasites, and can internally host over 60 types of infectious diseases. Among the most common are psittacocis, histoplasmosis, encephalitis, listeriosis, vibriosis, salmonellosis,

Figure 4.14 Pooling water mixed with bird droppings. Once the roof starts to leak, rain water contaminated with high amounts of pathogens may infiltrate to the inside of the food factory. Courtesy Reactive Group Services.

meningitis, candidiasis, toxoplasmosis, etc. Notice that dried bird dropping particles also can enter through ventilation systems and contaminate surfaces, equipment and food via that route. Usually with bird droppings, also a problem with insect infestation (e.g., bird mites, flies, etc.) arises, and even rodents are attracted to bird nests. Another problem is that bird droppings are acidic, eating through paint, roofing material, fabric, and soft metals like aluminum. Once roofs start to leak, heavily contaminated rain water may infiltrate to the inside of the food factory (Figure 4.14), creating severe unhygienic conditions in food preparation, processing, packaging, and storage areas. Finally, bird nests may clog exhaust vents and chimneys, hence increasing the risk of fire.

So as to prevent contamination, it is advisable to implement bird deterrent or bird repellent measures before evidence of contamination appears. Where bird control measures were not yet applied, it is important that pest control experts remove the droppings prior to installing any deterrents or antiperching systems. If the droppings are not removed, they still may contaminate the food products and as such bring the heath of the consumers at risk. Without resorting to bird poisons (avicides) that have very strict guidelines for use, there are a variety of effective and humane bird control measures one can use. In some circumstances, it may be necessary to have more than one type of bird deterrent depending on the target bird's behavior.

4.2.12.1 Visual Bird Deterrents

Visual bird deterrents are used to frighten birds from treated areas, triggering them to flee the area. Common scare triggers, such as predator features (decoys of owls, hawks, or falcons), reflective surfaces (e.g., aluminum plates, iridescent foil tape, tin foil) and scare eyes (e.g., scare diverter, scare balloons with predator eye pattern), are almost ineffective in protecting buildings because their effect only lasts a short while.

4.2.12.2 Laser Bird Repelling Devices and Power Strobe Lights

Laser bird repelling devices are an effective and humane way to repel pest birds. They send out bright, fat-beam lasers in multiple colors, scaring the birds day and night. They are effective over an area of 1000 m^2 in indoor and semienclosed areas. Birds also can be deterred using pulsing strobe light. The flashes of intense light (red, blue, and white) at a rate of 60–75 flashes/min and frightening shadows are intolerable and disorienting to birds (pigeons, sparrows, starlings, swallows, crows, blackbirds, and more), causing them to fly away from the source of annoyance. Pulsed strobe light is most effective in any dark or dimly-lit building where lights and shadows will be most visible.

4.2.12.3 Bird Spikes

Bird spikes are used to prevent birds, especially large birds, from landing. Available in galvanized steel, durable stainless steel, or unbreakable UV-protected polycarbonate, they can be quickly installed by nailing, screwing, or gluing down onto any surface. Some spikes may bend a full 360° to accommodate any contoured surface. If pest birds seem to gather in rain gutters, some spikes feature adjustable base clamps for easy attachment to the lip of the gutter. Birds may still nest and perch in unwanted spots if bird spikes are not properly installed.

4.2.12.4 Bird Coil (Spiral) and Bird Deterrent Wire

Bird coils are perfect for long, narrow, exposed ledges (Figure 4.15(a)). As birds like a solid landing platform, the bird coil destabilizes the ledge, where the unwanted pigeons and gulls like to roost and nest. Bird coils (available in high-grade stainless steel) have smooth curves which are completely harmless to both birds and unsuspecting workers needing access to ledges. Bird coils are adaptable to most architectural details. Somewhat similar bird deterrent wires (Figure 4.15(b)) also create an

(a) (b)

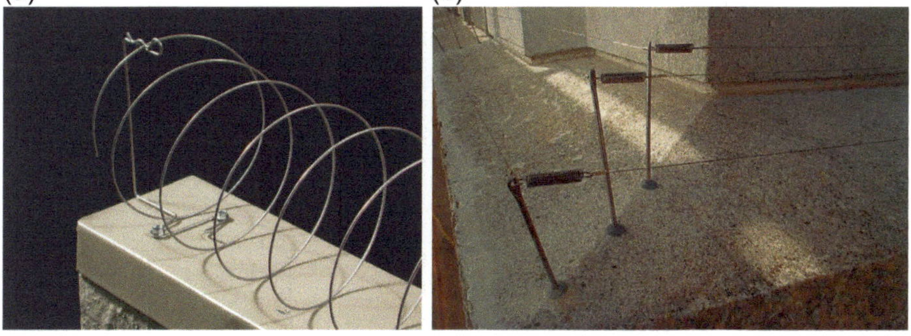

Figure 4.15 (a) Bird coil prohibits ledges from becoming a stable solid landing platform *(Courtesy of Bird Barrier America, Inc.)*. (b) Bird deterrent wires, arrayed in two or three parallel rows, create an unstable landing area preventing birds (e.g., pigeons) from attempting to perch *(Courtesy Reactive Group Services.)*.

unstable landing area, discouraging birds (e.g., pigeons) from attempting to perch on narrow building features such as window sills, stone, or brick ledges. They also work by preventing birds (e.g., pigeons) from looking over ledges into potential feeding sites. Nylon-coated wires (e.g., stainless steel) are secured and tensioned between stainless steel posts at differing heights and intervals (set at up to 1–1.3 m centers) and fixed into stone or brick using anchor plugs (e.g., in nylon). The supporting posts also can be welded to brackets or clamps for use on hard-to-protect and more fragile surfaces such as on cladding panels, lead flashing, rain gutters, piping and I-beams.

4.2.12.5 Bird Nets and Bird Wire Mesh

Bird netting is a well proven control measure to keep pest birds out of certain troublesome areas. Rot, water, and UV-proof netting should be used for extended outdoor use.

Solar panels are being installed not only on commercial and residential roofs but also on industrial facilities. Birds have caused many problems with solar panels, as these arrays provide perfect harborage for birds. Pigeon droppings and nesting is one of the many problems under solar panels. Wire mesh systems (Figure 4.16) are on the market that can keep all birds from getting under solar arrays, protecting the roof, wiring, and equipment from their damage. Special clips bind the mesh to the panels without drilling holes or damaging the system.

4.2.12.6 Bird Slides

Where netting is visually unacceptable or simply not possible, bird slides (Figure 4.17) should be used. Bird slides protect the ledge, eaves, angle irons, I-beams, and most L-shaped ledges against roosting and nesting of all bird species as they slide right off the ledge and cannot get grip.

Figure 4.16 Wire mesh systems can keep all birds from getting under solar arrays. Courtesy Bird B Gone, Inc.

Figure 4.17 With bird slides, birds slide right off the ledge, preventing them from getting grip. Courtesy of Bird Barrier America, Inc.

4.2.12.7 Electric Bird Deterrent Systems

Electric bird deterrent systems (electrified tracks) (Figure 4.18) are a new, valuable, and professional option for controlling birds alongside traditional bird spikes, bird nets, and bird wire. A harmless low level electric pulse (mild electric jolt) emitted by the electrified tracks on touch deters birds (pigeons, gulls) similar to a live electric fence that manages livestock. As such, it is not actually bird-proofing in the physical sense, but is designed to change the behavior of pest birds by preventing them from perching, loitering, roosting, and nesting on ledges, edges, parapet walls, beams, and roof peaks. Apart from flat surfaces, electropulse tracks also can follow shapes, curves, and contours in areas where other methods are not suitable.

Figure 4.18 Electric bird deterrent system.
Courtesy of Kusters Bird Control BVBA.

4.3 Interior Building Layout: Trends

4.3.1 Barrier and Zoning Concept

All food processing operations should be carried out in a way that the risk of contamination of the food product or packaging materials by any hazard (pathogens, foreign bodies, chemical contaminants, including other food ingredients/products that could act as allergens) is avoided. For that purpose, food factories have been segregated into separate areas or zones (the terms have identical meaning). Food production areas must be segregated from nonfood production areas, e.g., locker rooms, canteens/restaurants, smoking areas, boiler rooms, workshops, machinery rooms, laboratories, offices, meeting rooms, and living accommodation. Food production areas can be divided into basic, medium, and high hygiene areas, with further subdivisions such as dry versus wet processing, dry versus controlled wet, and wet cleaning depending on the identified food safety hazards. The barriers aim to limit the challenge of hazards on the factory and on subsequent processing zones. Notice that the zoning and barrier concept can be realized by horizontal segregation (factory with only one floor level) (Figure 4.19) and/or vertical segregation (multistorey factory) (Figure 4.20). Separation must occur by physical means (barriers, e.g., walls, floors), sufficient to prevent contamination of food production areas by microorganisms, pests, particulates, gases, and fumes.

4.3.2 Zoning for Microbial Contamination Control

4.3.2.1 Zone B, M, and H Areas

The concept of areas or zones within a food processing facility, and the barriers which separate them, are summarized in Table 4.1. The site reduces the challenge

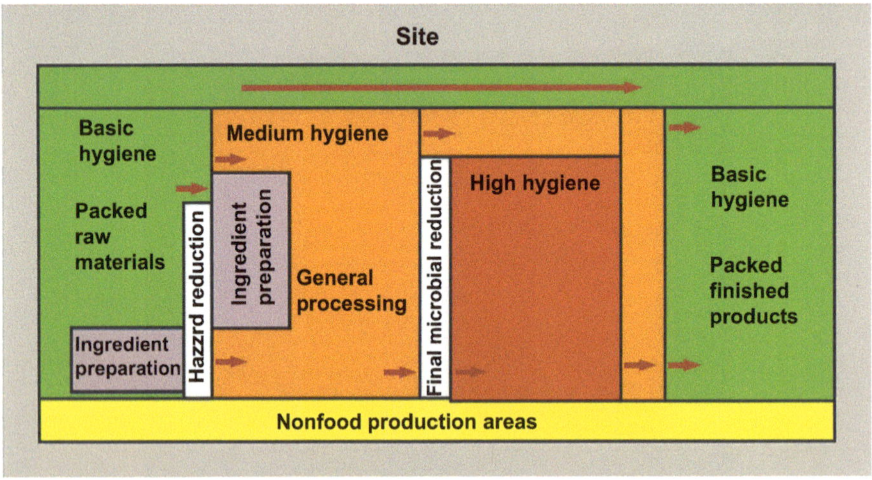

Figure 4.19 Products can be manufactured entirely within basic hygiene, may pass from basic to medium and then back to basic hygiene or pass from basic to medium to high and then, via medium, back to basic hygiene (Holah et al., 2014).

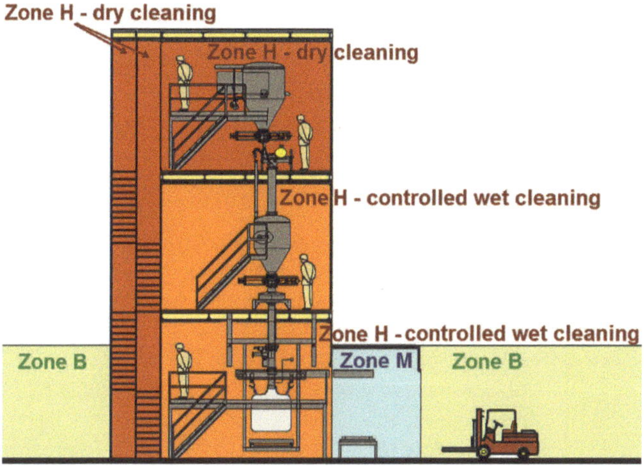

Figure 4.20 Also, a factory with two or more floors (multistorey design) provides the opportunity to segregate materials and different stages in the process. Using different floors for different stages in the process can provide an excellent barrier. Moreover, less piping, pumps, etc., are needed as many transfers can be done by gravity. Factories often use a floor for handling semifinished or finished products, so as to prevent contamination from dirty or unprocessed raw materials. Another floor may house all the packaging lines on one floor, although any filling or packaging could be conveyed or fed from a floor above or from a mezzanine level. Courtesy of J-tec, www.j-tec.com (Wallin, 2013).

of hazards on the factory envelope. The factory envelope reduces the challenge on the basic hygiene area. Segregation between basic and medium hygiene areas reduces the challenge into the medium hygiene area. The medium hygiene area reduces the challenge onto the high hygiene barrier. The high hygiene barrier reduces the challenge into the high hygiene area (CCFRA, 2003a; Moerman, 2010; Holah et al., 2014).

There is no perfect factory zoning solution, and a number of solutions may satisfy a food manufacturer's hazard analysis. For example, some food processing operations such as the sorting/grading of vegetables for size/weight could be undertaken in the field. Alternatively, such operations can be undertaken in basic hygiene areas in the factory. Basic hygiene areas also include warehousing in which ingredients and finished products are protected by (as a minimum) their primary packaging. Food processing that prepares products for cooking by the consumer (e.g., raw meats, fish, and vegetables) or produces RTE products that are effectively preserved (e.g., canned or baked products) will require processing in medium hygiene areas. RTE products that have undergone a decontamination process and in which spoilage and pathogenic microorganisms could survive or grow during the product's shelf-life will require processing in a high hygiene area. The degree of hygienic design applied to the high hygiene area will depend on the degree of microbial decontamination undertaken, the likelihood of spoilage and pathogenic microorganism growth or survival in the product, and the risk of cross-contamination from the external environment (Holah et al., 2014).

Table 4.1 **The Range of Potential Areas or Zones within a Food Manufacturing Facility and the Barriers between Them (Colored in Dark Grey). The Number of Manufacturing Zones Required Will be Dependent on the Food Product Produced (Holah et al., 2014)**

Site barrier	Security fencing, gatehouse.
Site	Grounds, driveways/boundary roads, effluent treatment facilities, rubbish tips/trash management areas. Maintenance workshops, storage tank areas, etc., located outside the factory buildings.
Building envelope	Foundations, floors, walls, roofs, air filtration.
Nonfood production area	Areas not associated with manufacturing processes (e.g., offices, utility rooms, locker rooms, toilets, rest rooms, canteens, laboratories).
Food production area barrier	Physical barrier from nonfood production area. Ideally one entrance to plant. Develop concept of "you are now entering a food production area." Food area protective clothing to be worn and handwashing to be undertaken.
Basic hygiene areas	Any manufacturing area in which raw materials and/or finished products (within their primary packaging) and packaging/labeling materials are received, sampled, or stored prior to dispatch. Areas in which raw agricultural products are sorted and cleaned. Areas in which waste is handled. Designed to minimize spillage, minimize product harborage and avoid cross-contamination. Basic GMPs and GHPs apply.
Medium hygiene barrier	Basic physical segregation by means of walls and doors. Handwashing is required on entry to medium hygiene area. Requirement for operative footwear or clothing change on a hazard analysis basis.
Ingredient preparation area	Area in which all legislated food handling activities are undertaken. Areas in which animals are slaughtered and dressed. Areas where products are subjected to further processing and are known to have the potential to be contaminated. All GMPs and GHPs apply. Food safety risk is low.
General processing area barrier	Basic physical segregation by means of walls and doors. Handwashing is required on entry to general processing area. Requirement for operative footwear or clothing change on a hazard analysis basis.
General processing area	Area in which raw materials are received, sorted, and sampled. Area within the plant where products susceptible to contamination and/or microbial growth are processed, treated, or handled. Areas in which animal carcasses are cut into fresh meat products. Areas in which vegetables and produce are packed. Further separation of ingredients and packaging stores, wet and dry processing areas and areas for specific ingredients, e.g., allergens may also be required. All GMPs and GHPs apply. These zones can also be the intermediate area to a high hygiene area. Food safety risk is medium.

Table 4.1 Continued

High hygiene barrier	Physical segregation via walls, floors, air, and drainage. Entrance of products via a decontamination step. Entrance of packaged, decontaminated ingredients via an outer packaging decontamination step. Controlled packaging entry procedures. Entrance of utensils, tools, equipment, etc., via a decontamination procedure. Personnel entry via a changing room, sluice, etc., where captive footwear and clothing is donned.
High hygiene area	Any manufacturing area, where microbiologically decontaminated ingredients or formulations are sampled, handled or further processed, and where such activities expose the product to microbiological contamination, and where such products may allow the survival or growth of spoilage and pathogenic microorganisms, and where these food products are intended to be consumed as supplied without a heat cooking step, e.g., ready to eat. The degree of hygienic design required for the high hygiene area is dependent on a risk analysis of the products to be manufactured in this zone. Food safety risk is high.
Aseptic barrier	Aseptic barriers prevent the ingress of all contamination into the aseptic area or machine.
Aseptic area or machine	A sterile filling zone in which sterile product can be filled into sterile containers.

4.3.2.2 Segregation Raw/Decontaminated RTE Products

When raw processing, preservation, and packaging of finished products take place in the same plant, design and construction of the facility should include a complete separation of production areas that house raw from decontaminated RTE products. Plant personnel can be separated by manufacturing function, keeping raw-side employees from the finished-product-side employees for food safety. These included separate parking areas for employees in raw and finished-product operations, as well as separate entryways, circulation areas, and welfare areas (e.g., employee locker rooms, support areas, break areas, and cafeterias within the building). To further prevent any risk of cross-contamination, food manufacturers may make use of color-coded gowning/uniforms and labels so that it is easier to identify if something is in the wrong place. Use of electronic access control can help further ensure that people—also equipment and materials—are only allowed to enter the facilities that they are allowed to enter (Wilkins, 2013; Holah et al., 2014; Labs, 2014).

4.3.2.3 Segregation Dry/Wet Processing

Different zones can be further classified according to wet and dry processing areas and the type of cleaning (Figures 4.21 and 4.22). Dry cleaning applies to areas with dry processing equipment where no cleaning liquids are used, and where cleaning is performed only by vacuum cleaners, dusting cloths, brooms, and brushes. Controlled wet cleaning applies to areas that are defined as dry during processing and normally also are cleaned dry, although some wet cleaning is permitted. However, full wet cleaning would impose too high a hygienic risk. Drying of all surfaces after

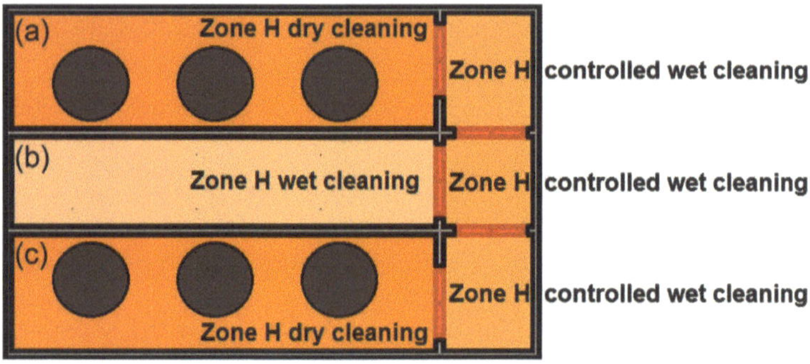

Figure 4.21 Factory with horizontal segregation between zone H areas subjected to, respectively, dry, controlled wet, and wet cleaning regimes. In the zone H areas (a, c), where food is prepared and undergoes a decontamination step, no water is allowed for cleaning as it creates too high of a hygienic risk. Notice that the same product must be produced in both critical zone H areas (a, c). In the intermediate zone H transfer station room (b) where wet cleaning is allowed, some operative work—relocated from the most critical zone H areas (a, c) to the zone H transfer station room (b)—is done. In this way, the space of the critical zone H areas (a, c) can be reduced, which allows that smaller air volumes must be treated and that air exchange rates can be decreased. This configuration is certainly not suitable when food with allergens is produced in one critical zone H area (a), while the other (c) houses allergen-free food processing operations.

Figure 4.22 Some of the operative work is relocated from the dry zone H preparation/decontamination areas to the zone H transfer station room where wet cleaning is applied. Courtesy of FRİGO MEKANİK & PANEL SİSTEM A.Ş.

controlled wet cleaning is essential. With controlled wet cleaning, ceilings and walls usually stay dry, which may influence the building design and plant layout to ensure these surfaces stay dry. Wet cleaning applies to areas where the entire room or zone is always cleaned wet. The contents (equipment, cable trays, ceilings, walls, etc.) are wet washed without restrictions on the amount of cleaning liquid used (Mager et al., 2003).

4.3.3 Zoning for Nonmicrobial Contamination Control

Nonmicrobial contamination, particularly allergens, may be reduced by manufacturing in separate locations/factories, by separation of operations and equipment within the same factory, by enclosed systems, by partition, by corridors, by airflow, by time with effective intermediate cleaning and—where appropriate—disinfection or other effective means. Facilities may be designed/redesigned to minimize nonessential employee and visitor traffic near key activities as defined by the FSMA and to make people or actions out of place easier to detect. Where clean corridors are used, the panel walls may be provided with windows, providing the staff walking through the corridor with an immediate view on the interior process operations. Moreover, corridors (Figure 4.23) also enable visitors (e.g., customers) to view all the operations without having to wear

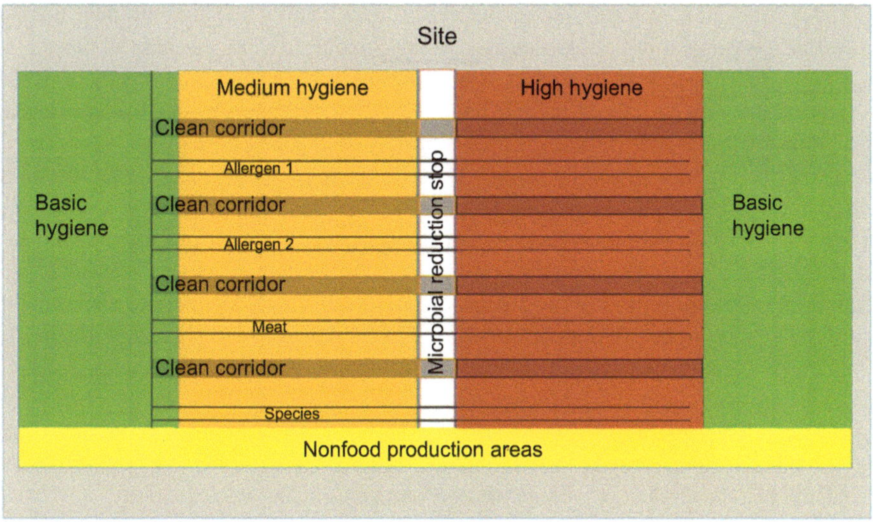

Figure 4.23 Schematic separation of nonmicrobiological hazards via horizontal segregation (in this case clean corridors) throughout as many factory zones as necessary to prevent product cross-contamination. Segregation is extended until the product is in its primary packaging. The corridor allows movement of clean materials, consumables, etc., to the manufacturing suites through unidirectional material airlocks. The airlocks prevent process cross-contaminant in the corridor and ensure no impact on the product manufactured on the different process lines.

protective clothing and encroaching on the factory floor. Windows between the corridors and suites, and suite to suite allow for visual communication.

When dealing with a separate building or a segregated area in a multiproduct site or building as the product/ingredient boundary, an assessment of the incoming and exhaust air as well as the pressure gradient is needed to ensure the risk of cross-contamination by airborne routes is controlled. The most common approach is to filter the incoming as well as the exhaust air. As these filters are considered critical controls for cross-contamination control, they should be monitored.

4.3.4 Reduction of the Size of the Production Area

By moving all dirty equipment components outside the process area (Figure 4.24), the size of the production area can be reduced, and the air quality can be better preserved. Moreover, installing heat producing components outside the process area has an advantage that heat no longer dissipates in the process area but is transferred to technical areas where it can be eliminated at a reduced cost, sometimes by natural ventilation. Therefore, moving dirty and heat producing components outside the process area allows for considerable savings on air handling because the heat load on the air handling system is reduced, and the smaller size of the process area makes that smaller air volumes must be treated and that air exchange rates can be decreased (Moerman, 2011c).

With open processing equipment, ingredient residues can be mechanically transferred to other products by materials and wastes that enter/exit the processing facility, and employees may even transfer residues outside the factory site and into their homes. Therefore, gowning must be disposed in a hygienic manner to limit mechanical transfer of ingredients to another product/process. A better solution is containing an ingredient within the process, because then the transfer to another product or process by mechanical transfer (e.g., airborne particulate transfer, people transiting the facility, employees' gowning) is reduced. This can be realized by using isolator technology, which means a hermetically sealed line separating the operator from the process area and bringing barriers very close to the product. An isolator is typically equipped with a system for biodecontamination and an air handling unit that ensures temperature control by heating or cooling, as well as permanent overpressure control of the process area compared with the operator environment in order to avoid ingress of contaminated air. The air can be dehumidified by a dryer to avoid condensation on equipment surfaces. By using isolator technology, the area around the enclosed process line can be further reduced and reclassified to a lower hygiene level. Therefore, isolator technology is more cost-effective than trying to create a complete clean room around a partially open line. Moreover, a hermetically closed line minimizes the likelihood that the product gets contaminated while optimizing its cleanability.

4.3.5 Auxiliary Area for the Cleaning of Movable Process Equipment

The formation of aerosols (liquid droplets, solid particles, or both types of matter) during hose-down operations of open process equipment is a serious problem, as they may redeposit on, and hence recontaminate, the process equipment afterward.

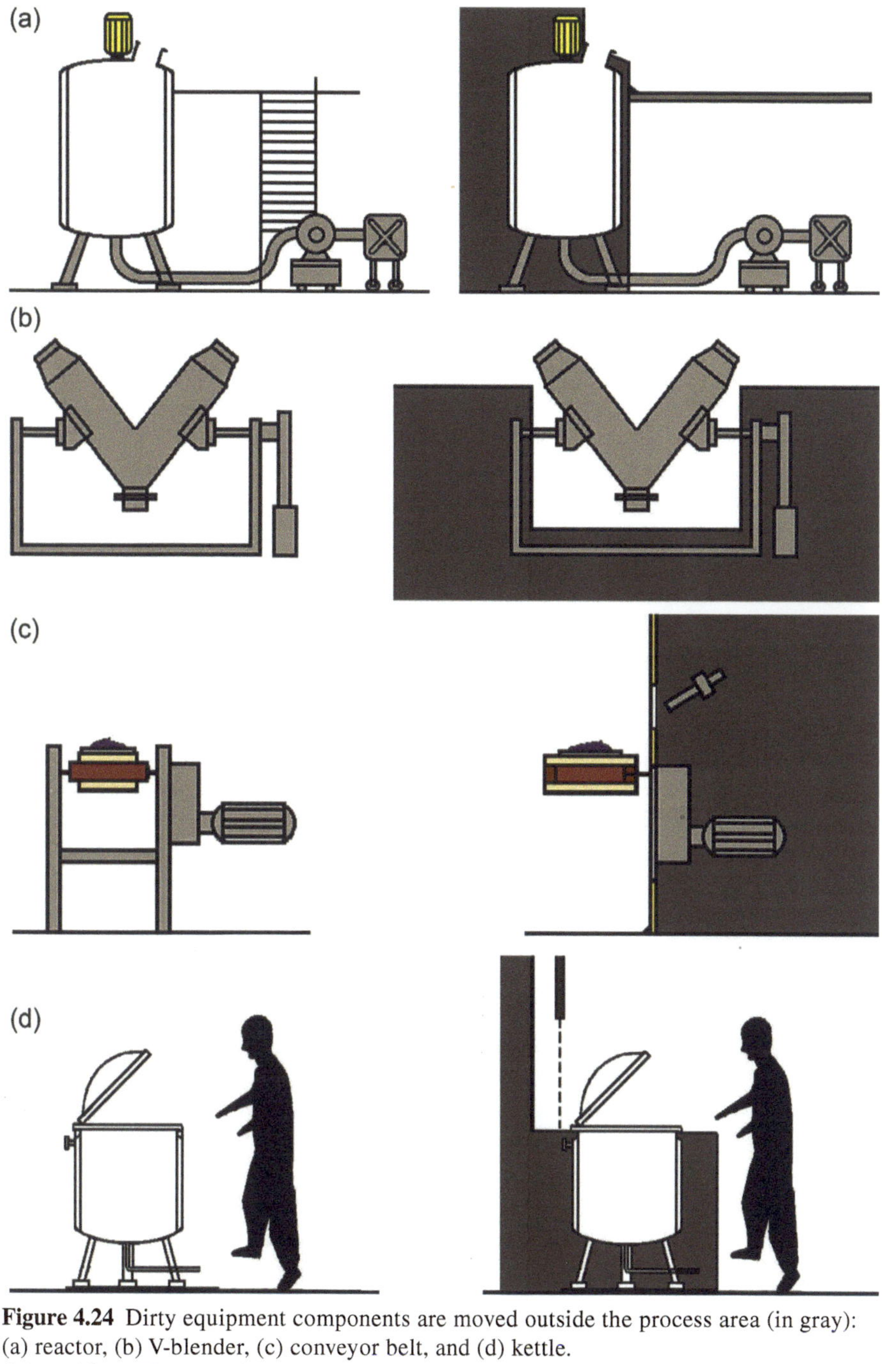

Figure 4.24 Dirty equipment components are moved outside the process area (in gray): (a) reactor, (b) V-blender, (c) conveyor belt, and (d) kettle.
Adapted from Lhoest (1991).

Burfoot (2003) have demonstrated that low-pressure hosing produces per cm^2 and per second a flux of about 144,000 particles with a diameter below 40 μm, and fewer than 900 particles with a diameter greater than 80 μm at a distance of 150 mm from the impact point of the water jet. Elsewhere in the room, the concentration of particles with a diameter larger than 1 μm was up to 1.4×10^7 per m^2. Particles in this range are particularly laden with microorganisms and, as they may travel many meters, they can recontaminate large parts of the process equipment in the surroundings. The smaller the particles, the greater the flight time and distance they may travel.

The use of wheeled process equipment and portable skid mounted unit operations and utilities not only allow fast adaptations for other manufacturing purposes; they also can be quickly transferred from the process area to an adjacent area dedicated to cleaning. The advantage is obvious; the formation of aerosols within the process area becomes minimal and recontamination of recently cleaned process equipment with dirt and microorganisms is excluded or reduced. It is also possible to keep the process area cleaner, as dirty cleaning solutions are no longer spilled within the process area but wasted to drains within the wash area. Moreover, removing process equipment from the process area also provides improved access for cleaning and disinfection of every spot in the process area. With obstructive fixed equipment, process rooms are just more difficult to clean and disinfect. Finally, production on adjacent open process lines within the same room may go on without any risk of contamination. In the pharmaceutical industry, many equipment parts are often cleaned out-of-place in wash cabinets, especially designed for that purpose. It must be possible to construct for the food industry completely enclosed wash cabinets that can automatically clean movable process equipment (e.g., belt conveyor systems) and food process containers with cleaning solutions at higher temperatures and detergent concentrations than normally would be the case with hose-down procedures. By means of cleaning nozzles strategically placed within the wash cabinet, it must be possible to spray cleaning solutions under medium/high pressure on the equipment to be cleaned, with emphasis on the difficult to clean places. In conclusion, cleaning-out-of-place rooms may reduce food contamination, but not without additional investment in an auxiliary room.

4.4 Air Handling Systems

Air handling systems control temperature and humidity (to avoid condensation) and the numbers of airborne dust particles and germs by means of filtration. Air handling recommendations for the different hygiene zones are defined in Table 4.2.

Airborne particles and viables can be reduced by increasing the air exchange rate (how many times per hour the total air volume of a room is conveyed through the air handling unit with filter). It is also possible to achieve different pressure levels among the process areas and pressure differences between a zone H area, and for example, a zone M area. In this system, the pressure levels of the different rooms are obtained by air flow control valves and the temperatures by heaters. Differences in pressure are required to ensure an optimal air balance within different sectors of the food factory, so as to ensure a correct flow of dust particles, viables, and contaminated air within the facility. No air exchange between rooms is allowed that might cause a risk of cross-contamination for product

Table 4.2 Filter Selection Dependent on the Food Product Risk (Brown et al., 2005)

Application	Designation	Final Filtration Level	General Air Filter Description	Approaching 100% Retention of Particles of Size
GMP area	General food production	Min. G4; ideal F5	Primary filters	>5 μm
Medium risk/High care	Processing of cooked/clean ingredients	F7 up to F9, depending on risk	Secondary filters	>2 μm up to >1 μm
High risk	Fully cooked ingredients	F9 up to H11, depending on risk	Secondary filters, semi HEPA	>1 μm up to >0.5 μm
Clean room	Aseptic food processing	H13	HEPA-filters	>0.3 μm

groups with differing requirements. As an example, air from kill floors and rendering areas in meat processing facilities (usually it contains dangerous airborne contaminants) should never flow to areas such as packaging, where airborne bacteria could contaminate the final product. So, the facility and heating, ventilation, and air conditioning (HVAC) systems need to be designed to provide the correct flow from the most clean to the less clean areas (Higgins, 2013; Labs, 2014):

- Production areas should be positively pressurized versus nonproduction areas, and the entire building should be positively pressurized versus the exterior, as outside airflow into production areas introduces outside contaminants that can jeopardize food safety. Badly placed or inadequate air handling systems may reentrain the contaminants released in the atmosphere back into the factory.
- High-speed, roll-up doors for fork truck traffic and automatic closing doors for people traffic can be used to minimize the amount of time that rooms and spaces are open to each other. In some situations, airlock vestibules and/or air curtains may be used to achieve the necessary air separation for the facility to function properly.
- Air locks that segregate air flows and maintain pressurization of the clean and dirty spaces, as well as upstream and downstream spaces.
- Rooftop air handling units must be sized and located properly, and the right filtration must be used.
- Well-thought-out pressurization schemes must be applied. HVAC systems must maintain positive air pressure in the cleanest compared with the dirtiest zones. As zoning will probably not work appropriately in a factory producing allergen and nonallergen ingredients, separate air systems with variable frequency drive sensors and controls for process rooms will be mandated in the future to provide and maintain the correct pressures within each space. Moreover, notice that with a central recirculation/mixed air conditioning system equipped with a common central air handling unit for all process areas, allergens from one process area may get in rooms where allergen-free products are produced. An individual local air handling unit features a cooler, filter, and humidifier, enabling temperature and humidity of the supply air to be controlled individually for each room and independently from internal thermal or humidity loads of other rooms.

4.5 Interior Construction Elements Reducing the Likeliness of Contamination

4.5.1 Construction of Beams and Columns for Basic and Medium Hygiene Areas

The advantages gained in plant layout, and in process and equipment selection, may be eroded if consideration of contaminants (dirt, dust, bioburden, etc.) is not carried through to interior building construction. Ideally, it should minimize horizontal ledges and surfaces that can collect contaminants. Also, potential building problems associated with water—such as condensation, freezing, and corrosion—need to be carefully considered. However, older food factories with a conventional braced-steel-frame construction contain a vast area of horizontal ledges, beams, duct, pipes and cable trays, etc., which become a catchment for dirt, liquids, bioburden, and dust. Figures 4.25 and 4.26 give an overview of some contamination problems related to the presence of steel members with exposed horizontal ledges, which are typical for a braced-steel-frame construction.

Exposed steel structures are only suitable for processes with lower hygiene requirements, but they must be protected by galvanizing or plating, e.g., with zinc, particularly in wet manufacturing areas. Solutions offered to relieve the contamination issues, as shown in Figures 4.25 and 4.26, include: avoiding the installation of beams in a way that they may form horizontal ledges, filling of I-beams outward sloping of horizontal ledges, ventilation openings, etc. Cross-bracing along walls (Figure 4.27) should be reduced, and if used, "breaks at the junction plates" should be provided to prevent build-up of dirt and liquids. To improve connections with masonry walls, steel columns can be embedded with concrete, but then the wall to column connections must be easy to clean. For basic hygiene areas, if used as columns, H or I type profiles should be filled at floor level with concrete (pitched at 60°) (Figure 4.28(b)). Medium hygiene area columns (Figure 4.28(c)) should be embedded in reinforced concrete to about 60 cm high, and slanted at the top to avoid build-up of dirt and liquids around the I-profile.

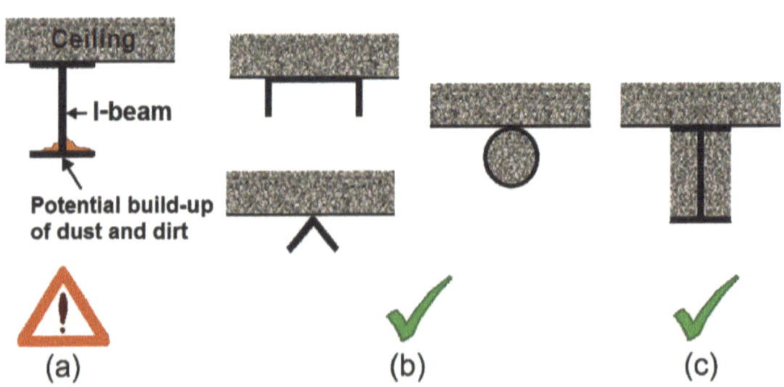

Figure 4.25 (a) Unhygienic steel structure, and hygienic steel structures in (b) low and (c) medium hygiene areas. Beams supporting the ceiling should not have horizontal ledges upon which dirt and dust may settle. Eventually fill the steel profiles, as exposed steel still may corrode (Holah et al., 2014).

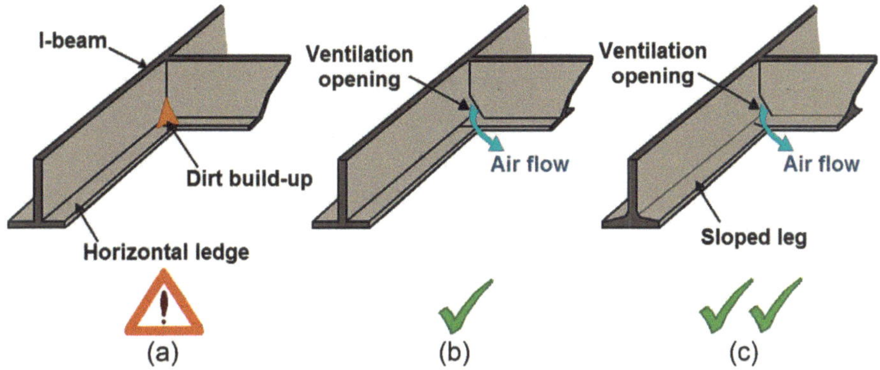

Figure 4.26 In older factories, (a) beams with horizontal ledges are often observed. These ledges are places upon which dirt, dust, etc., can deposit. Especially at the crossing of two beams, dirt will build up. (b) To prohibit the build-up of dirt at the crossing of two beams positioned in the horizontal plane, an opening to encourage air movement is recommended. However, the beams still have a horizontal ledge where contaminants may settle. (c) The most optimal construction is where the ledges are sloped outwards, and a ventilation opening is provided for air movement. Adapted from Corus (2004).

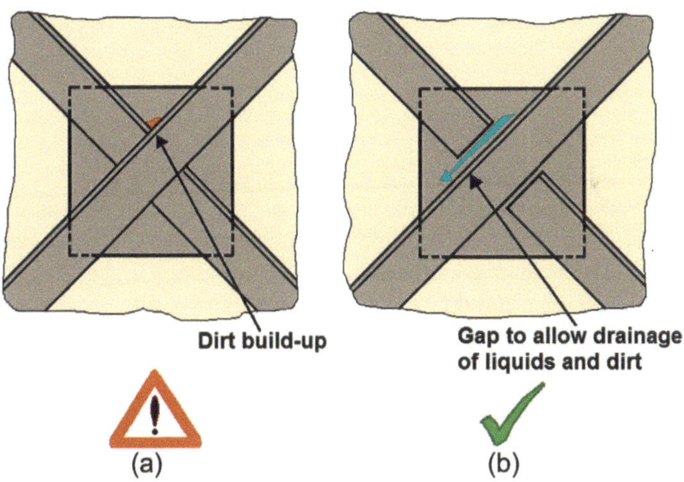

Figure 4.27 (a) Cross-bracing along walls should be reduced, and if used, (b) exclude the retention of water and dirt at junction plates by means of "breaks."
Adapted from Corus (2004).

4.5.2 Construction of Beams and Columns for High Hygiene Areas

Open steel structures with H or I type profiles in high hygiene areas are not acceptable. Structural steel should be concealed in concrete (e.g., roof beams, Figure 4.29) and walls, or can be avoided where stiffening of the building with stair towers, elevator, or stair shafts is possible (Goodfellow and Berry, 1986).

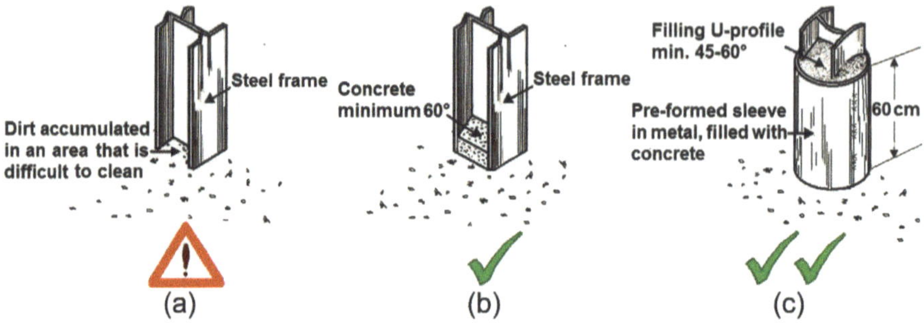

Figure 4.28 (a) Without any provisions, I- and H-profiles may collect dust and dirt. (b) In low-hygiene areas, at floor level, the gaps in the profiles can be filled with concrete slanted at the top with an angle of 60°. (c) In medium hygiene areas, columns should be embedded in reinforced concrete to about 60 cm high, again slanted at the top with an angle of 60° to avoid build-up of dirt and liquids around the I-profile.

Figure 4.29 In high hygiene areas, the beams should be full concrete (steel reinforced concrete, which means that the steel frame work, I-beam or steel mesh, is fully encapsulated in the concrete).

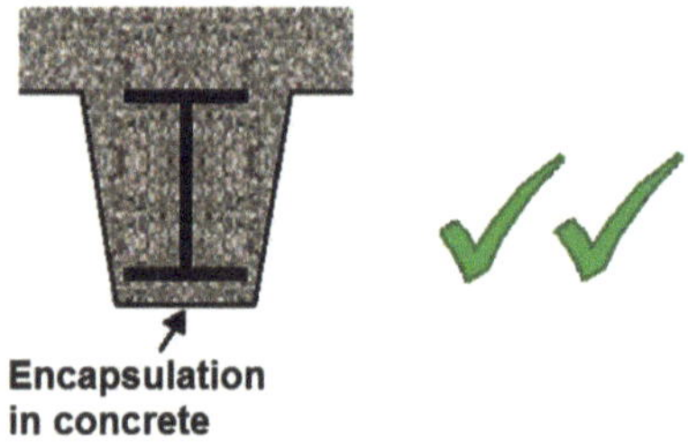

Beams take loads from floor slabs, equipment, and piping to further distribute them to columns or other compression-loaded vertical members. In high hygiene areas, columns should be of full concrete (steel reinforced concrete, which means that the steel frame work, I-beam, or steel mesh is fully encapsulated in the concrete). Where present, exterior columns should be flush with the interior side of adjacent walls. If not built into the walls, a gap of 15 cm should be provided between framing and walls.

4.5.3 Walls in Food Factories

Walls can have a load bearing and/or an enclosing function, and can be used to brace the building against horizontal loads (wind, horizontal impacts). To prevent broken edges from becoming a source of hygiene problems, especially in food processing areas, the corners of walls and columns should be protected with steel angles to a minimum height of 1.5 m, if exposed to heavy traffic. Columns or walls can be provided with block wall protection of about 60 cm high, slanted at the top (Figure 4.30).

4.5.4 Floor Slab and Floors

4.5.4.1 Floor Slab

The concrete floor slab must be properly designed by a structural engineer in accordance with building standards to meet the stresses of the in-service environment.

Figure 4.30 Columns or walls can be provided with block wall protection of about 60 cm high, slanted at the top with a slope of 45°–60°.

To protect the ground floor slabs from rising moisture, a damp-proof membrane should be installed below the concrete floor. Joint cracks must be avoided between the slab and building, especially where there is the potential for rising humidity and where residues could collect in the cracks. Expansion joints must have special attention, because they are maintenance items and weak points in floors. Ideally, they should be placed at the high points of the floor, and should not be made of material that cracks, accumulating and harboring residues. The insulation of floor slabs in freezers is essential to prevent freeze/thaw degradation of the slab (Holah et al., 2014).

4.5.4.2 Floor

Floors should be sloped to drains (Figure 4.31) and provided with curbed wall floor junctions, with the curbs having a 30° slope to prevent accumulation of water, dust, or soil. The following floor types are used (CIEH, 1997; CCFRA, 2002; Moerman, 2010):

- Concrete flooring, including the high-strength granolithic concrete finishes, are especially suitable in warehouses where its excellent resistance to heavy traffic is critical. Not normally used for open food preparation or handling areas, unless sealed with a light color

Figure 4.31 Floor sloped toward the drain.

epoxy resin floor paint to produce a finish more associated with a food preparation room. However, untreated concrete can be dusty if dry and highly susceptible to damage from water and especially acids when wet. Concrete floor is not recommended for high-care production areas, because it can spall (chip off) and because of its ability to absorb water and nutrients, allowing microbial growth below the surface.

- Resin flooring provides a durable, seamless, chemical-resistant, and readily cleanable surface; however, over time, the coating can crack and buckle due to attack from cleaning chemicals or wear by heavy traffic. Once this happens, moisture pockets under the coating can create a microbiological niche. As several types of resins are used, there may be big variations in chemical and temperature resistance. The choice of different aggregates will affect slip resistance, and the durability will be influenced by resin content, resilience, and thickness. In general, floors with larger aggregate have better scratch resistance and are more durable. The tougher the largest aggregate in a system, the more durable the floor. The more resilient the resin floor, the better the impact resistance. Brittle materials are more likely to crack under impact. Although suitable for all areas, resin flooring is rarely used for storage areas as there are cheaper alternatives that are almost as good. Resin floors are easily cleanable and promote plant worker safety with nonslip application.
- As regards epoxy resin flooring, the chemical resistance is limited especially to organic acids, severely limiting the life expectancy in many food environments. The temperature resistance is limited to <60 °C and the mechanical resistance can be limited especially in thinner systems. As an example, thin epoxy systems are not recommended in wet areas as long-term or frequent water exposure tends to result in delamination from the substrate. Generally, epoxy resin floors are appropriate for dry floors in light to medium duty environments (including warehousing) and floors where open food is handled or prepared.
- The chemical resistance of polyurethane floors is overall similar to epoxies, but they have better resistance to organic acids. Their temperature resistance is limited to <60 °C, and long-term/frequent water exposure in wet areas tends to result in delamination from the substrate. Generally, polyurethane floors are appropriate to dry floors in light to medium duty environments including warehousing. Heavy duty polyurethane floor systems (polyurethane–concrete or polyurethane–cement systems) are installed at thicknesses from 4 to 12 mm thick. The thermal shock resistance depends on the floor thickness. Floors of 9 mm

with temperature resistance up to 120 °C are available. The chemical resistance is generally good against all the chemicals, and heavy duty polyurethanes work well in medium to heavy duty and permanently wet environments, including warehousing.

- Polymethylmethacrylate (PMMA) floors are often hard, brittle floors with limited impact resistance. The temperature resistance is limited to 70 °C for thicker systems, but the chemical resistance is comparable with epoxy resin based systems. The limited resistance to organic acids reduces the life expectancy in many food environments. Hot vegetable oils can melt through PMMA floors. Generally polymethylmethacrylate floors are appropriate to floors in light to medium duty environments.
- Tile floor is an excellent surface for food plants. However, with heavy wear and in more aggressive cleaning environments, tiles may lose some of their grouting, which results in the penetration of water beneath them. Plastic or asphalt membranes may be laid between the underlying concrete surface and the tiles. Ceramic tiles have a glass-like (vitreous) surface, with a size of 15–30 cm square. A larger tile means there is less grouting to clean; however, unless it is a thick tile it will be more prone to cracking if heavy mechanical handling equipment moves over it, a heavy load is dropped on it, or a fixture/racking foot is placed on it. The grout should be brought near to level with the top of the tile, so as to avoid lips or ridges, and to provide a smooth surface. The presence of liquids including water on them may make them slippery. Quarry tiles are unglazed terra cotta tiles made of natural clay, which are cheaper than ceramic tiles while providing good impact resistance and chemical resistance. Quarry tiles of 1.3–1.9 cm thick only should be used in areas subjected to light and foot traffic. Quarry tiles that are 3 cm thick can be used in heavy traffic areas. Like any flooring material, they can get dirty, especially in high-traffic areas, but they are easy to clean. Terrazzo tiles consist of marble chips embedded in cement and are polished smooth. The grout comes level with the top of the tile to produce a completely smooth floor, thus avoiding the grooves of other tiled floors. However, they are slippery when wet. Spillages of alkali or acid products should be removed as soon as possible, as prolonged contact will discolour/bleach and then damage the surface of the floor. Treating with a chemical hardener or sealant after laying the tiles increases the resistance. Continuous running water, mixtures of heat and water, and water left for a very long time will also damage the surface of the tile. Hence, their use in very wet, humid areas is not recommended, as well as in areas with heavy mechanical equipment traffic which can cause significant wear and the tiles to crack. They can be used in preparation and storage rooms because of their attractive appearance and smooth, easy to clean surface (neutral detergent!). Vinyl tiles are cheap, easy to replace, and easy to clean. The appearance and life of the floor can be enhanced by sealing with a polish or sealant. The surface can be dented or scored by heavy weights. In heavy traffic areas, particularly where trolleys or wheeled cages are used, the surface can be damaged. Very hot items will damage the surface. Vinyl tiles are suitable for use in nonheavy traffic areas.
- Brick floors also may be satisfactory; however, they tend to be somewhat fragile, and—unless vitrified (glazed with heat during manufacture)—permit water penetration. Thicker brick should be used where loads are excessive, or where it is necessary to dissipate heat. Bricks are suitable for general use in most areas of a food plant.
- Vinyl sheeting, with any joints between polyvinyl chloride (PVC) sheets heat welded together to form a continuous easy to clean surface. It can be supplied with a nonslip surface, although this does inhibit cleaning. Vinyl sheeting has excellent chemical resistance and is often used in aseptic rooms. However, it is not suitable in hot and wet areas, and if a lot of heavy cart traffic is expected, the welded PVC may be damaged.

- Wood floors are satisfactory in packing and warehouse areas; however, the wood should be impregnated and coated with a durable plastic such as polyurethane. Generally, wood floors, while attractive, may become worn, porous, and absorbent requiring expensive maintenance. Therefore, they are not installed in modern food plants to a great extent.
- Metal sheeting is extremely hard, durable, and resistant but it can be slippery. Checker plates with a raised pattern are less slippery. The metal plates are normally bonded to a concrete subfloor, but the metal sheeting is difficult to bind to concrete. The epoxy must be formulated to allow movement caused by expansion of the steel tiles, and the plates are normally butted when they are installed to minimize exposure of the bonding agent. Notice that if the sheeting is not thick enough or properly fixed, it can buckle or turn up at the edges or corners. The metal sheeting also may corrode. Metal sheeting is used for extreme service such as refrigerators, walk-in chillers and freezers, shipping areas and loading bays, balconies, and walkways in the vicinity of the process.

4.5.5 Drains

4.5.5.1 Use and Location of Drains

Drains should have appropriate drain capacity to avoid "ponding" of water and hence contamination in the area to be drained. The drain bodies must be free of pockets that can hold food soil otherwise they will cause odor problems. The type of drainage selected—gully or channel—will depend on the nature of the operation and its requirements. Channels simplify floor gradients and offer better interception than gullies. Drains should be placed away from hygienic production areas but close enough to allow efficient drainage of liquids. Ideally, drains should be positioned near the wastewater/fluid source with space allowing access for inspection, cleaning, and maintenance. Where drainage is required from machinery, splashing can be minimized by use of funnels above the drain body. The air gap between the drain and the machine water outlet avoids cross-contamination and backflow from the drainage to the machine.

In dry areas with dry cleaning, gullies and channels should be avoided. If drainage is unavoidable, it should be a sealed, gas-tight system with a smooth surface which can be opened by a vacuum handle for access or disinfection. In dry areas with controlled wet cleaning, channels should be avoided. In these areas, gullies should be sealed and gas tight with a smooth surface, which can be opened by a vacuum handle for access and disinfection. In dry areas with wet cleaning, channels and gullies should have gratings of a high hygienic standard. Linear channel systems can be used to provide a physical barrier to fluids, and can be used to separate dry and wet areas. In cases where heavy traffic is unavoidable, care must be taken to ensure correct selection and fastening of gully, channel frame, and gratings to ensure stability for the specified load.

4.5.5.2 Gullies

The gully should have a round body, sloped toward the water trap, and must be able to completely drain. Removal of the water trap must allow the drain body to completely dry out. However, during operation, water traps must have their water level maintained

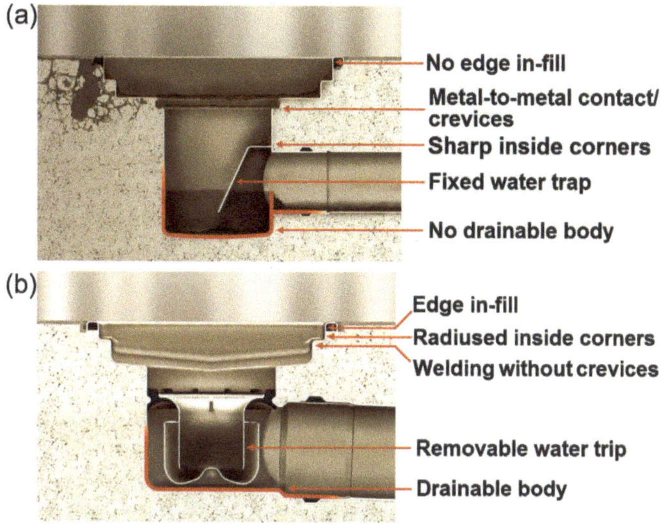

Figure 4.32 Gully of (a) poor and (b) good hygienic design.
Courtesy of ACO Technologies plc.

to prevent foul odors and possible pest entry. The typical characteristics of a hygienic gully are (Figure 4.32) (Holah et al., 2014):

- Continuous welded joints
- Radiused and rounded corners
- No crevices and dead spaces
- Water trap without seals under the waterline
- Removable water trap that allows full accessibility to the pipe system for jetting (cleaning with water jets) and rodding (cleaning via physical scrappers and brushes, both attached to flexible structures pushed or propelled down the pipes). A P-trap is not recommended: rodding and jetting via the P-trap is difficult, and the P-trap does not offer an adequate rodent and pest barrier and always retains polluted water.
- Built-in sediment basket to capture all solid material which otherwise would block the drain. The design should allow fluids to flow through the sides, and be hydraulically oversized to ensure the required design capacity of the system is maintained when sediments are present.

The correct installation of gullies in the flooring material is critical, as leakage between the floor and drainage element is a common problem, causing permanent contaminated moisture around the drain which itself is a source of bacteria.

4.5.5.3 Channels

Channels (Figure 4.33) should have removable gratings of a maximum length of 500 mm in order to facilitate removal and enable cleaning of the channel. Channels should have a lateral slope toward the outlet of the channel, forming a profile such as a V or U-shape in the base of the channel. Channels should have the same constant slope in the longitudinal direction toward the outlet of the channel, minimum 1% but preferably more. In areas of high solid waste, effective drainage requires a larger slope. During the installation, the

(a)

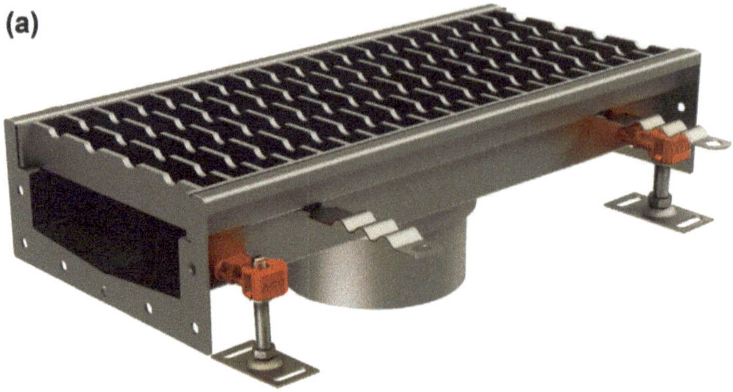

(b)

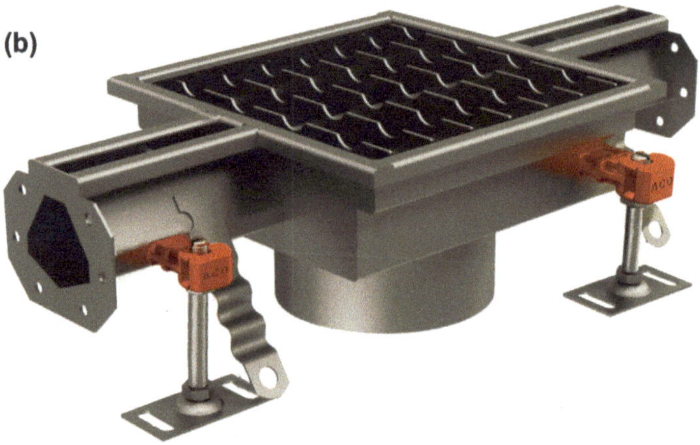

Figure 4.33 (a) Box channels should be sloped to the outlet in both the lateral and longitudinal direction, and have removable gratings of a maximum length of 500 mm in order to facilitate removal and enable cleaning. (b) Narrow slot channels should be avoided in areas with a high hygiene requirement, significant solid waste, or large flow. Courtesy of ACO Technologies plc.

frame profile should be filled to strengthen the frame and to avoid the potential for trapped water and bacteria in the profile. Proper sealant quality is important to separate effectively the surface of the channel and floor from the space below the tiles/resin and channel.

4.6 Integration of Process Piping and Utilities within Zone M Process Areas

4.6.1 Installation of Piping in Zone M Areas

In zone M areas, although piping shall preferably run in technical corridors, wall-compartments, or the ceiling, it also may run in open trays that are fixed to walls or columns (Figure 4.34), preferably close to the ceiling. Where fixations to walls or columns are not possible, piping trays can be fixed to the ceiling. Piping should be

Figure 4.34 Utility piping should be set off the wall for better cleaning. Pipe hanger suspension rods should be smooth and round, and suspended braces should be round tubing sealed at the ends. Cables drop from the ceiling, hence reducing the risk of contamination.
Courtesy of CSI Central States Industrial, www.csidesigns.com (Moerman, 2011c).

installed at least 6 cm from walls and floors to encourage thorough cleaning around it. Racks must be designed hygienically to minimize the presence of horizontal ledges, crevices, or gaps where inaccessible dirt can accumulate, and pipe connections between supporting racks and process equipment should be short. Pipes that pass through walls or ceilings should run through a protection pipe section at the point of traverse to allow for expansion or contraction. Holes in walls for pipe traverse must not be sealed water and airtight, when both sides of the wall are in rooms with the same hygienic requirements. However, if a wall separates rooms with different hygienic zone classification, all holes for pipe traverse must be sealed, e.g., by foaming-in-place. Finally, a plastic cap or stainless steel flashing around the piping and sealant may provide a water and airtight contact with the wall or ceiling (Figures 4.35 and 4.36).

Openings in floors for pipes should be guarded with a sleeve (Figure 4.37(b)) to avoid spill of cleaning solutions to a lower floor. It is even recommended that there should be no floor openings left. The use of a sleeve boot for single pipe floor penetrations belongs to the possibilities to solve that problem (CCFRA, 2003b) (Figure 4.37(c)).

When several pipes penetrate the floor (Figure 4.38), a larger curbed floor can circumvent several pipe sleeves to improve the cleanability of the surrounding process environment. However, that curbed floor may not create a large opening where pests may harbor and where dirt, water, etc., may accumulate. It must be a completely closed curb with a cover that leaves no gap around the penetrating piping (Moerman, 2011c).

4.6.2 Cable Routing in Zone M Areas

In medium hygienic areas, vertical wiring routes, cable trays, and cable ladders should be used instead of open rigid conduits as means to support current carriers over long

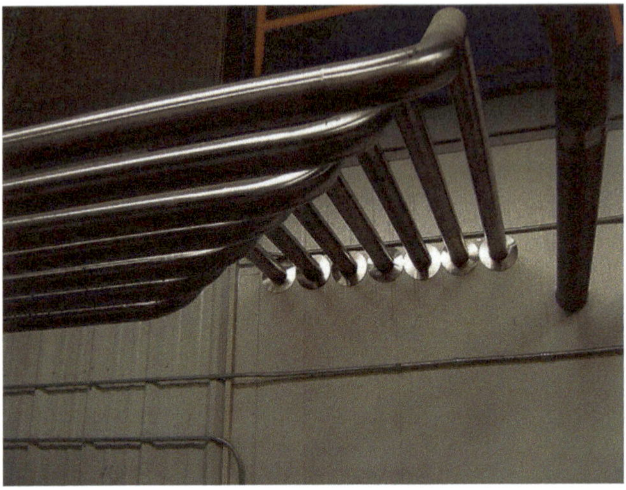

Figure 4.35 Stainless steel flashing around the piping and sealant provide a water and airtight contact with the wall (or ceiling). If one of the rooms is at positive pressure with respect to the adjacent room at the other side of the wall (or ceiling), excessive air volume losses will occur if the apertures through the wall or ceiling are not properly closed for air leakage.
Courtesy of CSI Central States Industrial, www.csidesigns.com (Moerman, 2011c; Holah et al., 2014).

Figure 4.36 Pipetite is a silicone wall or ceiling boot which attaches to the wall or ceiling and forms a flexible seal around the pipe.
Courtesy of CSI Central States Industrial, www.pipetite.us (Moerman, 2011c).

distances. As cables are exposed, they must withstand mechanical and chemical stress. Cable trays and cable ladders should allow for future expansion, which means that the trays and ladders must be wide enough and suitable to carry the additional load of new cables. They must follow the expected ways for the future installation, but may not block a place that is needed for equipment. On a cable tray, cable ladders and cable dust can accumulate. To make a system cleanable there should be a minimum distance between the cable tray/cable ladder and the cables. The accessibility around

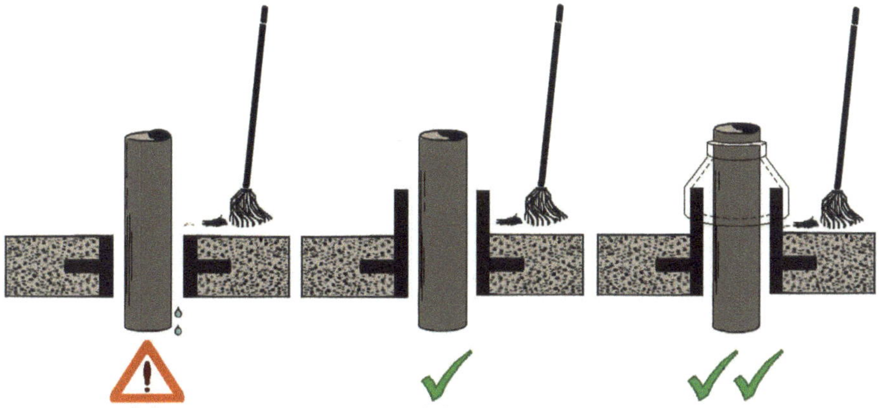

(a) (b) (c)

Figure 4.37 (a) Cleaning solutions may get spilled to a lower floor. (b) Openings in floors for pipes should be guarded with a sleeve. (c) With the use of a sleeve boot for single pipe floor penetrations there is no floor opening left (Moerman, 2011c).

Figure 4.38 When several pipes penetrate the floor, a larger curbed floor can circumvent several pipe sleeves to improve the cleanability of the surrounding process environment. However, that curbed floor may not create a large opening, where pests may harbor and where dirt, water, etc., may accumulate. It must be a completely closed curb with a cover that leaves no gap around the penetrating piping. When installed somewhere in the middle of a food factory, preference should be given to a gable roof construction that it is less sensitive to damage. Along a wall, a shed roof construction is more preferable (Moerman, 2011c).

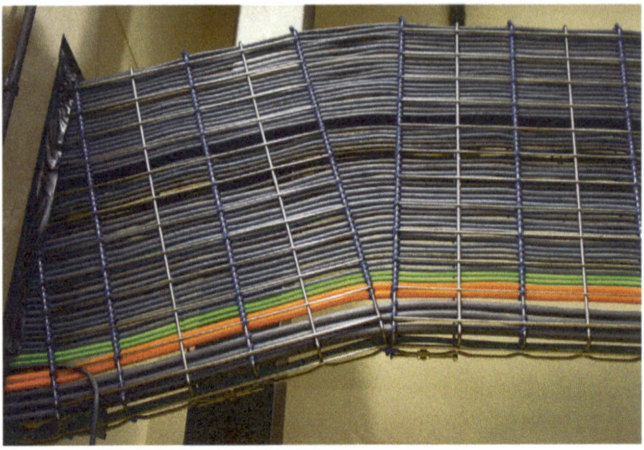

Figure 4.39 It is recommended to lay the cables separated from each other to make cleaning operations easier. When cables pass from one hygiene zone to another, care should be taken to avoid the formation of pest ingress channels. Cable binders do not facilitate the cleanability. Courtesy VEZOR media, publisher Solids Processing Benelux.

the cable tray/cable ladder is very important. However, the cable trays/cable ladders are often placed too close to the wall for proper cleaning of the backside.

Horizontal cable ways can be installed vertically (on their side) (Figure 4.39) to avoid the creation of a horizontal surface, but this has no use, because the total surface where dust can collect is the same for horizontal and vertical cable ways with cables and wires laid on a minimum distance. The electrical cables must be routed and fastened individually in a distance from each other to improve their cleaning, but cable binders are not facilitating the cleanability and should be avoided where possible. Therefore, a new type of horizontal cable support system is recommended, which at present seems the only system that is fully cleanable dry, as well as controlled wet and wet. Figure 4.40 demonstrates the new concept to lay cables in a separate way. Another system (Figure 4.41) allows cables to be fastened in a horizontal and vertical plane perpendicular to the wall without the use of cable binders. In parallel with the wall the backside of cable support systems is difficult to clean, especially when installed too close to this wall. Where cable trays and cable ladders enter the medium hygiene production area, the opening remaining after the passage of the cable way should be sealed with fire-resistant material so as to maintain the degree of fire resistance as well as the hygienic standard of the respective element (e.g., wall, ceiling) (Moerman, 2011a).

There are now flexible conduits (Figure 4.42) on the market providing a permanent and fluid-tight protection tube for multiple electrical cables and wires. These flexible conduits have stainless steel AISI 316L fittings at their ends to be connected to the equipment that must be powered. The fittings have a smooth surface without visible thread in the screwed state. Hygienically designed conduits of this type may reduce the number of cables that must be individually laid on cable ways, hence increasing the hygiene within food factories by decreasing the amount of contaminants built up on current carriers and by enhancing the cleanability of the cable transport system.

Figure 4.40 A new hygienically designed cable support system with springs (Streamline HD) keeps the cables separated at a correct distance from each other (developed by NIEDAX Gouda Holland B.V. in collaboration with SafeFoodFactory.com).
Courtesy of Burggraaf and Partners.

Figure 4.41 Cable support systems preferably should be installed perpendicular to the wall instead of parallel with the wall. In parallel with the wall the backside of cable support systems is difficult to clean, especially when installed too close to this wall. The ladder-like cable support systems shown allow cables to be fastened without the use of cable binders. Maximum cleanabilty of the cables in the horizontal cable support systems can be achieved when the distance between the crossbars amounts to 100–160 cm. A distance of 50 cm between the crossbars should be applied for the vertical cable support systems.
Courtesy of SafeFoodFactory.com and Burggraaf and Partners.

4.7 Relocation of Process Piping and Utilities Outside Zone H Process Areas

4.7.1 Arguments to Relocate Process Piping and Utilities

Ceilings uncluttered with piping and services (e.g., electrical cable runs, HVAC ducts, etc.) reduce the risk of contaminants falling on exposed food products. Moreover, piping

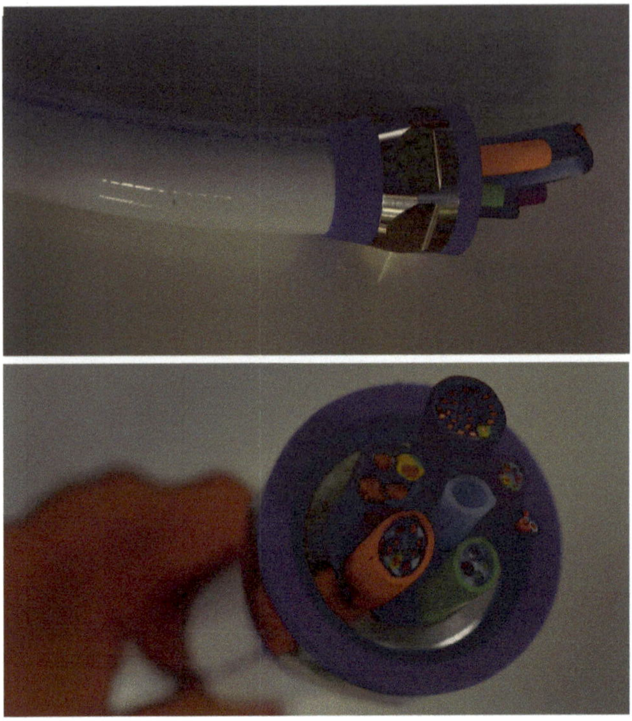

Figure 4.42 Multiple cables and wires can be combined in a flexible conduit with a hygienic connector. Connections between cable support racks and process equipment must be short and must be made via the bottom side of the equipment to be powered.
Courtesy ANAMET Europe B.V. (Moerman, 2011a; Holah et al., 2014).

and utilities running below lighting leads to the formation of shadows which will make cleaning and inspection of floor, walls, or ceilings difficult. Piping or utilities fixed to walls by supports or running in the vicinity of walls make the cleaning of both the walls and piping and services more difficult. Moreover, piping and utilities installed in the horizontal plane create a horizontal ledge upon which dirt may accumulate. There are a number of ways to reduce the number of piping and utilities within process areas so as to keep walls and ceilings uncluttered (Moerman, 2011b,c):

- Make process piping and services an integral part of the building structure.
- Relocate piping and utilities to false ceilings and mezzanines.
- Make use of pipe trains, preferably down from the ceiling installed.
- Relocate piping and utilities to false floors, technical rooms, and corridors.
- Use wireless technology where possible.
- Make use of ceiling mounted utility columns, standardized to provide electricity, pressurized air, process gases, cooling water, etc., for diverse manufacturing skids in flexible plug and play setups.
- Position equipment in adjacent areas along the same wall, back-to-back.
- Make use of gravity flow by applying multilevel factory concepts.

4.7.2 Methods to Keep Walls and Ceilings Uncluttered

4.7.2.1 Making Services an Integral Part of the Building Structure

Process piping and services can be integrated in wall-compartments, more specifically a shaft or utility duct chase (Figure 4.43). Piping and utility drops integrated in wall-compartments usually end in piping hook-up and/or utility panels (Figure 4.44), which allow that connections to equipment can be done neatly. Technical service shafts should be well ventilated to prevent accumulation of dust, as this integration concept is usually applied in high hygiene areas (in food the equivalent of clean rooms in the pharmaceutical industry) operating under overpressure relative to adjacent areas. Process and service pipe infrastructure integrated in walls must be controlled for leaks on a regular base, to avoid contamination of certain process aids like process water, food gases, compressed air, etc. (Moerman, 2010, 2011c).

The utility connections can be enclosed in a wall compartment with panel door and lower free space area (Figure 4.44). That free space permits hoses and/or cables to leave that wall compartment, when the plug-in connections are made (Moerman, 2011a).

4.7.2.2 Relocate Piping and Utilities to False Ceilings and Mezzanines

Services can be positioned on a mezzanine floor (Figure 4.45) or in a roof void. The interstitial space above a process area may house lighting, piping, utilities, large volume HVAC ducts, etc. With a larger clear false ceiling space, layering of utilities, duct

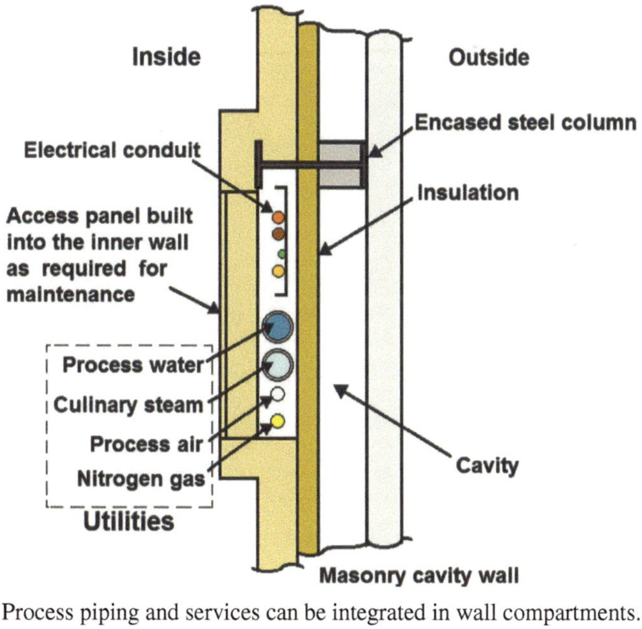

Figure 4.43 Process piping and services can be integrated in wall compartments. All built-in cabinets, closets, and similar casement must be sealed leaving no gaps for pests to hide in. Adapted from Goodfellow and Berry (1986).

Figure 4.44 Utility connections can be enclosed in a wall compartment with panel door and lower free space area.
Courtesy of CSI Central States Industrial, www.csidesigns.com (Moerman, 2011a).

Figure 4.45 By relocating piping utilities, large volume heating, ventilation and air conditioning (HVAC ducts), etc., on a mezzanine floor, maintenance may occur without disruption of the cleanliness of the space just below. A mezzanine floor is an intermediate and partial storey, usually between the ground floor and the first floor, and commonly fully or partially open on one or more sides (Moerman et al., 2014).

Figure 4.46 The use of false ceilings for utility piping (like HVAC ducts) improves the hygiene in the process areas below, as they reduce maintenance costs by elimination of horizontal surfaces.
Courtesy of AES Clean Technology, Inc. (Moerman, 2011c).

crossings and installation of large horizontal HVAC and piped service distribution systems (Figure 4.46) are possible, as such minimizing the number of vertical shafts and the floor space they require. The floor-to-floor height should be a minimum of 4.2 m, so as to allow a false ceiling space of at least 0.9–1.2 m high. If the clear false ceiling space is small, utility duct diameter sizes are obliged to be much smaller, requiring shorter multiple duct runs and the need to enlarge the vertical shaft space. Concomitantly, the amount of available space for production activities will decrease (Moerman, 2011c).

Walkable ceilings (Figure 4.47) are becoming a standard feature in new food plant construction, with the interstitial area above the ceiling and below the roof providing a convenient workspace for easy routine maintenance operations, while isolating utility piping and equipment from process areas. The change-out of HEPA filters and the service of piping and valves can occur without disruption of the cleanliness of the high hygiene space below. The lamps of lighting, which may be built-in hermetically closed into the false ceiling (Figure 4.48(b)) and protected by a transparent polycarbonate glass, also can be changed via the technical area (Moerman, 2011b,c; Higgins, 2013).

4.7.2.3 Make Use of Pipe Trains, Preferably Down from the Ceiling Installed

Pipes running in a roof void logically enter the process area through the ceiling. As they drop from the ceiling, they should be grouped together in easily accessible pipe trains whenever possible. In an attempt to minimize individual ceiling drops, the points of use also should be grouped (Figure 4.49). Vertical entrance of piping into the equipment or equipment jacket is more hygienic than horizontal utility piping runs (Moerman, 2010).

Figure 4.47 The cost of structural steel needed to build a walkable ceiling where personnel can stand erect for easy maintenance discourages some companies from including that feature, but the long-term efficiencies will make it standard within 10 years, at least for pumpable foods. A walkable ceiling has as advantage that maintenance personnel can access the technical area without special gowning.
Courtesy of BNP Media, publisher Food Engineering, Richard Zerby (Moerman, 2011c; Higgins, 2013).

(a) (b)

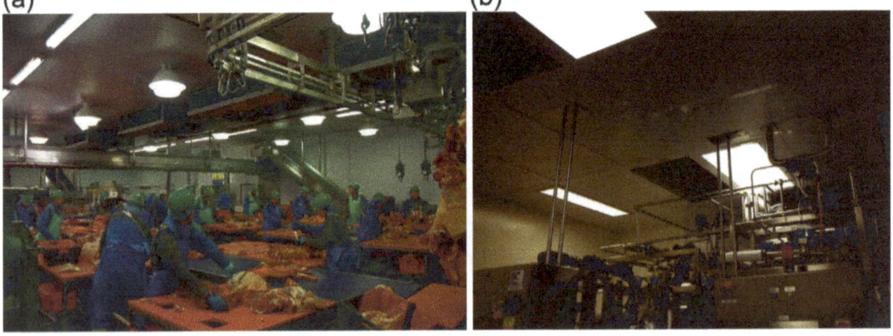

Figure 4.48 (a) Uncluttered ceiling with lighting having a sloped body with the lower part provided with a polycarbonate glass *(Courtesy of Veelite Technologies, Ltd.)*. (b) Unclutted ceiling. Lighting can be built in into the false ceiling with a hermetically closed seal and further separated with a transparent polycarbonate glass from the process area just below. It is a procedure typically for high hygiene areas where lamps are changed via the technical area *(Courtesy of Environmental Systems Corporation, ESC Cleanrooms (Moerman, 2011b))*.

4.7.2.4 Relocation of Piping and Utilities in a Technical Room or Technical Corridor

It is the general philosophy of good manufacturing practice (GMP) to relocate all process utility services to a technical location with minor hygienic requirements. Mechanical, electrical, pneumatic, hydraulic and electronic components, together with distribution conduits, valves, pumps, pressure reducers, gas cylinders, vacuum sources, compressors, etc., should be isolated in a technical room (Figure 4.50)

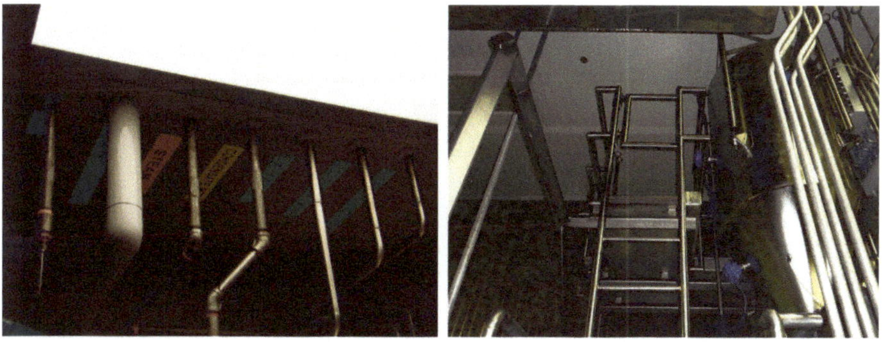

Figure 4.49 As pipes drop from the ceiling, they should be grouped together in easily accessible pipe trains whenever possible.
Photo left: courtesy of Environmental Systems Corporation, ESC Cleanrooms; photo right: courtesy of The Stellar Group.

Figure 4.50 Utilities should be placed in a technical room.
Courtesy of Bay State Industrial Welding and Fabrication, Inc.

or technical corridor (Figure 4.51) adjacent to the production room, with an easy, well-designed access for maintenance personnel (Moerman, 2011c).

Process support systems like the plant steam equipment and the infrastructure to heat thermal oil can be placed in a building adjacent to the food production plant, while utilities like water cooling towers, chillers, condensers, and air handling systems can be placed in respectively open air in the neighborhood or on the roof of the food production plant. Other utilities like water treatment systems, electrical cabinets, vacuum pumps, compressors, etc., are usually placed in-service rooms, close to the point-of-use to keep the length of piping short.

Figure 4.51 Utility piping placed in a technical corridor.
Photo right: courtesy of Environmental Systems Corporation, ESC Cleanrooms.

4.7.2.5 Reducing the Number of Cables by Using Wireless Measurement, Communication, and Control Techniques

The number of cables can be reduced by appropriate installation of process equipment or by making use of remote input/output and/or bus systems. The design of a hygienic plant can be performed in such a way that only a single air and electrical connection is required. Wireless transfer of data between instruments and control equipment, and (wireless) battery supplied low-energy sensors and actuators are techniques that little by little find their way in industry (Moerman, 2011a).

4.7.2.6 Ceiling Mounted Utility Columns

In high hygiene rooms (in food, the equivalent of clean rooms), utilities also can be supplied via pendent services (Figure 4.52). A pendent is a column that brings various services (water piping, chilled water piping, steam piping, electrical cabling, nitrogen gas, and compressed air lines) in a grouped and hygienic manner down from the ceiling in a zone H area. To prevent cross-contamination between various services, they may be housed in different segments of the pendent, isolated and independent of each other. In a housing that is completely sealed, utility pipes, lines, and cabling going through the outer shell of the service shaft are no longer exposed to the contaminants present in the process area, and as such they no longer will accumulate dirt at their exterior. Hence, in reverse, in a closed configuration, the utility infrastructure also will not act as a source of contaminants toward the process environment.

4.7.2.7 Position Equipment in Adjacent Areas Along the Same Wall, Back-to-Back

The amount of piping in process areas also can be reduced by installing process equipment in adjacent areas back-to-back (Figure 4.53).

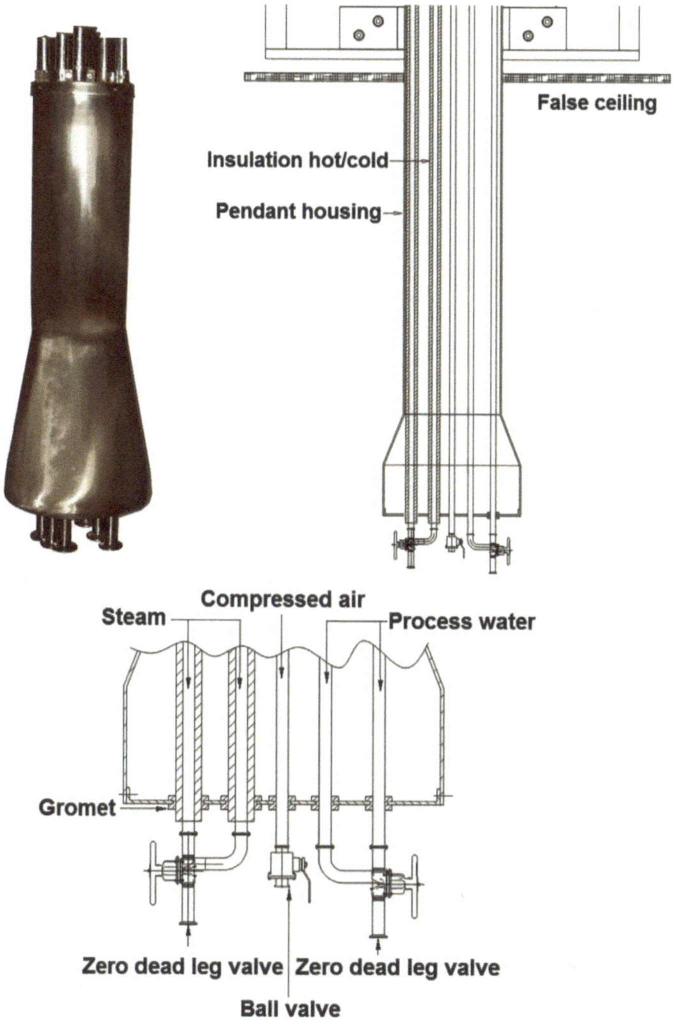

Figure 4.52 Utilities can also be supplied via pendant services.
Courtesy of Industrial Equip Wash Inc.

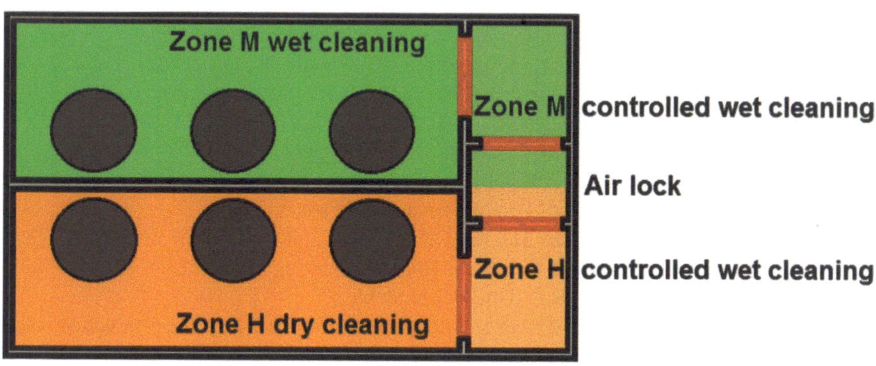

Figure 4.53 By positioning process equipment back-to-back, the number of pipings in the process area can be reduced.

4.7.2.8 *Make Use of Gravity Flow by Applying Multilevel Concepts*

Some processes may be better suited in a multistorey plant (Figure 4.20). The product can be transferred by gravity rather than conveyors, elevators, or other powered mechanical transport systems that cause inherent hygiene problems.

4.8 Conclusion

Since 1989, the European Hygienic Engineering and Design Group has largely demonstrated the value of hygienic engineering and design of food factories as a tool to enhance the safety of the food produced. Hygienic food factory design starts with the selection of an appropriate site and the application of hygienic building concepts that must prevent the entry of pests and the build-up of dirt. The factory layout must permit the correct flow of materials, waste, air, and personnel without compromising food safety, and the implementation of hygiene zones must offer maximal protection to the food produced. Zoning and the establishment of barriers, to ensure that product of acceptable hygienic quality is produced, should only be applied where their use will significantly help to protect products. To avoid unnecessary investment, zoning and barrier technology must be applied in an appropriate and consistent way based on the risks to be managed for the food products produced. Therefore, putting the whole factory into a complete clean room concept cannot be the purpose of food area segregation to protect product and consumer. By relocation of piping, dirty process, and utility components to technical areas, it is possible to further reduce the likeliness of product contamination, and at the same time the size of the process area. The result is that smaller amounts of air must be treated to maintain the required air quality, reducing the size of the air conditioning installation needed and the operational costs for air handling. By applying isolator technology, it is possible to even further reduce the size of the production while allowing for faster decontamination and increasing flexibility.

References

Brown, K.L., Aichinger, K., Burfoot, D., Fahldieck, A., Freund, M., Germain, J.P., Hamer, A., Klemetti, I., Lacroix, N., Nakatani, S., Sasia, G., Sun, X., Van Beek, Ph., Wray, S., 2005. Guidelines on air handling in the food industry. In: EHEDG Guideline N° 30. EHEDG working group 'Air handling', EHEDG, Frankfurt, Germany, pp. 1–43.

Burfoot, D., 2003. Generation and dispersion of droplets from cleaning equipment used in the chilled food industry. Journal of Food Engineering 58 (4), 343–353.

Chartered Institute of Environmental Health (CIEH), 1997. 'Industry Guide to Good Hygiene Practice: Baking Guide', Food Safety on Hygiene Working Group of the Baking Guide. Chadwick House Group, London, United Kingdom, 107 p.

CCFRA, 2002. Guidelines for the design and construction of floors for food production areas. In: Timperley, A. (Ed.), Guideline N° 40 of Campden & Chorleywood Food Research Association, second ed. Chipping Campden, United Kingdom, 67 p.

CCFRA, 2003a. Guidelines for the hygienic design, construction and layout of food processing factories. In: Holah, J. (Ed.), Guideline N° 39 of Campden & Chorleywood Food Research Association, second ed. Chipping Campden, United Kingdom, 94 p.

CCFRA, 2003b. Guidelines for the design and construction of walls, ceilings and services for food production areas. In: Guideline N° 41 of Campden & Chorleywood Food Research Association, second ed. Chipping Campden, United Kingdom, 93 p.

Corus, 2004. 'A corrosion protection guide, for steelwork exposed to atmospheric environments', document CC&I: JP:C:2000:UK:05/2004, Corus Construction & Industrial, Barnetby, North Lincoln-shire, United Kingdom, 5 p.

Goodfellow, H.D., Berry, J., 1986. Clean-plant design. Chemical Engineering 6, 55–61.

Higgins, K.T., 2013. Food Plant 2023. Food Engineering 89–100. January magazine.

Holah, et al., 2014. Hygienic design principles for food factories. In: EHEDG guideline N° 44. EHEDG subgroup 'Building design', EHEDG, Frankfurt, Germany, pp. 1–115.

Labs, W., 2014. A clean slate: use it wisely. Food Engineering 49–56. December magazine.

Lelieveld, H.L.M., 2011. Aspects to be considered when selecting a site for a food factory. Ch. 11. In: Lelieveld, H.L.M., Holah, J. (Eds.), Hygienic Design of Food Factories, N° 216. Woodhead Publishing, Cambridge, United Kingdom, pp. 203–216.

Lhoest, W.J., 1991. Design and selection of pharmaceutical production equipment in the scope of modern automated plants. Pharmaceutical Engineering 11 (2), 46–57.

Mager, K., Duffey, J.L., Hauser, G., Hutten, H., Masters, K., Meesters, G.M.H., Oosterom, J., Rumpf, W., Schleining, G., Roels, J., 2003. Hygienic engineering of plants for the processing of dry particulate materials. In: EHEDG Guideline N° 26. EHEDG subgroup Dry Materials Handling, EHEDG, Frankfurt, Germany, pp. 1–28.

Moerman, F., 2010. Hygienic design of food processing facilities. Food Safety Magazine 16 (5), 50–54.

Moerman, F., 2011a. Hygienic supply of electricity in food factories. Ch. 19. In: Lelieveld, H.L.M., Holah, J. (Eds.), Hygienic Design of Food Factories, N° 216. Woodhead Publishing, Cambridge, United Kingdom, pp. 369–411.

Moerman, F., 2011b. Hygienic and innovative energy-saving lighting systems for the food industry. Ch. 20. In: Lelieveld, H.L.M., Holah, J. (Eds.), Hygienic Design of Food Factories, N° 216. Woodhead Publishing, Cambridge, United Kingdom, pp. 412–470.

Moerman, F., 2011c. Hygienic integration of process, process support and utility systems in the Food Factory. Ch. 21. In: Lelieveld, H.L.M., Holah, J. (Eds.), Hygienic Design of Food Factories, N° 216. Woodhead Publishing, Cambridge, United Kingdom, pp. 471–493.

Moerman, F., Holah, J., Steenaard, P., 2014. Hygiene practice during Maintenance Operations in the Food industry. Ch. 11. In: Lelieveld, H.L.M., Holah, J., Napper, D. (Eds.), Hygiene in Food Processing: principles and practice, N° 258, second ed. Woodhead Publishing, Cambridge, United Kingdom, pp. 384–407.

Moerman, F., Fikiin, K., 2015. Effect of hygienic design and operational parameters on frosting and defrosting of evaporators in refrigerated food processing and storage facilities. Chapter 11. In: Gaspar, P.D., da Silva, P.D. (Eds.), Handbook of Research on Advances and Applications in Refrigeration Systems and Technologies. IGI Global, Hershey, Pennsylvania, United States, pp. 646–707.

Sutton, K.P., 2013. Site Considerations. Ch. 11. In: Baker, C.G.J. (Ed.), Handbook of Food Factory Design. Springer Science and business Media, New York, United States, pp. 283–296.

Wallin, P.J., 2013. Design principles. Ch. 12. In: Baker, C.G.J. (Ed.), Handbook of Food Factory Design. Springer Science and business Media, New York, United States, pp. 297–324.

Wilkins, S., 2013. Are you controlling your boundary? Pharmaceutical Engineering 33 (5), 86–89.

Factory-Friendly Approaches to Applying Multivariate Analytics for Productivity and Yield Gains

5

Sean Robinson
Kerrco Automation Ltd, Glasgow, UK

5.1 Introduction

The food and beverage industries are uniquely positioned between a heritage of artisanal production and an increasingly engineering-driven future. Different regions around the world are driven by specific local issues, but the response to those pressures is in almost every case a trend toward increasing use of science and automation in the pursuit of optimized operations. Evolution in both computing hardware and software is creating the conditions for the food industry to embed advanced analytical capacity into a range of production systems and supporting functions. Even in 2014, paper-based methods or simple spreadsheets are all-too-commonly used for tracking everything from material usage to food-safety-related quality parameters. Two important trends are disrupting the entrenchment of these simple tools: the increasing level of processing power available in forms that can be incorporated alongside common process automation platforms; and the adoption of descriptive standards such as ISA-95 as the basis for modeling the data involved in production operations. The emergence of this combination means that the food factory of the future will be characterized by highly automated flows of data between master system and adaptive system "nodes" that support key operations with real-time measurement of, and responsiveness to, complex local (process- or machine-level) sets of data. The economy and ease-of-use of these tools will underpin a shift in thinking on the part of food enterprise leaders, bringing operational effectiveness into focus as a source of competitive advantage.

In developed economies, production has lagged as an area of focus until relatively recently; brand owners have historically prioritized product development and marketing over the more mundane areas of manufacturing and facility management. A combination of factors—the rise of retailers' own brands, increasing fluctuation in material, energy, and water costs, and the increasing pace of product innovation—have driven producers to recognize the need for finer, predictable control of production processes. In less advanced economies, traditionally localized food supply networks are making room for an increasing proportion of processed foods, partly due to changes in migratory and living patterns, and partly due to an increasing focus on maximizing the utility of harvested crops, rather than risking spoilage of

Innovation and Future Trends in Food Manufacturing and Supply Chain Technologies
http://dx.doi.org/10.1016/B978-1-78242-447-5.00005-8

fresh foods and grains in a world with increasing food insecurity. In both cases, it is now understood that opportunities to reduce waste, whether for competitive reasons or to stretch stressed primary crop supplies (and other resources) must see producers take a more rigorous approach to tuning processes and equipment for optimum performance.

In this challenging environment, improving the analytical capability of food processors offers one route toward addressing these needs. Because it has historically been difficult to collect, correlate, and interpret complex data sets from production, quality, and material supply operations, applications of statistical analysis in food and beverage production operations have traditionally been limited both in terms of scope and in terms of style of application. For the most part, statistical tools have been used to address a specific factor in need of control (e.g., biological load on a raw material or a quality-critical characteristic managed within a process), thus use of sampling or measurement regimens have been mainly to establish proof of adherence to previously defined control measures or to validate performance characteristics in single-variable analysis. To the degree that multiple critical factors related to a particular material batch or process have been subject to statistical analysis, the most common approach has been to execute the measurement and related data interpretation of each factor in relative isolation, trusting to an understanding of the process engineering and food science involved in the process design to guide the prioritization of the analysis and control effort. This reliance on single-factor approaches can result in producers missing significant opportunities to improve performance in areas such as material usage, utility consumption, and capacity utilization; at the same time the efficacy of root-cause analysis efforts in relation to quality and food safety incidents may also be lessened.

Single-variable focused analysis and control efforts can also be stymied by slowly changing conditions or that are not anticipated in the original development of products or design of processing equipment. Examples of this include

1. Process and equipment characteristics can change over time. Factors ranging from wear on parts to investments in improved componentry or food contact materials can shift operating characteristics in unexpected ways
2. Processes and equipment may suffer perturbations with extended periodicity, making it difficult to determine whether an issue is unique/short term, or requires a systemic approach to "engineering it out"
3. Ingredient characteristics may have periodic (seasonal) or seasonal variations, or long-cycle shifts not tracked in existing quality regimens
4. Poorly understood or mischaracterized relationships may exist between ingredients or between ingredients and the processing conditions and equipment
5. Combinations of the above may be simultaneously present.

Further complicating these issues is the reality that many processing operations have grown in place over decades with inconsistent automation and sensory array present on different generations of equipment. This combination of difficulty in measurement and cumbersome analytical tools can deceive producers into believing that worthwhile process changes would be difficult to achieve. But it is often the case that the analytical process itself is the greater barrier and that relatively small changes to controllable factors such as process set points or material specifications can deliver high impacts. The greatest gains to be realized from next-generation systems with embedded analytics are the speed with which outcomes can be related to controllable factors, and the precision with which the sensitivity of those factors can be characterized.

Multivariate approaches that measure causal factors both within and external to process execution and equipment (e.g., addressing material characteristics, equipment characteristics, ambient conditions, etc.) offer a more robust framework for ensuring a true understanding of how outcomes are achieved. A multivariate approach based on deriving insight from historical data ensures that knowledge extracted from the effort will definitely be applicable to the real-world environment in which future production will occur, offering a higher predictive value than exercises conducted solely on the basis of initial equipment, product, and process design. This in turn allows a producer to not only drive remediation and reengineering insights based on that extracted knowledge, but to use the models derived from past "good" performance as a baseline against which future operating signatures can be compared, with controllable factors adjusted to drive convergence to the models.

The application of analytical methods and statistical modeling in food and beverage is thoroughly treated in a mix of academic and private sector works. The focus of this chapter is on the practical aspects of embedding multivariate methods into the common, "live" factory environment, exploring the nature of problem sets that are commonly present but less commonly "solved," then describing challenges typically confronted in reaching a state of readiness to analyze, and finally developing an understanding of how a project flows from analysis through to deployment of intervention techniques. Throughout, the capacity of modern plant systems to enable this mix of capabilities is also explored; too often analytical control methods are seen as something that must be applied in a slow process of offline iterative analysis followed by disruptive reengineering, whereas the evolution of factory control systems, data collection tools, and interactive process visualization and control technologies allows a much faster and less intrusive approach.

5.2 Areas of Opportunity for the Typical Producer

Though improvements in tools available to assist with data collection, management, and interpretation have made them easier to use than in the past, multivariate analysis still demands an investment in time and often in tools. Fortunately, many opportunities to find significant payback will present themselves. Some modes of processing (and their respective product categories) are likely to be easily singled out as having thorny multivariate problems that relate to core processes and the ingredients that pass through. But by assessing operations with a broader perspective, a producer will often find that data from ancillary operations can be as important—both for providing improvement opportunities in their own right and for properly characterizing the optimization potential within core processes. In general, most firms will find they can make gains in these areas:

- Material usage/yield—with a direct impact on cost-of-good/unit sold
- Capacity recovery—leading to either or both of improved revenue opportunities (if capacity constrained) or more efficient use of labor (if not)
- Energy intensity—though especially impactful in thermal processes such as cooking, baking chilling or other obviously energy-intensive operations, surprising gains can be made when one considers asset categories such as building and environmental systems and cleaning systems.

With these in mind, we could consider the following to represent scenarios with a high potential for returns on the investment in analytics.

5.2.1 Complex Products and/or Complex Processes

These combine a high quantity of ingredients, many process steps, many recipe stages and related set points. If ingredients are also prone to variability, this will compound all of the above factors.

Optimization potential could be explored for:

- Constraining use of most expensive ingredients or additives, in relation to desired final product attributes or yield curves. Examples include dosing flavorings and seasonings on snacks; determining proportions of emulsifiers or stabilizers in confectionery.
- Understanding shortest possible cycle times at different stages to ensure desired attributes while recovering capacity (and related changes to operating profiles for supporting processes/assets). Appropriate for: doughs, pastes, sauces/soups, blended/cooked confectionery, and similar products.
- Using indirect measures to infer effects of variability in upstream processes or ingredients. Examples include color measurement of baked goods to infer effects of variability in moisture content through mixing and baking, and pH and conductivity analysis of wastewater in dairy processing to characterize milk fat yields in various liquid milk processing stages

5.2.2 Processes with Significant Inertia

Whether considering thermal inertia, or slowly-developing chemical or biological processes, these scenarios tend to have hidden sensitivities that are effectively impossible to control for "in the moment," but which can be addressed with refined recipe/process control engineering, and with improved understanding of ideal ingredient characteristics.

Optimization potential can be explored for:

- Constraining use of expensive additives or ingredients through improved understanding of ideal dosing profiles (solving for quantity and timing). Examples include evaluating yeast and enzyme activation additives in fermentation processes relative to other ingredient characteristics and processing conditions; and determining recipe and control adaptations needed to adjust for variance in bulk ingredients or ingredient water.
- Balancing utility demands using refined models of processes' needs for heating or cooling in relation to chemical or biological processes. Examples include optimizing boiler burn and compressor pump cycles in brewing and distilling; tuning compressor control and air movement in cold storage or freezer chambers.

5.2.3 *Packaging Processes with Significant Mechanical Complexity*

An increased pace of packaging material innovation can expose new failure modes—especially on high-speed lines where even subtle changes in product and package characteristics can cause capacity losses due to material jams and other equipment breakdowns.

Optimization potential can be explored for:

- Reducing downtime by developing properly characterized matrices of products and their optimal start-up sequences and run-rates through various packaging stages. Factors to evaluate include material rigidity, coefficients of friction, shifts in center of gravity from different package designs, or product densities. As sudden stops on high-speed lines tend to involve jams that are destructive to work in process, material savings generally go hand-in-hand with capacity recovery here.
- Reducing material waste by more finely tuning mechanical processes in packaging—both to optimize packaging weight control and to determine more optimal sequences for minimizing changeover losses in capacity, materials, and cleaning inputs. Factors to evaluate include methods and specifications for material transfer relative to different compositions and characteristics (variations in fattiness, viscosity, particle size distribution, etc.); methods related to the composition of packaging materials themselves—consider specifications for sealing process variables such as pressures, dwell time, temperatures, treatment factors etc. to develop understanding of packaging material-related performance variation.

The improvement opportunities noted above are not a comprehensive list, but should serve to illustrate the range of potential improvements to be pursued, and to illustrate the mix of data types involved in supporting the analytical and remedial work. In almost every case, a mix of ingredient characteristics, process variable data and outcome data (quality or effectiveness) is required to first characterize current performance and then expose the opportunities locked therein. Establishing a state of readiness, where such information is at hand in a form useful for analysis, is the first step in moving from recognition of opportunity to realization.

5.3 Using Existing Data Sources or Establishing the Data Sets

A robust approach requires a broad scope of data collection and analysis—after all the idea is to look for relationships that not yet identified or well characterized. That does not necessarily mean that "new" data must be collected. Most production environments are data rich; a review of data sets currently captured or maintained will generally show many if not all of the categories noted in Table 5.1 (below).

For the best chance to derive new insight into a process or product, it is important that as many of these elements as possible are made available for analysis. Often, remediation efforts founder when the approach taken is simply to apply more powerful tools to data sets that have been limited in scope either as a matter of perceived practicality, or as a matter of methodology (restricting analysis efforts to combinations of factors already believed to be at issue).

Table 5.1 **Types of Data Sets Useful for Multivariate Analysis**

Data Types and Sources		
Category	**Comment**	**Common Source**
Reference data	These data elements generally describe the standards against which material and process characteristics should be measured. A broader definition also encompasses equipment capabilities and capacities	Recipes Performance standards Quality specifications Equipment specifications Ingredient/material characteristic data
Process data	These data sets should encompass not only the details of key process variables "as run," but also data to check adherence to set points, as well as event/fault occurrences in core processing and related ancillary equipment	Equipment sensory array Automation such as Programmable Logic Controller (PLC) systems Supervisory Control and Data Acquisition (SCADA) systems Human–Machine Interface (HMI) units
Laboratory/test data	Encompassing material or product characteristic data captured through a mix of observational and technical methods	Laboratory Information Management Systems (LIMS) Supplier provided Certificate of Analysis (COA) Line-side logs
Ancillary data	Information about the consistency and quality of how power, steam, compressed air, or other critical inputs are delivered to equipment and processes	Equipment sensory array PLC systems SCADA systems HMI devices
Ambient data	Often tracked, but not commonly incorporated into productivity or quality analysis	Building sensors/building management systems Weather services (historical)

Note: As product categories and related process technologies vary widely, general terminology is used here.

5.3.1 Challenges in Preparing Data Sets

In an ideal world, all of the data elements described above would be readily available in a common database format, structured to make correlation easy. Realistically, the typical factory will possess partial data sets, with the most complete data generally being logged from processing equipment and key ancillary items such as boilers, chillers, and environmental systems, as these are likely the most automated systems. Regardless of the relative accessibility of data, however, it is important to make an initial mapping of the relationships between the different data elements available, as this will ensure that subsequent analytical work is properly focused. Figure 5.1 provides a

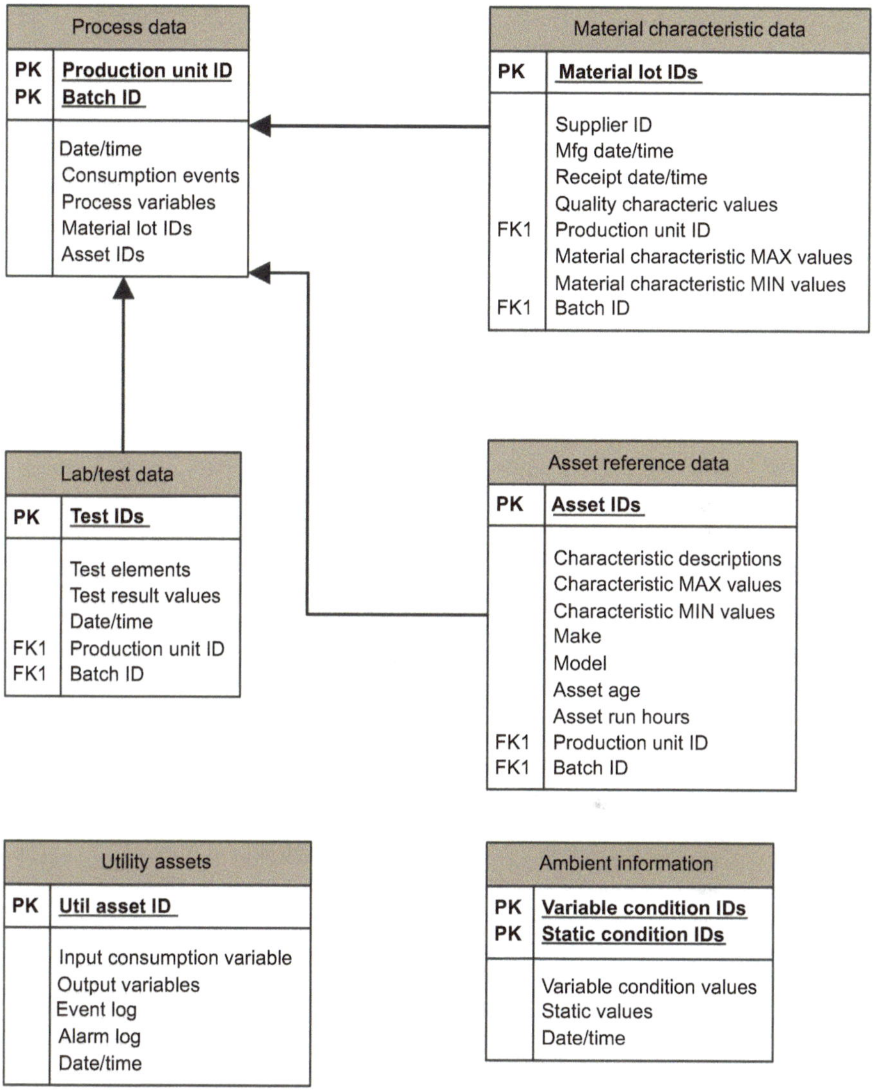

Figure 5.1 Generalized mapping of data sets commonly used for multivariate analysis. Note: The model presented here has been simplified, so should be considered a conceptual guideline rather than a true data model.

simplified overview of how disparate data sources might have their basic relationships mapped in rough model. Note that while there are natural links between several data sets, based on specific identifiers such as batch or material identification, and date and time stamps are often the only way to ensure that some data types are correctly associated with others during an analytical exercise. Considering that each of the material, asset, recipe, and process data sets for each combination of product and process could carry many individual data elements, the scope of the challenge in developing process effectiveness models through manual means becomes clear!

The initial mapping exercise typically reveals several immediate challenges:

- Some, but not all, potentially useful data sets are captured—though most commonly, the required data is collected, but manually or in formats that are not convenient for correlative analysis
- Inconsistencies in naming conventions and data structures for elements that are otherwise describing similar or identical things (especially in cases where multiple production lines of different age and engineering provenance are present)
- Data quality is inconsistent, with resulting potential to introduce "noise" into analysis efforts

5.3.2 Attributes of a Suitably Robust Data Collection Infrastructure

Each of the issues noted above lends itself to straightforward one-time remedies but cumbersome manual interventions cannot form the basis of an optimized production infrastructure when processes and equipment are increasingly complex, and generating data that might be useful, if more easily delivered to analytical and operational environments. To ensure that silos of data are not maintained even as individual process points or operations are laden with increasing automation and sensory array, best practices are evolving to underpin the data management strategy of future factories. Forward looking organizations will:

- Implement a data collection backbone that is designed to handle the high-volume flows of event and parametric data generated by equipment and sensors. And collect all data streams that "may" be useful. The advent of relatively cheap processing power and data storage removes constraints that used to limit such efforts when many production lines were initially deployed. Many "process historians" are available that offer a combination of high-speed/capacity (outstripping most general "enterprise/corporate" database systems), fault-tolerance, and connectivity to many types of plant automation and devices.
- Select a data collection backbone that allows raw data streams to be correlated "on-the-fly" to models that reflect asset structures, material definitions, and event descriptions rather than through periodic transformations. Industrial software vendors are increasingly making use of standards such as the ANSI/ISA-95 set (parts 1–5) to develop data models that provide consistent descriptors and structures for production data types, so producers can take advantage of a commonly used approach, rather than being forced to develop their own unifying structure and schema for their disparate data sources.
- Embed data quality management into the collection infrastructure data quality problems can be rooted in matters ranging from appropriate selection of sensory array and automation, through choice of communications topologies and protocols, to techniques to maintain the health of that communications infrastructure. Good data preparation tools or analytical tools have the capacity to identify and set aside portions of data sets whose accuracy is in question for issues relating to any of these reasons.

With suitable data collection infrastructure in place, and a methodology established to keep data sets and linkages current (or at least reproducible), knowledge extraction will no longer be hampered by issues of completeness and fidelity that are prevalent today.

5.4 Deriving Reliable Process Models from Historical Performance

The key element in this stage is summed up by the word "historical". The approach we propose is based on extracting hidden rules that govern the interaction of machines, materials, and process decisions as they have been run, rather than depending primarily on initial process engineering and food science to guide investigations into optimization strategies. Because of that, its value is governed by the availability of data to inform the analysis and conclusions, hence the intense focus on accessing and linking data sets as described above. But with that in place, a processor can characterize the behavior of their processes based on past experiences and results. Generally, the number of variables under consideration will outstrip the capacity to perform meaningful analysis unaided, so this section assumes the use of appropriate software tools for data processing. Generally, to derive a reliable model, a producer will follow these steps.

1. Determine factors correlating strongly to desired outcomes
2. Evaluate individual controllable factors for consistent behavior (reducing noise in subsequent optimization efforts)
3. Characterize adjustments to controllable factors that will reinforce the desirable outcomes, focusing on quantifying the sensitivity of the process to changes
4. Predict outcomes through "what if" exercises that use the initially derived correlative model as the basis for extrapolating outcomes as input values for controllable factors are manipulated in simulation exercises
5. Test the validity of the model by executing changes in controllable factors in a series of test production runs
6. Repeat and refine

For this process of learning, the author's experience has primarily been with tools that make use of artificial neural network (ANN) technologies to model the relationships between factors and their relative weighting and sensitivity. For readers unfamiliar with ANNs, a general overview, with a focus on food industry applications, is presented by Huang et al. (2007). The simplest summary of the approach might be to say that an ANN characterizes the relationships between different data elements (nodes in the network) by rapidly applying pattern matching techniques in conjunction with mathematical transformations to slowly "weight" or reinforce the emerging correlations of different data elements present in the sets released to the network for processing. Figure 5.2 presents the output of such an exercise—having identified a range of values for a quality-critical characteristic (in this case, bulk density of a powder), as the desirable outcome, application of an ANN has yielded a ranking of other factors according to the range of how they correlate (strongly positive to weakly through strongly negative).

This initial view makes it possible to identify the factors that deserve closer analysis, yielding a list of individual factors that may be well served by traditional statistical process analysis/control evaluations (performance within control limits, etc.), to ensure that their performance is within expectations. Once their consistent behavior

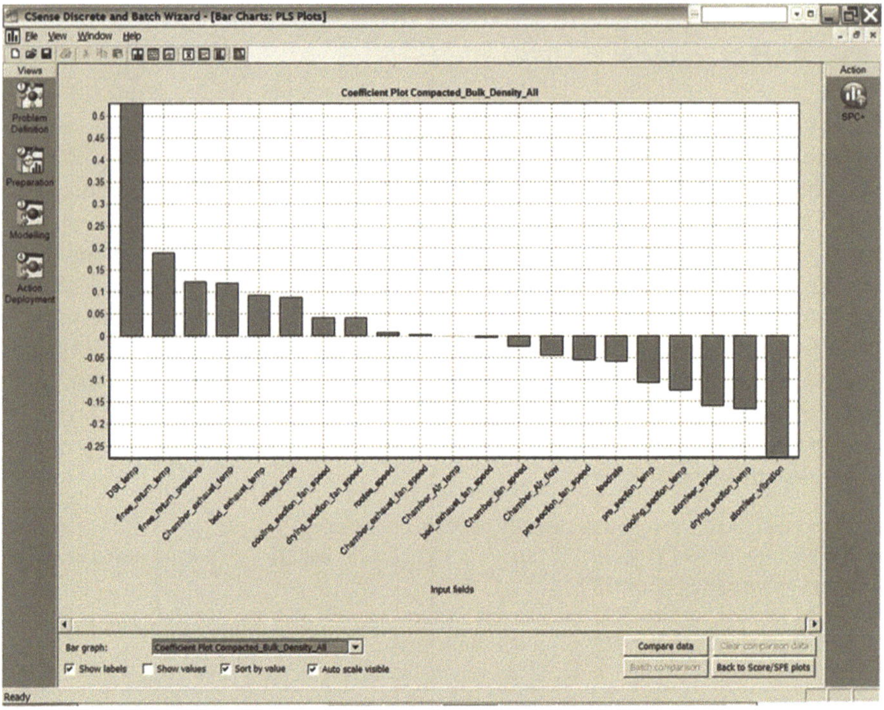

Figure 5.2 Evaluating a range of material characteristics and control recipe factors against quality outcomes (in this example, bulk density of a powder). Note: Use of appropriate software tools speeds modeling correlation of outcomes to potentially related factors. Art to Science, presentation by Steve Diggines, Factora.

has been validated, then each can be tested for the sensitivity of their relationship with desired outcomes. Figure 5.3 provides an illustration of a simple trending tool overlaying steam injection temperature and bulk density outcomes (y-axis values) for different sampling periods (x-axis values). This sort of view quickly provides an intuitive understanding of the relationship between the variables measured, while the underlying data will yield the precise matches between good outcomes and the range of values into which the controllable variable fell at the same time.

A similar set of views can be useful when predicting future performance. As a controllable factor's values are held steady in a range that correlates with good outcomes, a feed-forward simulation can illustrate how other factors must either be held (if controllable) or how they are likely to develop (if not controllable). This combination of history-based insight and predictive modeling becomes the basis for a variety of remediation efforts, including but not limited to:

- Recipe adjustments—addressing both how set points are managed for process variables and material quantities/ratios. Though our example figures focused on a machine setting, ANNs can be configured to evaluate outcomes in relation to variances in feed quantities at various stages as well

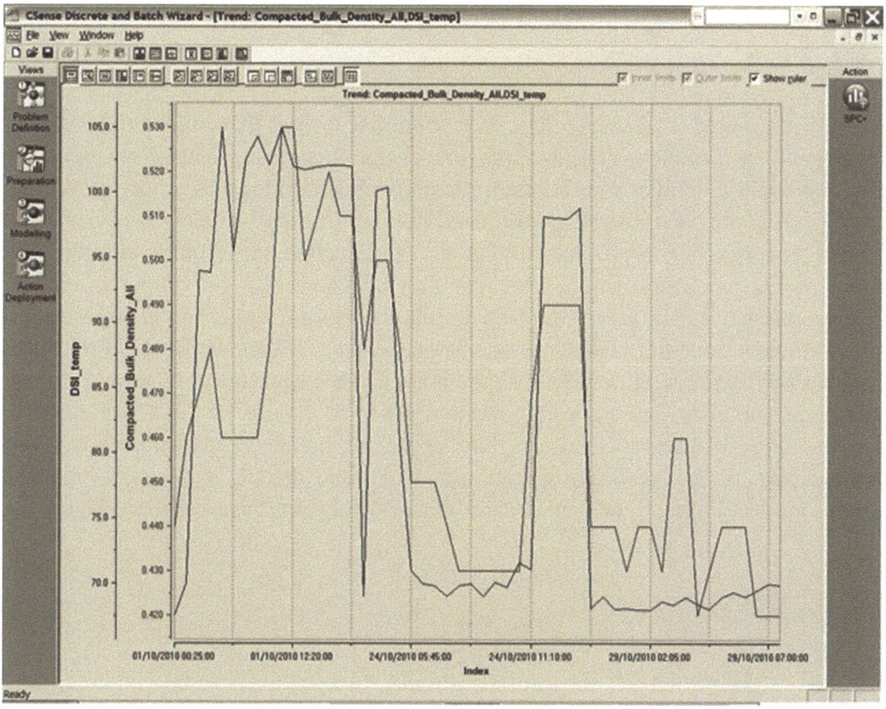

Figure 5.3 Evaluating individual controllable factors. Note: Reviewing historical outcomes against the pattern of the controllable factors enables testing for desirable ranges as well as for and understanding of the sensitivity of the relationship.
Art to Science, presentation by Steve Diggines, Factora.

- Equipment reengineering—comparing outcomes across different asset bases can yield new insight into the physical equipment that enables a process. Factors such as vessel sizes, contact surface materials, sizing/scaling of agitators, steam valves, and so on can be considered within a model, often helping to prove the need for capital investment
- Changes in material specification—often, assumptions about ideal ingredient characteristics are proven to have been true in relation to initially installed equipment (or to pilot scale production assets), but not to the production assets currently being run.

As remediation efforts are executed, the in-place data collection and analysis infrastructure allows their effectiveness to be rapidly assessed. As the results are developed over a statistically significant set of production event, the level of further refinement needed will become evident.

While in some cases remediation efforts will be self-contained in nature, in others it may be appropriate to develop the capability to drive adaptations on a per-batch or per-run basis, accommodating variability in raw materials or ambient conditions. Thus, it can be advantageous to be able to present incoming data streams to a tool that can compare them to predicted values, based on the trusted models developed in the exercises described here.

5.5 Embedding Predictive Capability within Common Plant Systems

As with the initial analysis project, it is common to find that multivariate models require different implementation approaches when it comes to real-time monitoring and intervention. While a traditional process control optimization effort may focus entirely on a core control strategy expressed through a primary automation system and related sensory array, in-process multivariate optimization may require monitoring of a mix of real-time information, while running calculations or derivations in relation to data sets not easily accessible to a traditional control system (e.g., information about ambient conditions, material characteristic data, machine/process event histories, etc.). With the work to define a data map, and the subsequent relationship analysis done, capturing and sustaining gains will see adaptation of familiar plant-level systems to incorporate mechanisms (think of them as model comparison engines) that can compare trusted predictive operational and process models to process events and conditions developing in real-time—and to guide interventions as needed. Figure 5.4

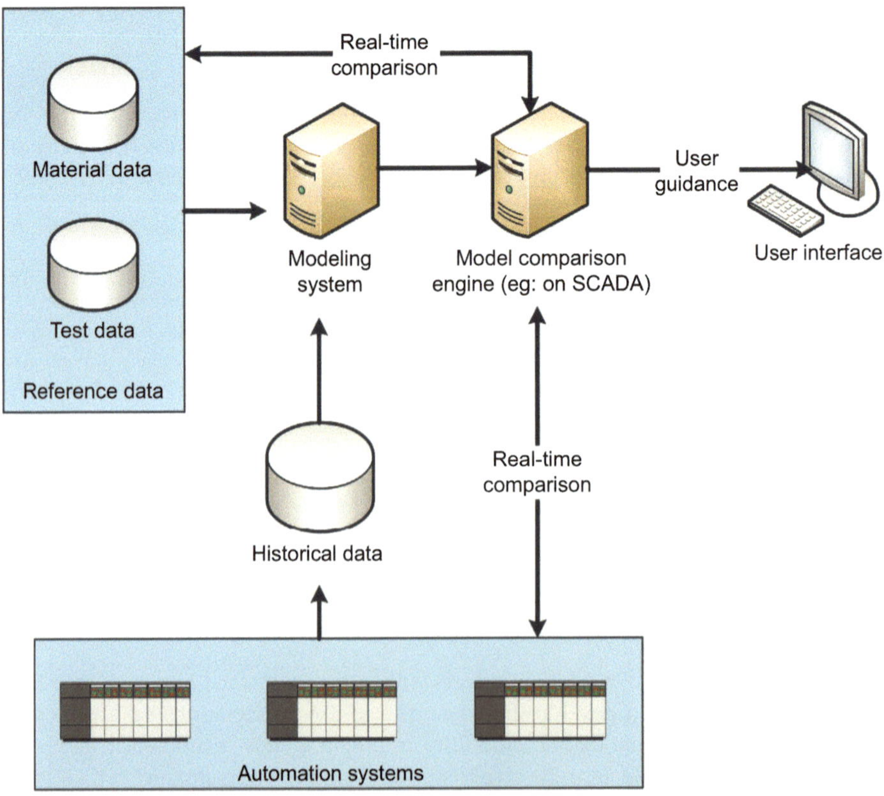

Figure 5.4 Logical overview of systems architecture with a model comparison engine.
Note: A real-time service that monitors developing events and parametric data can trigger interventions based on the process rules derived through the modeling effort.

provides an example of how the logical architecture of the different systems can be deployed.

Commercial software tools are evolving that can be used to facilitate this integration directly with standard automation equipment; when transitioning from modeling and analysis to shop floor activities, it is generally best to look for ways to integrate the outputs of the comparison engine with existing tools, rather than introduce additional systems into the production environment. Factors to consider include

- User training and adoption—it is ideal if an existing user environment can host and present the outputs of the comparison engine (alarms, reports, work instructions, etc.) in a form and framework that will require minimal additional training
- Integration pathways to automation—if established means to gather data from, and pass set points or variables to controls systems, etc. are in place, making use of those paths will reduce the risks associated with making changes in those underlying systems

Currently in common use for straightforward process visualization, Supervisory Control and Data Acquisition (SCADA) systems often provide the best bridge between real-time process execution, reference and ambient data sets, and a model comparison engine. Aside from the fact that they are a broadly used and accepted class of technology, they provide specific capabilities useful for turning analysis to real-time optimization:

- They have been designed to work in conjunction with real-time control systems, thus have the capacity to keep up as data is generated at that layer
- When intervention needs to be channeled through a lineside operator, or directly to one or more subsystems in an automation platform, SCADA provides both the user environment for manual notification/interaction and the native communications capability to drive changes in recipes or set points directly to the necessary control elements
- SCADA's native monitoring and alarming functions provide an ideal vehicle for triggering interventions when the model comparison engine detects drifts in controllable factors away from desirable ranges
- SCADA provides a useful means to visualize not only alarms and work instructions, but also underlying data—trends, values, and such—this allows operators and supervisors to understand the conditions against which intervention is directed, and apply additional judgment as required

By taking advantage of a well-accepted plant floor technology, adoption and usage risks can also be minimized; routing the outputs of complex analysis tools through modernized yet familiar shop floor systems such as SCADA will ensure that user environments remain simple and task-directed, even as the level of analytical capacity applied to operations increases.

5.6 Developing Trends

Multivariate analysis and process intervention clearly offer high value to producers, but it only in recent years that the conditions have become ripe for increased adoption of these tools and methodologies. To date, it has been the continuously decreasing cost of processing power for both offline analysis and lineside process control, data storage and in-plant, high-speed communications that have allowed richer and richer data sets to be captured and made available for analysis. However, as illustrated in Section 5.3 above, significant work still remains simply to establish a state of readiness for analysis! But with the basic infrastructure costs and complexities having largely been tamed, it is likely that two specific trends will accelerate the use of multivariate analytics in the food and beverage industries:

1. Increasing prevalence of standard-based data models being delivered in common shop floor systems. Bodies such as the Manufacturing Enterprise Solution Association, the Instrumentation, Systems and Analysis society (ISA) and the Packaging Workgroup within the Organization for Machine Automation and Control (OMAC) have been building consensus among producers, machine builders, information technology (IT) providers and service provider on the need for common data models and descriptors across asset and operational models. As engineering and IT tools are modernized, it becomes easier and more cost-effective to apply such standards for naming conventions and structures in ways that make it easier to establish the linkages and correlations described herein—even across the boundaries between systems delivered by different technology vendors.
2. Increasing development of autonomous, agent-based technologies that will manage individual processes and drive interactions between systems—at the very least for integration, and even for execution. The Intelligent Manufacturing Systems consortium has pioneered work in this field, moving beyond theoretical exploration to pilot systems based on the principles of distributing decision processes among independent nodes in a production network. The next logical step will be for relatively automated agents to exploit the metadata around core data elements (noting their descriptions/classifications as things such as "quality variables" with "targets"), and the linkages between recipes, production routes, and inputs, allowing autonomous agents to begin constructing models of process behavior an interaction as an automatic function.

The degree to which industry collaborative bodies such as Campden Brewing Research Institute (BRI) or the Grocery Manufacturers of America, or even regulators, are able to foster intercompany collaboration will also be enhanced. In a world where the assets, materials and general process models for a given product are described in standardized terms, or where analysis tools can more easily impose common definitions across like data elements, it will be easier than ever to combine data sets from different sites and firms to support analysis and resolution of problems for the common good of producers, consumers, and policy makers in different jurisdictions.

5.7 Sources of Further Information and Advice

There is a growing body of research detailing how multivariate techniques and supporting tools such as ANNs can be applied against specific product categories or process types. However, much of the work being done at a data modeling level is not as well known, as often these are seen as "IT issues" rather than enablers for embedding analytical capabilities that support a mix of product and process development as well as furthering applied food science in situ. With that in mind, the work of the groups below merits a deeper look, as their projects serve to reduce the technical difficulty, and by extension the cost and time required, to execute modeling efforts and to deploy real-time multivariate optimization engines.

Industry Groups

Instrumentation, Systems and Analysis society (ISA): https://www.isa.org/isa95/

Manufacturing Enterprise Solutions Association: www.mesa.org.

Intelligent Manufacturing Systems consortium, Holonic Manufacturing Systems projects: http://www.ims.org/2011/11/hms-%E2%80%93-phase-i-and-ii-holonic-manufacturing-systems/

Organization for Machine Automation and Control, Packaging Workgroup: http://www.omac.org/content/packaging-workgroup.

Reference

Huang, Y., Kangas, L., Rasco, B., 2007. Applications of artificial neural networks (ANNs) in food science. Critical Reviews in Food Science and Nutrition 47 (2), 113–126.

Emerging Methods and Principles in Food Contact Surface Decontamination/Prevention

6

Torstein Skåra, Jan T. Rosnes
Nofima, Stavanger, Norway

6.1 Introduction

Food contact surfaces comprise all surfaces that may come into contact with food products during production, processing, and packaging. These surfaces are typically made of stainless steel or some kind of plastic material, but contact surfaces may also consist of other materials like wood, rubber, ceramics, or glass. They should not "leak" any substances to the food product, and they should be easy to clean and maintain. This implies that they should have an even surface without cuts and crevices that could support build-up or trapping of biological material, and sometimes specific material characteristics like hydrophobicity.

Microorganisms have a natural affinity to surfaces (Faille et al., 2001; Zhao et al., 2007). When attached, cells are still readily removed by mild rinsing, but when left to proliferate they can also adsorb irreversibly and form layers of bacteria and extracellular products (Hood and Zottola, 1995). These layers, called biofilms, are more difficult to remove from a surface (Jefferson, 2004).

There are several definitions of decontamination, e.g., the inactivation of microorganisms by disinfection or sterilization. In this setting, disinfection implies destroying or removing vegetative forms of harmful microorganisms, whereas sterilization destroys fungi, viruses, as well as bacteria and their spores. Disinfection and sterilization are closely related to cleaning. Cleaning, however, is a large area in itself and will not be covered in this chapter. Nor will it cover specific aspects of biofilms and their removal, common currently used disinfectants and packaging materials.

Sanitizers reduce, but do not necessarily eliminate the microorganisms they are brought in contact with. A good sanitizer should have a broad spectrum of activity toward bacteria, fungi, and viruses and be cost-effective. It should be easy to apply, have low toxicity and be noncorrosive, but have good penetrative power and be fast acting. It should be stable in the concentrated form and active after dilution in hard water, but not be adversely affected by organic matter. Finally, it should not have harmful or offensive odors, nor should it be incompatible with other chemicals and materials of construction or persistent in the environment (Gaulin et al., 2011), but (according to Environmental Protection Agency, United States specifications) kill or irreversibly inactivate at least 99.9% of all bacteria, fungi, and viruses present on a surface.

Innovation and Future Trends in Food Manufacturing and Supply Chain Technologies
http://dx.doi.org/10.1016/B978-1-78242-447-5.00006-X

These are numerous demands that the currently available agents and methods fulfill to varying degrees. Hence, there is continuous development, which this chapter seeks to summarize and evaluate.

6.2 Methods and Principles

There are a number of ways to achieve some form of decontamination. The most-used methods involve the use of some kind of chemical agent. In addition, there are a number of physical methods available, and even surface and material modifications can be used to achieve some degree of decontamination effect (Figure 6.1).

Common to many emerging methods is the combination of two or more agents or technologies to achieve synergistic effects.

The most appealing principle, however, is the prevention of recontamination, which can lead to reduced use of disinfectants altogether. This is not a new method, but knowledge is continuously generated on harbors and contamination routes. Combined with smart use of technology through hygienic design, use of barriers and control of airflows, significant advances can be made.

6.2.1 Decontamination

6.2.1.1 Nonthermal Methods

6.2.1.1.1 Chemical Methods

A chemical disinfectant is a chemical agent used on food contact surfaces and objects to destroy infectious fungi and bacteria. Sporicidal and antiviral agents are often considered a special class of disinfectants. Only chemical sanitizers that have been approved for use in food facilities can be used, and specific requirements or regulations may vary between countries. For disinfectants to be effective, proper cleaning and rinsing must be completed before sanitizers are applied.

Some agents are widely used and described in textbooks and literature. These include chlorine, iodine, quaternary ammonium compounds (QAC), carboxylic acid, and acid–anionic sanitizers. Their use is not considered emerging and hence they will not be covered in this chapter. In addition, some of these have been coupled to the formation of unwanted, potentially toxic substances (see section Future Trends). This is also a market driver that may influence the future of food contact surface decontamination agents and technologies. It follows from these considerations that methods and technologies used for food surfaces can also be applied for food contact surfaces. In this chapter, some methods and agents are listed for which the effects have been documented for both, but some have only been documented for foods. Despite that, they have been included in this chapter as further developments may lead to their future use for food contact surfaces.

Peroxy compounds are on the list of commonly used disinfectants, but possess qualities that are unique both with respect to efficiency and toxicity after treatment. Hydrogen peroxide has been used to decontaminate a number of foods (Joshi et al., 2013;

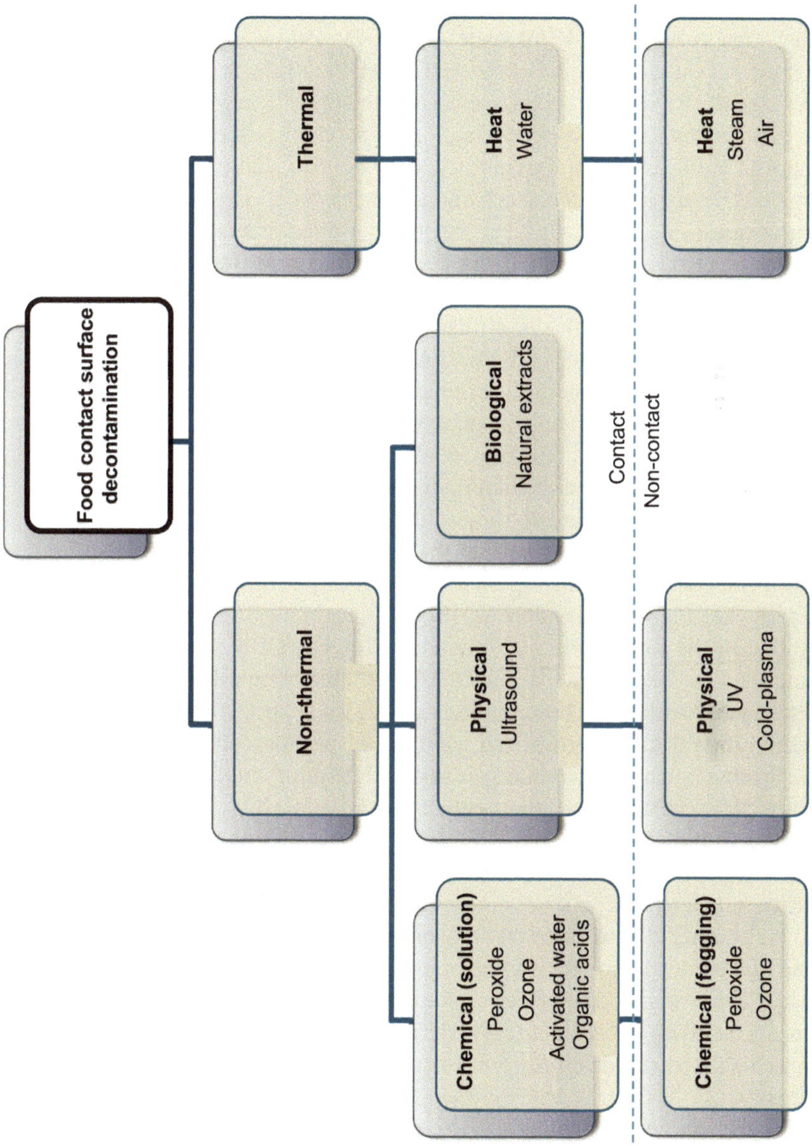

Figure 6.1 Examples of food contact surface decontamination principles. Adapted from Warriner (2011).

Lopez-Galvez et al., 2013; Upadhyay et al., 2013) as well as packaging in aseptic filling lines (Ansari and Datta, 2003; Cardoso et al., 2011). It removes electrons from susceptible chemical groups and becomes itself reduced in the process (Finnegan et al., 2010). Hydrogen peroxide does not react with organic compounds to produce carcinogenic compounds. It breaks down into water and oxygen and has been given the status of Generally Regarded as Safe in 1986 for some food commodities. Hence, there is sustained interest in this agent, and modifications (Hassan et al., 2013), as well as combinations with other agents like organic acids have been tested (Martin and Maris, 2012a,b).

Most disinfectants are delivered through direct contact, either in the form of foam or liquid. Again, this is within the currently used domain, and will not be explored in this chapter. Delivery of disinfectants by noncontact systems is, of course, a somewhat inaccurate expression. Eventually there has to be a direct interaction by the disinfectant on the target organism, but in this context noncontact refers to no immediate contact. The active component(s) or destructive energy is applied in the vicinity of the surface to be treated and eventually diffuses to or "lands" on the bacteria.

Aerosolization, or fogging, is dispersion of a liquid or a solution in air, in the form of fine mist. Distribution of disinfectants by fogging was found to be effective in reducing the number of organisms on upward-facing surfaces, but not on vertical or downward-facing surfaces (Burfoot et al., 1999). These authors also found fogs to be most effective when the median diameter of the fog droplets lies between 10 and 20 μm; dispersing well and settling within about 45 min, they give good coverage. Bagge-Ravn et al. (2003) compared the use of sodium hypochlorite-based foam and a peroxyacetic acid-based fog sanitizing procedure in a salmon smokehouse. Interestingly, the fogging procedure was found to improve the hygienic level, compared to that of the foam sanitizing procedure. In contrast to Burfoot et al. (1999), these authors also found that fogging improved the hygienic levels of vertical and downward-facing surfaces. These findings were also supported by Oh et al. (2005), who studied the effect of aerosolized sanitizer of peroxyacetic acid and hydrogen peroxide against inoculated target organisms on laboratory media. It should be noted that the droplet size was 2 μm, the treatment time was 1 h at ambient temperature, and the reduction of *Bacillus cereus*, *Listeria innocua*, *Staphylococcus aureus*, and *Salmonella* Typhimurium were in the range between 3 and 8 log units, with no significant differences for each organism, independent of height and orientation. This led the authors to conclude that aerosolized sanitizers diffuse like gaseous sanitizers and that the technology has great potential for use in commercial applications. Park et al. (2012) studied the effect of aerosolized sanitizers (sodium hypochlorite and peracetic acid) on *Escherichia coli* E157 and *S.* Typhimurium biofilms, concluding that aerosolized sanitizers have a potential as a biofilm control method in the food industry.

Fogging technology and sanitizing procedures have many additional applications including medical, e.g., for room disinfection (Thorn et al., 2013) that can be relevant also for food applications. The technology, however, demands a certain size of the room/chamber to be disinfected or adaptation of delivery systems, and there may be limitations to its uses caused by the oxidative nature of the disinfectants used, which may cause problems related to corrosion.

Ozone is a colorless gas with a distinctive odor, produced industrially by bombarding oxygen with ultraviolet (UV) radiation or by passing air through a high-voltage alternating current electrical discharge (Anon., 2014). It has a very high oxidation potential and leaves no toxic residues because it decomposes into oxygen (Nicholas et al., 2013). Its powerful antimicrobial properties are related to the reaction with cell membranes, causing lysis and death. Ozone has several food applications, e.g., water treatment and disinfection of equipment surfaces as well as food (Akbas and Ozdemir, 2006; Beltran et al., 2005; Crapo et al., 2004; Gelman et al., 2005; Kleiner, 2001; Manousaridis et al., 2005; Simons and Sanguansri, 1997). Its low toxicity makes it an agent of continued focus and interest. There are several studies available of its use for decontamination of food contact surfaces (Khadre and Yousef, 2001; Martin et al., 2009; Nicholas et al., 2013; Wirtanen et al., 2002; Zoutman et al., 2011). The efficiency of ozone has also been tested for use in whole room disinfection (Staniforth, 2014), on *Listeria monocytogenes*, *Pseudomonas aeruginosa*, and *S. aureus*. As with other chemical agents there was a relationship between concentration, contact time, and log reduction. Sensitivity was also an issue, with *S. aureus* being the least, and *L. monocytogenes*, the most sensitive. For the latter, a reduction of >4 logs was achieved for an ozone concentration of 20 ppm within a contact or dwell time of 1 h. Interestingly, there was also little practical difference in log reductions achieved between orientations. Ozone can effectively penetrate all areas of a room, including those that are difficult to access with conventional liquids and disinfection procedures.

The major disadvantage of using aerosols and gases such as ozone is the potential toxicity which precludes their use in areas where people are present. Hence, they can only be used in areas that can be isolated during the decontamination process and well ventilated after, or where the gas can be allowed enough time to break down.

Acidic electrolyzed water and neutral electrolyzed water (NEW) are generated from the electrolysis of a dilute sodium chloride solution. Both have shown strong antimicrobial activity against many pathogens from food surfaces (Guentzel et al., 2008; Koseki et al., 2001; Park et al., 2002) as well as food contact surfaces (Arevalos-Sanchez et al., 2013; Deza et al., 2007; Monnin et al., 2012). NEW does not cause irritation of skin or corrosion of processing equipment and is stable because chlorine loss is significantly reduced at pH values of 6–9 (Len et al., 2002).

Natural extracts like pomegranate peel extract and cinnamon bark extract have been tested as biopreservatives to beef trimmings (Tayel and El-Tras, 2012) and combined with lemongrass leaves extract these have resulted in complete bacterial inhibition for 2 days when applied on meat steak surfaces (Tayel et al., 2012). The effect of natural extracts incorporated in food packaging material, on the growth of *L. innocua* and spoilage bacteria have also been studied (Iturriaga et al., 2012), but no studies on the use of plant extracts on food contact surfaces have been found.

Antimicrobial enzymes aimed at the disruption of bacterial cellular machinery and biofilm formation have been commercialized for application in the healthcare, food, and biomedical industries. Advances in synthetic biology, enzyme engineering, and whole DNA sequencing technologies are expected to bring the development of these enzymes further and increase their efficiency (Thallinger et al., 2013).

Nanomaterials are defined as particles less than 100 nm (Meyer and Persson, 1998). The use of nanoparticles may be an emerging technology for decontamination. Thus, too few studies are available that utilize this technology in food preservation, but zinc oxide (ZnO) nanoparticles have been shown to exhibit much stronger antibacterial activity than ZnO powder. This effect may be attributed to the larger surface to volume ratio of the nano particles and also the generation of hydrogen peroxide from the surface of ZnO (Tayel et al., 2011).

6.2.1.1.2 Physical Methods

The effect of ultrasound on the inactivation of microorganisms in foods has most often been demonstrated in combination with e.g., thermal processes. The inactivation itself is a complex process, and the efficiency is governed by a number of factors (Piyasena et al., 2003) as well the size and the nature of the microorganism (Goode et al., 2013). When it comes to food contact surfaces, however, the use of ultrasound is most often seen as a means to aid cleaning (Tolvanen et al., 2009) and to remove biofilm (Assere et al., 2008; Berrang et al., 2008; Martin et al., 2009).

Cold plasma is a relatively new antimicrobial process operating at or near room temperature being developed for applications in the food industry (Niemira, 2012). Plasma treatments use no chemicals and are contact free. Gas molecules are ionized by high-voltage electricity, thereby generating a rich abundance of highly reactive, short-lived chemical species (ions, electrons, excited atoms, UV photons, radicals, and reactive neutral species, e.g., oxygen and nitrogen) that are capable of inactivating a wide range of microorganisms. This technology has shown promise as a direct treatment for fresh and freshly cut fruits (Critzer et al., 2007) and vegetables (Fernandez et al., 2013), as well as for nuts and other foods.

The effect of cold plasma technology has also been investigated for food contact surfaces. Small volumes of water have been seen to increase the inactivation, presumably due to the plasma treatment reducing the pH of the liquid. At conditions with water present, as is often the case in food industry, both *S.* Typhimurium and *L. monocytogenes* were reduced by approximately 4 log units after 2 min, whereas it took 5 min to achieve the same effect on *E. coli*. *S. aureus* was even less sensitive to plasma treatment (Bayliss et al., 2014). These authors also found that the food contact surface impacts the effect of plasma treatment. Reductions for *L. monocytogenes*, *E. coli*, *S. aureus*, and *S.* Typhimurium were higher on stainless steel disks compared with polyurethane conveyor belts. Niemira et al. (2014) studied the inactivation of *Salmonella* biofilms on glass, and found reductions in the range between 1 to 2 log units.

UV is electromagnetic radiation with a wavelength of between 100 and 400 nm, classified into three general areas by wavelength: UV-A, UV-B, and UV-C. UV-C, at 100–280 nm, has the shortest wavelength range of the three and is the type applied to food and beverages, although the technology has long been known and used for other industrial applications. For example, UV-C has been used for many years to disinfect or sanitize drinking water, wastewater, air, and food contact surfaces. UV light (254 nm UV-C) is bactericidal and can be used for decontamination of juice in the United States (FDA, 2000). Exposure to UV-C irradiation leads to changes in the bacterial chromosome (Reardon and Sancar, 2005), which kill the bacteria or make them unable to reproduce.

Sommers et al. (2010) studied the inactivation of *Salmonella* spp., *L. monocytogenes*, and *S. aureus* on a number of food products, including raw meat, precooked sausages, shell eggs, and fruits and vegetables, as well as stainless steel surfaces. Of all the surfaces tested, the effect of UV-C was best on stainless steel, where >5 log reduction of all three pathogens was observed at a UV-C dose of 0.4 J/cm^2, an inactivation rate that is also in accordance with the findings of Kim et al. (2002).

6.2.1.2 Thermal Methods

Traditional thermal treatments with hot water or water vapor demand high energy input and long treatment times and also that all parts subjected to the treatment are made of materials with adequate thermal resistance (Otto et al., 2011). Elevated temperatures, however, often increase the reaction rates and efficiency of many products, and hence combinations of heat and other agents can be favorable.

Among food grade chemicals without corrosive or toxic potential for use as bactericidal and sporicidal agents, esters of fatty acids acting as both solvents and detergents have shown spore inactivation potential. Lee and Pascall (2012) investigated the use of an agent consisting of selected fatty acids and lactate esters. At room temperature, this gave a 3 log reduction in number of viable *Clostridium sporogenes* spores. Combined with heat, the agent (10%) showed 5 log reduction after 10 min at 68 °C.

The use of steam to decontaminate food contact surfaces is quite common in the food industry. Steam is an efficient energy carrier when condensed on a surface, but it can be difficult to monitor contact time and temperature. Additionally, one should be aware of some issues related to the use of steam on surfaces. The inactivation kinetics of surface attached cells may differ from those of free floating. Valdramidis et al. (2008) found that this was the case with surface bound *L. innocua*. *Listeria innocua* cells attached to abiotic Teflon had a much larger heat resistance than free floating cells. In addition, substantial tailing and erratic inactivation data made it difficult to calculate basic inactivation parameters, like D-value and z-value for surface bound *L. innocua*. Jermann et al. (2014) studied the effect of steam pasteurization on the inactivation of *E. coli* O157:H7, *S.* Typhimurium, and *L. monocytogenes* biofilms on stainless steel and polyvinyl chloride (PVC). They found that biofilm cells of *E. coli* O157:H7, *S.* Typhimurium, and *L. monocytogenes* on stainless steel were reduced by more than 6 log colony forming units (CFU)/coupon after exposure to steam at 75 °C for 30–40 s, and at 85 °C for 20–30 s. On PVC coupons only 1 to 2 log reductions were seen on these organisms after 50 s of exposure to steam at 75 °C. Exposure to steam at 85 °C for 50 s increased the efficiency somewhat, to 2 to 3 log units. Differences in microbial reduction patterns between stainless steel and PVC may be due to different thermal conductivity. When stainless steel and PVC coupons were treated with steam at 75 °C, the temperature of stainless steel coupons reached 75 °C within 5 s, while PVC coupons reached this temperature within 15 s. Higher efficiencies can be achieved by increasing steam temperature. One should also be aware of the potential effects of the surface material properties (like thermal conductivity) when setting up steam based decontamination procedures.

Table 6.1 summarizes a number of available decontamination studies for the reduction of *Listeria* spp.

Table 6.1 Recent Studies of Food Contact Surface Inactivation on Stainless Steel, Plastic Materials, and Glass

Food Contact Surface	Organism	Surface Decontamination Treatment	Dose (Concentration, Temperature, Energy)/ Treatment Time	Bacterial Reduction/ Unit	References
Teflon	*Listeria innocua*	Steam	58 °C/20–300 s	1–2 log CFU	Valdramidis, et al. (2008)
PP	*Listeria monocytogenes (surface attached)*	Ozone	45 ppm/1 h	1.1 CFU/cm^2	Nicholas et al. (2013)
PP	*Listeria monocytogenes (biofilm)*	Ozone	45 ppm/1 h	0.9 CFU/cm^2	Nicholas et al. (2013)
PP	*Listeria monocytogenes (surface attached)*	Open air factor	1 h	ND	Nicholas et al. (2013)
PP	*Listeria monocytogenes (biofilm)*	Open air factor	1 h	1.8 CFU/cm^2	Nicholas et al. (2013)
PVC	*Listeria monocytogenes* (biofilm)	Steam	75 °C/>50 s	1–2 log CFU/coupon	Jermann et al. (2014)
PVC	*Listeria monocytogenes* (biofilm)	Steam	85 °C/50 s	2–3 log CFU/coupon	Jermann et al. (2014)
Stainless steel	*Listeria monocytogenes* (biofilm)	Steam	75 °C/30–40 s	>6 log CFU/coupon	Jermann et al. (2014)
Stainless steel	*Listeria monocytogenes* (biofilm)	Steam	85 °C/20–30 s	>6 log CFU/coupon	Jermann et al. (2014)
Stainless steel	*Listeria monocytogenes*	UV-C	0.4 J/cm^2	>5 log CFU	Sommers et al. (2010)
Stainless steel	*Listeria monocytogenes* Scott A (biofilm)	Neutral activated water	65 ppm/>600 s	>6 log CFU	Arevalos-Sanchez et al. (2012)
Stainless steel	*Listeria monocytogenes*, *Listeria innocua*, and *Listeria ivanovii*	Moist heat and silver dihydrogen citrate sanitizer	66 °C/>20% Relative Humidity/7 h	>6 log CFU/coupon	Masuku et al. (2014)

Stainless steel	*Listeria monocytogenes, Listeria innocua*, and *Listeria ivanovii*	Silver dihydrogen citrate sanitizer		4.9 log CFU/coupon	Masuku et al. (2014)
Stainless steel	*Listeria monocytogenes, Listeria innocua*, and *Listeria ivanovii*	Moist heat	66 °C/<20% RH/7 h	<5 log CFU/coupon	Masuku et al. (2014)
Stainless steel	*Listeria innocua*	Dry heat	66 °C/0,5–15 h	1.5–3 log CFU/coupon	Crandall et al. (2010)
Stainless steel	*Listeria innocua*	Dry heat	80 °C/>15 h	>4 log CFU/coupon	Crandall et al. (2010)
Stainless steel	*Listeria monocytogenes*	Bacteriophages	1 h	5 log CFU/cm2	Arachchi et al. (2013)
Stainless steel	*Listeria monocytogenes*	Electrochemicallly activated water (ECAW)	100 mg/L ECAW	>54 log CFU/coupon	Yang et al. (2013)
Stainless steel	*Listeria monocytogenes* (surface attached)	Ozone	45 ppm/1 h	3.4 CFU/cm^2	Nicholas et al. (2013)
Stainless steel	*Listeria monocytogenes* (biofilm)	Ozone	45 ppm/1 h	0.6 CFU/cm^2	Nicholas et al. (2013)
Stainless steel	*Listeria monocytogenes* (surface attached)	Open air factor	1 h	1.9 CFU/cm^2	Nicholas et al. (2013)
Stainless steel	*Listeria monocytogenes* (biofilm)	Open air factor	1 h	1.4 CFU/cm^2	Nicholas et al. (2013)
Glass	*Listeria monocytogenes* Scott A (biofilm)	Neutral activated water	65 ppm/>600 s	>6 log CFU	Arevalos-Sanchez et al. (2012)

6.2.2 Other Measures

Antimicrobial surfaces is a term used to describe surfaces that can improve the cleanability and hence reduce the build-up of biofilm, as well as surfaces that have a lasting lethal effect on microorganisms. Antimicrobial surfaces cannot replace regular cleaning and should only be used as an additional control for food hygiene.

A strong, even and easily cleanable surface can be obtained by the use of parylene. Similarly, as well, hydrophobic surface properties can be obtained by the use of liquid glass coatings. Both these compounds are commercially available.

Surfaces that prevent adhesion of microorganisms are either superhydrophobic or superhydrophilic. A superhydrophobic surface has unique self-cleaning attributes through outstanding water repellency and antifouling properties, but it requires a flow stream of liquid (Patel et al., 2010). On a superhydrophilic surface, antifouling is achieved by tightly water stretching reactions, which are able to form a stable bacteria-free hydration layer on the surface of metal substrate (Yoon et al., 2014). These are useful properties for many food production applications. Both superhydrophobic and superhydrophilic surface properties can be achieved by the use of nanocomposite coating. For example, can silver nanoparticles be used to make a stainless steel surface more hydrophobic (Araujo et al., 2013)? Wear resistance and toxicological safety, however, still remains to be investigated (Yoon et al., 2014).

There are a number of elements and compounds that can be used to actively reduce the number of microorganisms on a food contact surface through a lasting lethal effect. Jermann et al. (2014) listed nine different agents, their modes of action, their current use and legislative status, as well as the advantages and limitations. They are presented, together with their main legislation status (in the European Union), in Table 6.2.

Moretro and Langsrud (2011) reviewed the effects of a number of materials containing antimicrobial compounds, silver compounds, and triclosan in particular. The antibacterial activity of these compounds have also been studies in cutting boards (Møretro et al., 2011; Møretrø et al., 2012). This concept often suffers from too rapid release of the active compound, leading to a short product lifetime, or too slow release, which leads to a lack of antibacterial effect.

6.2.3 Prevention

Sources of recontamination can be removed or prevented from being established by reducing the capability of the growth of microorganisms on processing equipment and in the environment. In addition, routes of recontamination should be broken by reducing direct or indirect contact between potential sources of contamination and the product. In order to achieve this, hygienic design, appropriate use of materials, proper maintenance, cleaning, and production routines are all key elements.

It may vary between different microorganisms, but in general one sees increased presence of microorganisms in areas where organic material tends to build up, areas that are difficult to clean, areas that are not cleaned, or areas that tend to stay moist and rarely dry up. Such areas should be identified in each plant, and measures should be taken to remove them. Some examples are drains, which are a huge challenge in most plants, and have repeatedly been documented to harbor pathogenic bacteria. Wheels of trolleys,

Table 6.2 **Products for Active Reduction of Microorganisms**

Agent/Product	Legislation
Copper	Cannot be incorporated into a food contact material as a biocide unless approved for this purpose under the biocides regulation. If the contact material was a plastic, then copper would also need to be listed as an authorized substance in Annex I of Commission Regulation (EU) No 10/2011.
Triclosan	Currently used in plastic food contact materials. Has been assessed by EFSA but their use in plastics has not been agreed and authorized at European Union level and to date they remain on the "provisional list," which allows their use at national level.
Silver	EFSA (2004 and 2005) has allocated a group specific migration limit for various silver releasing biocides in food contact materials of 0.05 mg Ag/kg food.
Zinc	If intentionally incorporated into a food contact material, zinc is subject to the biocide controls including the requirement for the active substance to be approved for any given product type. In the case of biocides that are used to impregnate materials which may enter into contact with food, this would require an approval under product type 4 (PT4). In the European Union, the biocidal product must also be authorized.
N-halamines	Chlorine reacts with organic matter to produce chloramine. Antimicrobial substances such as monochloramine incorporated into a food contact material is subject to the biocide controls including the requirement for the active substance to be approved for any given product type. In the case of biocides that are used to impregnate materials, which may enter into contact with food, this would require an approval under product type 4 (PT4).
Light-activated antimicrobial agents	Titanium dioxide that has been intentionally incorporated into the surface of a contact material and is working as a surface biocide is subject to the biocide controls and to Regulation (EC) No 1935/2004 on food contact materials.
Essential oils	Essential oils that have been added to foods have been incorporated to perform a flavoring function. Regulation (EC) No 1334/2008 lays down rules for flavorings, source materials for flavorings, and foods containing flavorings. It also covers certain food ingredients with flavoring properties which are naturally occurring undesirable substances, their source material and foods containing them.
Bacteriolytic enzymes	Enzymes may be working as an active material. Commission regulation (EC) No 450/2009 on active and intelligent materials and articles intended to come into contact with food controls only the components responsible for creating the active and/or intelligent parts, not the passive base materials. Base materials will still need to comply with European Union food contact material controls, for example, the plastics controls.
AEGIS Microbe Shield®	Does not seem to be food contact approved.

Adapted from Jermann et al. (2014).

trucks, and equipment are also very important in that they can contribute to transporting contamination within the production area. Condensates from cooler, roof, or tubes may drip onto the food contact surfaces or products, but this problem can be alleviated by the use of ventilation or dehumidifiers. Attachments of bacteria to conveyor belts can also be reduced through the use of conveyor belts made of stainless steel, as these show less attachment to bacteria than plastic belts and are also less prone to wear and tear that can harbor niches for bacteria.

Risk assessment studies increasingly found the basis of European Union regulations (Anon., 2004). den Aantrekker et al. (2003) state that recontamination should be included in quantitative microbial risk assessment studies and should be quantified using predictive models. Key elements of a model for recontamination via factory environment are the source of recontamination (harbor), the transfer rate to some kind of intermediate phase (surface, hands, or air) and the transfer rate to the product. Carrasco et al. (2012) reviewed the cross-contamination and recontamination routes and sources for poultry and other meats, eggs and egg products, low-moisture foods, fresh produce, and spices, milk, dairy and milk products, and seafood. They point to the model for the transfer of *Salmonella* during small-scale grinding of pork, which was developed by Møller et al. (2011). The model predicted the observed behavior of *Salmonella* during its cross-contamination in the grinding of up to 110 pork slices. It can be considered a useful tool to examine the effect of cross-contamination in microbial risk assessment. Chaitiemwong et al. (2014) quantified the transfer of *L. monocytogenes* between cooked ham and slicing machine surfaces, and found that the transfer rate was highest to the table, followed by the handle board and plate, the guard and front and back of the slicer blade. When more contamination models become available, one can expect more accurate exposure assessments and risk characterization to be developed.

A series of papers reviewing outbreaks where food workers have been implicated in the spread of food-borne disease have been published (Greig et al., 2007; Todd et al., 2007a,b, 2008a,b, 2009, 2010a,b,c,d,e). Many of these are focused on hand hygiene, which indeed is a major food contact surface and source of contamination (Michaels, 2002), although not covered in this chapter. Gloves can be considered a barrier to prevent contamination (Todd et al., 2010b), and proper use of gloves can reduce contamination. However, gloves have limitations and may become a source of contamination if punctured or used improperly. With inadequate training, glove use can create a false sense of security, resulting in more high-risk behaviors that can lead to cross-contamination. The developments in computer vision and robot technology, however, provide increasingly sophisticated and efficient solutions that can minimize manual contact, at least in industrial production.

Other physical barriers include properly engineered building walls and doors to minimize the flow of outside particles and pests to food storage and food preparation areas, food shields to prevent aerosol contamination of displayed food by customers and workers, and work clothing designated strictly for work and utensils such as spoons, etc., to prevent direct contact between hands and the food product.

The most critical area is called a primary pathogen control area, a high-risk area where the food is not subjected to subsequent microbial reduction steps. Examples are postcooking areas like filling and packaging, where the food is at risk of exposure to the environment. In general it is advisable to minimize the volume and area of these zones, in order to better control both the surface hygiene and the air quality. Use of computational fluid dynamics enables modeling of airflows and support for equipment design and placement of air ventilation outlets. With improving software and availability of computational power, this is becoming an increasingly useful tool for the food industry (Chorel et al., 2010; Kubiak and Jakubowski, 2013; Norton, 2013). The use of localized (refrigerated) air delivery can minimize airborne recontamination (Burfoot et al., 2000).

Chemical barriers include sanitizing solutions used to remove microorganisms from objects or materials used during food production and preparation, and also to launder work clothes. Many of these have already been described.

6.3 Future Trends

6.3.1 Increased Focus on Decontamination

With the emerging presence of multiresistant microorganisms, it is likely that the emphasis on decontamination and prevention of recontamination will increase. The use of antimicrobial agents in humans and food-producing animals has important consequences for human and animal health, as it can lead to development of resistant bacteria (pathogens and/or commensals with resistance genes), (Blaak et al., 2014; Guerra et al., 2014).

6.3.2 Reduced Susceptibility of Antibiotic-Resistant Bacteria to Disinfectants

The ability of bacteria to withstand disinfectants through resistance is well documented (Langsrud et al., 2003) and subject of ongoing research. Moreover, resistant bacteria in animals can be transferred to humans—usually through the consumption of food, but also through direct contact with food-producing animals or through environmental spread. Ultimately, this can result in human infections with bacteria that are resistant to antimicrobial agents and that can therefore be difficult or impossible to cure (Aidara-Kane, 2012).

Within the hospital environment, contaminated surfaces have been demonstrated to play an important role in the transmission of microorganisms causing health care-associated infections. The bacterial infections associated with primary surface colonization include methicillin-resistant *S. aureus*, vancomycin-resistant enterococci, and extended-spectrum-beta-lactamase-producing gram-negative organisms, such as *E. coli* and *Acinetobacter baumannii*, which prevail in the hospital environment for extended periods, i.e., months, in viable form.

Growing concern over the development of resistance to certain therapeutic drugs has led to questions over microorganisms developing resistance to sanitizers. As with antibiotics, reduced susceptibility (or acquired "resistance") of bacteria to disinfectants can arise by either chromosomal gene mutation or acquisition of genetic material in the form of plasmids or transposons (Gilbert and McBain, 2003). As most sanitizers are nonspecific, the development of resistance is mostly caused by innate factors (Russell, 1997) that are chromosomally controlled properties naturally associated with the organism.

When changes occur in bacterial susceptibility that renders an antibiotic ineffective against an infection previously treatable by that antibiotic, the bacteria are referred to as "resistant." In contrast, reduced susceptibility to disinfectants does not correlate with failure of the disinfectant because concentrations used in disinfection still greatly exceed the cidal level. Thus, in accordance to (Russell, 1999) the word "resistance" when applied to these changes is incorrect, and the preferred term is "reduced susceptibility" or "increased tolerance." No data are available that show that antibiotic-resistant bacteria are less sensitive to the liquid chemical germicides than antibiotic-sensitive bacteria at currently used germicide contact conditions and concentrations. Inadequate disinfection procedures in livestock production and food processing facilities, however, may contribute to the selection of biocide resistant isolates as a result of exposure to sublethal biocide concentration. Therefore, the potential for development of resistant strains must continue to be evaluated (Davidson and Harrison, 2002).

6.3.3 Less use of Traditional Compounds due to Potential Toxicity

The awareness and demand of the consumers may limit the use of e.g., halogenated compounds, e.g., by minimizing concerns about the formation of carcinogenic compounds like trihalomethanes resulting from the use of chloride compounds (Platikanov et al., 2010). Among the most used disinfectants are QACs that are cationic surfactants containing a tetrasubstituted ammonium salt and characterized by a positively charged quaternary nitrogen atom. These compounds strongly adsorb to negatively charged surfaces and bind to the fatty acids of cell membranes of organisms, which makes them useful as biocides. Ferk et al. (2007) found two common QACs, namely benzalkonium chloride (BAC) and dimethyldioctadecyl-ammonium bromide, to cause genotoxic effects in mammalian and plant cells. There is currently concern over potentially high residue levels of these and similar compounds, which in the European Union are covered by the regulation on "maximum residue levels of pesticides in or on food and feed …" (Anon., 2005), in which all foods and feeds are subject to a maximum residue level (MRL) of 0.01 mg/kg. A problem has arisen with this in that analyses have shown residue levels greater than 0.01 mg/kg to be present in some foods in the supply chain. It is thought that, apart from uses in agriculture and animal husbandry, these foods may be picking up these QACs from food contact surfaces, where they have been used as disinfectants. The situation has been discussed at European Union level and in 2012, a temporary guideline MRL of 0.5 mg/kg has been set for BAC and didecyl dimethyl ammonium chloride by the Standing Committee on the Food Chain and Animal Health for all food and animal feed of plant and animal origin, while investigations to develop more safety data and a monitoring program are carried out. The outcome of these investigation remains to be seen, but it may have great impact on the future use of QACs.

6.3.4 More use of Natural Alternatives

The use of biocontrollers may become an alternative treatment method for the control of *L. monocytogenes* in the food processing environment, particularly in difficult-to-reach locations. Schobitz et al. (2014) tested a biocontroller that consisted of thermally treated fermentate from three lactic acid bacteria combined with nisin. This biocontroller was effective in reducing *L. monocytogenes* counts in biofilms grown on stainless steel chips to undetectable levels ($<\log 1.0\,CFU/cm^2$) during 48 h of contact time.

With a number of natural extracts and agents showing promising results, significant progress will be seen in the application of these (Powitz, 2014).

6.3.5 Specific Challenges

Salmonella poses a substantial challenge to the food industry, with strains persisting in food processing environments for 10 years or more (Nesse et al., 2003; Russo et al., 2013). Studies have documented the failure of conventional cleaning and sanitation to eradicate *Salmonella* from surfaces used for handling raw meat (Arguello et al., 2012; McKee et al., 2008). Corcoran et al. (2014) found that neither sodium hypochlorite, benzalconium chloride nor sodium hydroxide could eradicate mature *Salmonella* biofilm, even after a 90-min contact time, leading them to emphasize the priority of preventing access of *Salmonella* to postcook areas of food processing facilities. Dehydration also affects the resistance of *Salmonella* to many stressors including heat, alcohol, sodium hypochlorite, and hydrogen peroxide (Gruzdev et al., 2011).

6.4 Sources of Further Information and Advice

The first information source to consult is the food legislators. These have information on what is legal and required with respect to agents, procedures and processes, but most also contain additional information which can be of great use. United States Food and Drug Administration (FDA, http://www.fda.gov) provides legislation, but also a number of guidance documents and regulatory information which are very useful.

The European Food Safety Authority (EFSA, http://www.efsa.europa.eu) is an independent European agency that operates separately from the European Commission, Parliament, and European Union member states. Currently available EFSA publications relevant to surface decontamination are mainly focused on carcass decontamination.

A very active consortium of equipment manufacturers, food industries, research institutes, and public health authorities play a leading role in Europe. The European Hygienic Engineering and Design Group (EHEDG, http://www.ehedg.org) aims to promote hygiene in food production by improving hygienic engineering and design in all aspects of food manufacture. EHEDG actively supports European legislation and has international experts organized working groups, which develop guidelines in most areas of food processing, including cleaning validation and air handling as well as building design and hygienic systems integration.

Campden BRI (http://www.campdenbri.co.uk) is an independent food research institute that offers a combination of scientific expertise and food and drink factory experience, supported by a portfolio of food hygiene guidelines, which detail good hygiene practices and factory design. They also work as an independent cleaning chemical and disinfectant testing center, run a wide range of hygiene training courses, and undertake factory sampling and audits.

The most comprehensive text book in this area is the (Handbook of Hygiene Control in the Food Industry, 2005), edited by H.L.M. Lelieveld, M.A. Mostert, and J. Holah. This book covers most aspects of hygiene, including issues like factory planning, sampling procedures and auditing.

References

Aidara-Kane, A., 2012. Containment of antimicrobial resistance due to use of antimicrobial agents in animals intended for food: WHO perspective. Revue Scientifique Et Technique-Office International Des Epizooties 31 (1), 277–287.

Akbas, M.Y., Ozdemir, M., 2006. Effectiveness of ozone for inactivation of *Escherichia coli* and *Bacillus cereus* in pistachios. International Journal of Food Science and Technology 41 (5), 513–519.

Anon, 2004. Regulation (EC) No 853/2004 of the European Parliament and of the Council of 29 April 2004 Laying down Specific Hygiene Rules for Food of Animal Origin. In, vol. 853/2004.

Anon, 2005. Regulation (EC) NO 396/2005 of the European Parliament and of the Council of 23 February 2005 on Maximum Residue Levels of Pesticides in or on Food and Feed of Plant and Animal Origin and Amending Council Directive 91/414/EEC. In, vol. 396/2005.

Anon, 2014. Ozone: health hazards and control measures. In: Health and Safety Executive (Ed.), Guidance Note EH38, third ed., pp. 1–10.

Ansari, M.I.A., Datta, A.K., 2003. An overview of sterilization methods for packaging materials used in aseptic packaging systems. Food and Bioproducts Processing 81 (C1), 57–65.

den Aantrekker, E.D., Boom, R.M., Zwietering, M.H., van Schothorst, M., 2003. Quantifying recontamination through factory environments–a review. International Journal of Food Microbiology 80 (2), 117–130.

Arachchi, G.J.G., Cridge, A.G., Dias-Wanigasekera, B.M., Cruz, C.D., McIntyre, L., Liu, R., Flint, S.H., Mutukumira, A.N., 2013. Effectiveness of phages in the decontamination of *Listeria monocytogenes* adhered to clean stainless steel, stainless steel coated with fish protein, and as a biofilm. Journal of Industrial Microbiology and Biotechnology 40 (10), 1105–1116.

Araujo, E.A., de Andrade, N.J., da Silva, L.H.M., Bernardes, P.C., Teixeira, A.V.N.D., Fialho, J.F.Q., de Sa, J.P.N., Fernandes, P.E., 2013. Modification of stainless steel surface hydrophobicity by silver nanoparticles: strategies to prevent bacterial adhesion in the food processing. Journal of Adhesion Science and Technology 27 (24), 2686–2695.

Arevalos-Sanchez, M., Regalado, C., Martin, S.E., Dominguez-Dominguez, J., Garcia-Almendarez, B.E., 2012. Effect of neutral electrolyzed water and nisin on *Listeria monocytogenes* biofilms, and on listeriolysin O activity. Food Control 24 (1-2), 116–122.

Arevalos-Sanchez, M., Regalado, C., Martin, S.E., Meas-Vong, Y., Cadena-Moreno, E., Garcia-Almendarez, B.E., 2013. Effect of neutral electrolyzed water on lux-tagged *Listeria monocytogenes* EGDe biofilms adhered to stainless steel and visualization with destructive and non-destructive microscopy techniques. Food Control 34 (2), 472–477.

Arguello, H., Carvajal, A., Collazos, J.A., Garcia-Feliz, C., Rubio, P., 2012. Prevalence and serovars of *Salmonella enterica* on pig carcasses, slaughtered pigs and the environment of four Spanish slaughterhouses. Food Research International 45 (2), 905–912.
Assere, A., Oulahal, N., Carpentier, B., 2008. Comparative evaluation of methods for counting surviving biofilm cells adhering to a polyvinyl chloride surface exposed to chlorine or drying. Journal of Applied Microbiology 104 (6), 1692–1702.
Bagge-Ravn, D., Gardshodn, K., Gram, L., Vogel, B.F., 2003. Comparison of sodium hypochlorite-based foam and peroxyacetic acid-based fog sanitizing procedures in a salmon smokehouse: survival of the general microflora and *Listeria monocytogenes*. Journal of Food Protection 66 (4), 592–598.
Bayliss, D., Snipe, T., Bennington, M., Staniforth, L., Everis, L., Leadley, C., 2014. Cold Atmospheric Plasma: A Feasibility Study. Campden BRI, Chipping Campden.
Beltran, D., Selma, M.V., Marin, A., Gil, M.I., 2005. Ozonated water extends the shelf life of fresh-cut lettuce. Journal of Agricultural and Food Chemistry 53 (14), 5654–5663.
Berrang, M.E., Frank, J.F., Meinersmann, R.J., 2008. Effect of chemical sanitizers with and without ultrasonication on *Listeria monocytogenes* as a biofilm within polyvinyl chloride drain pipes. Journal of Food Protection 71 (1), 66–69.
Blaak, H., van Hoek, A.H.A.M., Veenman, C., van Leeuwen, A.E.D., Lynch, G., van Overbeek, W.M., Husman, A.M.D., 2014. Extended spectrum beta-lactamase- and constitutively AmpC-producing Enterobacteriaceae on fresh produce and in the agricultural environment. International Journal of Food Microbiology 168, 8–16.
Burfoot, D., Brown, K., Xu, Y., Reavell, S.V., Hall, K., 2000. Localised air delivery systems in the food industry. Trends in Food Science and Technology 11 (11), 410–418.
Burfoot, D., Hall, K., Brown, K., Xu, Y., 1999. Fogging for the disinfection of food processing factories and equipment. Trends in Food Science and Technology 10 (6–7), 205–210.
Cardoso, C.F., Faria, J.D.F., Walter, E.H.M., 2011. Modeling of sporicidal effect of hydrogen peroxide in the sterilization of low density polyethylene film inoculated with *Bacillus subtilis* spores. Food Control 22 (10), 1559–1564.
Carrasco, E., Morales-Rueda, A., Garcia-Gimeno, R.M., 2012. Cross-contamination and recontamination by Salmonella in foods: A review. Food Research International 45 (2), 545–556.
Chaitiemwong, N., Hazeleger, W.C., Beumer, R.R., Zwietering, M.H., 2014. Quantification of transfer of *Listeria monocytogenes* between cooked ham and slicing machine surfaces. Food Control 44, 177–184.
Chorel, F., Kondjoyan, A., Mirade, P.S., 2010. Toward quantitative CFD prediction of contaminant particle deposition against surfaces in large forced-ventilation food plants. Aerosol Science and Technology 44 (1), 10–28.
Corcoran, M., Morris, D., De Lappe, N., O'Connor, J., Lalor, P., Dockery, P., Cormican, M., 2014. Commonly used disinfectants fail to eradicate *Salmonella enterica* biofilms from food contact surface materials. Applied and Environmental Microbiology 80 (4), 1507–1514.
Crandall, P.G., O'Bryan, C.A., Martin, E.M., Pendleton, S., Shannon, E., Marcy, J., Ricke, S.C., 2010. Dry heat thermal inactivation of Listeria innocua on deli slicer components. Food Protection Trends 30 (9), 588–592.
Crapo, C., Himelbloom, B., Vitt, S., Pedersen, L., 2004. Ozone efficacy as a bactericide in seafood processing. Journal of Aquatic Food Product Technology 13 (1), 111–123.
Critzer, F.J., Kelly-Wintenberg, K., South, S.L., Golden, D.A., 2007. Atmospheric plasma inactivation of foodborne pathogens on fresh produce surfaces. Journal of Food Protection 70 (10), 2290–2296.
Davidson, P.M., Harrison, M.A., 2002. Resistance and adaptation to food antimicrobials, sanitizers, and other process controls. Food Technology 56 (11), 69–78.

Deza, M.A., Araujo, M., Garrido, M.J., 2007. Efficacy of neutral electrolyzed water to inactivate *Escherichia coli, Listeria monocytogenes, Pseudomonas aeruginosa*, and *Staphylococcus aureus* on plastic and wooden kitchen cutting boards. Journal of Food Protection 70 (1), 102–108.

Faille, C., Fontaine, F., Benezech, T., 2001. Potential occurrence of adhering living Bacillus spores in milk product processing lines. Journal of Applied Microbiology 90 (6), 892–900.

FDA, 2000. In: CFR Part 179. Irradiation in the Production, Processing and Handling of Food, vol. 65, pp. 71056–71058 Federal Register.

Ferk, F., Misik, M., Hoelzl, C., Uhl, M., Fuerhacker, M., Grillitsch, B., Parzefall, W., Nersesyan, A., Micieta, K., Grummt, T., Ehrlich, V., Knasmuller, S., 2007. Benzalkonium chloride (BAC) and dimethyldioctadecyl-ammonium bromide (DDAB), two common quaternary ammonium compounds, cause genotoxic effects in mammalian and plant cells at environmentally relevant concentrations. Mutagenesis 22 (6), 363–370.

Fernandez, A., Noriega, E., Thompson, A., 2013. Inactivation of *Salmonella enterica* serovar Typhimurium on fresh produce by cold atmospheric gas plasma technology. Food Microbiology 33 (1), 24–29.

Finnegan, M., Linley, E., Denyer, S.P., McDonnell, G., Simons, C., Maillard, J.Y., 2010. Mode of action of hydrogen peroxide and other oxidizing agents: differences between liquid and gas forms. Journal of Antimicrobial Chemotherapy 65 (10), 2108–2115.

Gaulin, C., Lê, M.L., Shum, M., Fong, D., 2011. Disinfectants and Sanitizers for Use on Food Contact Surfaces (pp. 1–15). National Collaborating Centre for Environmental Health.

Gelman, A., Sachs, O., Khanin, Y., Drabkin, V., Glatman, L., 2005. Effect of ozone pretreatment on fish storage life at low temperatures. Journal of Food Protection 68 (4), 778–784.

Gilbert, P., McBain, A.J., 2003. Potential impact of increased use of biocides in consumer products on prevalence of antibiotic resistance. Clinical Microbiology Reviews 16 (2), 189.

Goode, K.R., Asteriadou, K., Robbins, P.T., Fryer, P.J., 2013. Fouling and cleaning studies in the food and beverage industry classified by cleaning type. Comprehensive Reviews in Food Science and Food Safety 12 (2), 121–143.

Greig, J.D., Todd, E.C.D., Bartleson, C.A., Michaels, B.S., 2007. Outbreaks where food workers have been implicated in the spread of foodborne disease. Part 1. Description of the problem, methods, and agents involved. Journal of Food Protection 70 (7), 1752–1761.

Gruzdev, N., Pinto, R., Sela, S., 2011. Effect of desiccation on tolerance of *Salmonella enterica* to multiple stresses. Applied and Environmental Microbiology 77 (5), 1667–1673.

Guentzel, J.L., Lam, K.L., Callan, M.A., Emmons, S.A., Dunham, V.L., 2008. Reduction of bacteria on spinach, lettuce, and surfaces in food service areas using neutral electrolyzed oxidizing water. Food Microbiology 25 (1), 36–41.

Guerra, B., Fischer, J., Helmuth, R., 2014. An emerging public health problem: Acquired carbapenemase-producing microorganisms are present in food-producing animals, their environment, companion animals and wild birds. Veterinary Microbiology 171 (3-4), 290–297.

Handbook of Hygiene Control in the Food Industry, 2005. Woodhead Publishing, Cambridge.

Hassan, F., Lakshmanan, P.T., Geethalakshmi, V., Mukundan, M.K., 2013. Evaluation of stabilised hydrogen peroxide as sanitiser in seafood processing industry. Indian Journal of Fisheries 60 (2), 145–149.

Hood, S.K., Zottola, E.A., 1995. Biofilms in food-processing. Food Control 6 (1), 9–18.

Iturriaga, L., Olabarrieta, I., de Maranon, I.M., 2012. Antimicrobial assays of natural extracts and their inhibitory effect against *Listeria innocua* and fish spoilage bacteria, after incorporation into biopolymer edible films. International Journal of Food Microbiology 158 (1), 58–64.

Jefferson, K.K., 2004. What drives bacteria to produce a biofilm? Fems Microbiology Letters 236 (2), 163–173.

Jermann, C., Price, R.J., Staniforth, L., 2014. Antimicrobial Surfaces and Coatings: How they work, Their Advantages, Limitations and Legal Status. Campden BRI, Chipping Campden, UK.
Joshi, K., Mahendran, R., Alagusundaram, K., Norton, T., Tiwari, B.K., 2013. Novel disinfectants for fresh produce. Trends in Food Science and Technology 34 (1), 54–61.
Khadre, M.A., Yousef, A.E., 2001. Decontamination of a multilaminated aseptic food packaging material and stainless steel by ozone. Journal of Food Safety 21 (1), 1–13.
Kim, T., Silva, J.L., Chen, T.C., 2002. Effects of UV irradiation on selected pathogens in peptone water and on stainless steel and chicken meat. Journal of Food Protection 65 (7), 1142–1145.
Kleiner, U., 2001. Ozone applications in the food industry. Fleischwirtschaft 81, 88–90.
Koseki, S., Yoshida, K., Isobe, S., Itoh, K., 2001. Decontamination of lettuce using acidic electrolyzed water. Journal of Food Protection, 64 (5), 652–658.
Kubiak, M.S., Jakubowski, M., 2013. Cfd simulations as a supporting tool for process and construction optimization in food industry production practice: a case study of a single truck smoking chamber. Italian Journal of Food Science 25 (3), 251–254.
Langsrud, S., Sidhu, M.S., Heir, E., Holck, A.L., 2003. Bacterial disinfectant resistance - a challenge for the food industry. International Biodeterioration & Biodegradation, 51 (4), 283–290.
Lee, J., Pascall, M.A., 2012. Inactivation of *Clostridium sporogenes* spores on stainless-steel using heat and an organic acidic chemical agent. Journal of Food Engineering 110 (3), 493–496.
Len, S.V., Hung, Y.C., Chung, D., Anderson, J.L., Erickson, M.C., Morita, K., 2002. Effects of storage conditions and pH on chlorine loss in electrolyzed oxidizing (EO) water. Journal of Agricultural and Food Chemistry 50 (1), 209–212.
Lopez-Galvez, F., Ragaert, P., Palermo, L.A., Eriksson, M., Devlieghere, F., 2013. Effect of new sanitizing formulations on quality of fresh-cut iceberg lettuce. Postharvest Biology and Technology 85, 102–108.
Manousaridis, G., Nerantzaki, A., Paleologos, E.K., Tsiotsias, A., Savvaidis, I.N., Kontominas, M.G., 2005. Effect of ozone on microbial, chemical and sensory attributes of shucked mussels. Food Microbiology 22 (1), 1–9.
Martin, H., Maris, P., 2012a. Synergism between hydrogen peroxide and seventeen acids against five agri-food-borne fungi and one yeast strain. Journal of Applied Microbiology 113 (6), 1451–1460.
Martin, H., Maris, P., 2012b. Synergism between hydrogen peroxide and seventeen acids against six bacterial strains. Journal of Applied Microbiology 113 (3), 578–590.
Martin, S.E., Baumann, A.R., Hao, F., 2009. Removal of *Listeria monocytogenes* Biofilms from stainless steel by use of ultrasound and ozone. Journal of Food Protection 72 (6), 1306–1309.
Masuku, S.M., Babu, D., Martin, E.M., O'Bryan, C.A., Crandall, P.G., Ricke, S.C., 2014. Lethality of moist heat and silver dihydrogen citrate sanitizer combinations on *Listeria* spp. adhered to components of a deli meat slicer. Food Control 44, 227–232.
McKee, S.R., Townsend, J.C., Bilgili, S.F., 2008. Use of a scald additive to reduce levels of *Salmonella* Typhimurium during poultry processing. Poultry Science 87 (8), 1672–1677.
Meyer, M., Persson, O., 1998. Nanotechnology–interdisciplinarity, patterns of collaboration and differences in application. Scientometrics 42 (2), 195–205.
Michaels, B., 2002. Handwashing: an effective tool in the food safety arsenal. In: Food Safety, pp. 45–53.
Monnin, A., Lee, J., Pascall, M.A., 2012. Efficacy of neutral electrolyzed water for sanitization of cutting boards used in the preparation of foods. Journal of Food Engineering 110 (4), 541–546.

Møretrø, T., Høiby-Pettersen, G.S., Habimana, O., Heir, E., Langsrud, S., 2011. Assessment of the antibacterial activity of a triclosan-containing cutting board. International Journal of Food Microbiology 146 (2), 157–162.

Møretrø, T., Høiby-Pettersen, G.S., Halvorsen, C.K., Langsrud, S., 2012. Antibacterial activity of cutting boards containing silver. Food Control 28 (1), 118–121.

Møretrø, T., Langsrud, S., 2011. Effects of materials containing antimicrobial compounds on food hygiene. Journal of Food Protection 74 (7), 1200–1211.

Møller, C.O.A., Nauta, M.J., Christensen, B.B., Dalgaard, P., Hansen, T.B., 2011. Modelling transfer of *Salmonella* Typhimurium DT104 during the grinding of pork. In: Cummins, E., Frías, J.M., Valdramidis, V.P. (Eds.), Predictive Modelling of Food Quality and Safety—Conference Proceedings. UCD, DIT, Teagasc, Dublin, IE, pp. 37–40.

Nesse, L.L., Nordby, K., Heir, E., Bergsjoe, B., Vardund, T., Nygaard, H., Holstad, G., 2003. Molecular analyses of *Salmonella enterica* isolates from fish feed factories and fish feed ingredients. Applied and Environmental Microbiology 69 (2), 1075–1081.

Nicholas, R., Dunton, P., Tatham, A., Fielding, L., 2013. The effect of ozone and open air factor on surface-attached and biofilm environmental *Listeria monocytogenes*. Journal of Applied Microbiology 115 (2), 555–564.

Niemira, B.A., 2012. Cold plasma decontamination of foods. Annual Review of Food Science and Technology 3, 125–142.

Niemira, B.A., Boyd, G., Sites, J., 2014. Cold plasma rapid decontamination of food contact surfaces contaminated with Salmonella biofilms. Journal of Food Science 79 (5), M917–M922.

Norton, T., 2013. CFD in the agri-food industry: a maturing engineering design tool. Computers and Electronics in Agriculture 93, 149–150.

Oh, S.W., Gray, P.M., Dougherty, R.H., Kang, D.H., 2005. Aerosolization as novel sanitizer delivery system to reduce food-borne pathogens. Letters in Applied Microbiology 41 (1), 56–60.

Otto, C., Zahn, S., Rost, F., Zahn, P., Jaros, D., Rohm, H., 2011. Physical methods for cleaning and disinfection of surfaces. Food Engineering Reviews 3 (3-4), 171–188.

Park, H., Hung, Y.C., Brackett, R.E., 2002. Antimicrobial effect of electrolyzed water for inactivating *Campylobacter jejuni* during poultry washing. International Journal of Food Microbiology 72 (1-2), 77–83.

Park, S.H., Cheon, H.L., Park, K.H., Chung, M.S., Choi, S.H., Ryu, S., Kang, D.H., 2012. Inactivation of biofilm cells of foodborne pathogen by aerosolized sanitizers. International Journal of Food Microbiology 154 (3), 130–134.

Patel, P., Choi, C.K., Meng, D.D., 2010. Superhydrophilic surfaces for antifogging and antifouling microfluidic devices. JALA 15 (2), 114–119.

Piyasena, P., Mohareb, E., McKellar, R.C., 2003. Inactivation of microbes using ultrasound: a review. International Journal of Food Microbiology 87 (3), 207–216.

Platikanov, S., Tauler, R., Rodrigues, P.M., Antunes, M.C., Pereira, D., Esteves da Silva, J.C., 2010. Factorial analysis of the trihalomethane formation in the reaction of colloidal, hydrophobic, and transphilic fractions of DOM with free chlorine. Environmental Science and Pollution Research International 17 (8), 1389–1400.

Powitz, R.W., 2014. Chemical-free cleaning: revisited. In: Food Safety Magazine.

Reardon, J.T., Sancar, A., 2005. Nucleotide excision repair. Progress in Nucleic Acid Research and Molecular Biology 79, 183–235 79.

Russell, A.D., 1997. Plasmids and bacterial resistance to biocides. Journal of Applied Microbiology 83 (2), 155–165.

Russell, A.D., 1999. Bacterial resistance to disinfectants: present knowledge and future problems. Journal of Hospital Infection 43, S57–S68.

Russo, E.T., Biggerstaff, G., Hoekstra, R.M., Meyer, S., Patel, N., Miller, B., Quick, R., S.A.O.I.T., 2013. A recurrent, multistate outbreak of salmonella serotype Agona infections associated with dry, unsweetened cereal consumption, United States, 2008. Journal of Food Protection 76 (2), 227–230.

Schobitz, R., Gonzalez, C., Villarreal, K., Horzella, M., Nahuelquin, Y., Fuentes, R., 2014. A biocontroller to eliminate *Listeria monocytogenes* from the food processing environment. Food Control 36 (1), 217–223.

Simons, L.K., Sanguansri, P., 1997. Advances in the washing of minimally processed vegetables. Food Australia 49, 75–80.

Sommers, C.H., Sites, J.E., Musgrove, M., 2010. Ultraviolet light (254 Nm) inactivation of pathogens on foods and stainless steel surfaces. Journal of Food Safety 30 (2), 470–479.

Staniforth, L., 2014. (pp. 1–24). In: The Use of Ozone in Whole Room Disinfection. Campden BRI, Chipping Campden, UK.

Tayel, A.A., El-Tras, W.F., 2012. Plant extracts as potent biopreservatives for *Salmonella* Typhimurium control and quality enhancement in ground beef. Journal of Food Safety 32 (1), 115–121.

Tayel, A.A., El-Tras, W.F., Moussa, S., El-Baz, A.F., Mahrous, H., Salem, M.F., Brimer, L., 2011. Antibacterial action of zinc oxide nanoparticles against foodborne pathogens. Journal of Food Safety 31 (2), 211–218.

Tayel, A.A., El-Tras, W.F., Moussa, S.H., El-Sabbagh, S.M., 2012. Surface Decontamination and quality enhancement in meat steaks using plant extracts as natural biopreservatives. Foodborne Pathogens and Disease 9 (8), 755–761.

Thallinger, B., Prasetyo, E.N., Nyanhongo, G.S., Guebitz, G.M., 2013. Antimicrobial enzymes: an emerging strategy to fight microbes and microbial biofilms. Biotechnology Journal 8 (1), 97.

Thorn, R.M.S., Robinson, G.M., Reynolds, D.M., 2013. Comparative antimicrobial activities of aerosolized sodium hypochlorite, chlorine dioxide, and electrochemically activated solutions evaluated using a novel standardized assay. Antimicrobial Agents and Chemotherapy 57 (5), 2216–2225.

Todd, E.C.D., Greig, J.D., Bartleson, C.A., Michaels, B.S., 2007a. Outbreaks where food workers have been implicated in the spread of foodborne disease. Part 2. Description of outbreaks by size, severity, and settings. Journal of Food Protection 70 (8), 1975–1993.

Todd, E.C.D., Greig, J.D., Bartleson, C.A., Michaels, B.S., 2007b. Outbreaks where food workers have been implicated in the spread of foodborne disease. Part 3. Factors contributing to outbreaks and description of outbreak categories. Journal of Food Protection 70 (9), 2199–2217.

Todd, E.C.D., Greig, J.D., Bartleson, C.A., Michaels, B.S., 2008a. Outbreaks where food workers have been implicated in the spread of foodborne disease. Part 5. Sources of contamination and pathogen excretion from infected persons. Journal of Food Protection 71 (12), 2582–2595.

Todd, E.C.D., Greig, J.D., Bartleson, C.A., Michaels, B.S., 2009. Outbreaks where food workers have been implicated in the spread of foodborne disease. Part 6. Transmission and survival of pathogens in the food processing and preparation environment. Journal of Food Protection 72 (1), 202–219.

Todd, E.C.D., Greig, J.D., Michaels, B.S., Bartleson, C.A., Smith, D., Holah, J., 2010a. Outbreaks where food workers have been implicated in the spread of foodborne disease. Part 11. Use of antiseptics and sanitizers in community settings and issues of hand hygiene compliance in health care and food industries. Journal of Food Protection 73 (12), 2306–2320.

Todd, E.C.D., Greig, T.D., Bartleson, C.A., Michaels, B.S., 2008b. Outbreaks where food workers have been implicated in the spread of foodborne disease. Part 4. Infective doses and pathogen carriage. Journal of Food Protection 71 (11), 2339–2373.

Todd, E.C.D., Michaels, B.S., Greig, J.D., Smith, D., Bartleson, C.A., 2010b. Outbreaks where food workers have been implicated in the spread of foodborne disease. Part 8. Gloves as barriers to prevent contamination of food by workers. Journal of Food Protection 73 (9), 1762–1773.

Todd, E.C.D., Michaels, B.S., Greig, J.D., Smith, D., Holah, J., Bartleson, C.A., 2010c. Outbreaks where food workers have been implicated in the spread of foodborne disease. Part 7. Barriers to reduce contamination of food by workers. Journal of Food Protection 73 (8), 1552–1565.

Todd, E.C.D., Michaels, B.S., Holah, J., Smith, D., Greig, J.D., Bartleson, C.A., 2010d. Outbreaks where food workers have been implicated in the spread of foodborne disease. Part 10. Alcohol-based antiseptics for hand disinfection and a comparison of their effectiveness with soaps. Journal of Food Protection 73 (11), 2128–2140.

Todd, E.C.D., Michaels, B.S., Smith, D., Greig, J.D., Bartleson, C.A., 2010e. Outbreaks where food workers have been implicated in the spread of foodborne disease. Part 9. Washing and drying of hands to reduce microbial contamination. Journal of Food Protection 73 (10), 1937–1955.

Tolvanen, R., Lunden, J., Horman, A., Korkeala, H., 2009. Pilot-scale continuous ultrasonic cleaning equipment reduces *Listeria monocytogenes* levels on conveyor belts. Journal of Food Protection 72 (2), 408–411.

Upadhyay, A., Upadhyaya, I., Kollanoor-Johny, A., Baskaran, S.A., Mooyottu, S., Karumathil, D., Venkitanarayanan, K., 2013. Inactivation of *Listeria monocytogenes* on frankfurters by plant-derived antimicrobials alone or in combination with hydrogen peroxide. International Journal of Food Microbiology 163 (2–3), 114–118.

Valdramidis, V.P., Peroval, C., Portanguen, S., Verhulst, A., Van Impe, J.F., Geeraerd, A.H., Kondjoyan, A., 2008. Quantitative evaluation of thermal inactivation kinetics of free-floating versus surface-attached *Listeria innocua* cells. Food and Bioprocess Technology 1 (3), 285–296.

Warriner, K., 2011. Surface contamination of foods: reduction. In: Encyclopedia of Agricultural, Food and Biological Engineering, second ed. Taylor & Francis, pp. 1–5.

Wirtanen, G., Langsrud, S., Salo, S., Olofson, U., Alnås, H., Neuman, M., Homleid, J.P., Mattila-Sandholm, T., 2002. Evaluation of Sanitation Procedures for Use in Dairies. VTT, VTT, Finland.

Yang, H.S., Feirtag, J., Diez-Gonzalez, F., 2013. Sanitizing effectiveness of commercial "active water" technologies on *Escherichia coli* O157:H7, *Salmonella enterica* and *Listeria monocytogenes*. Food Control 33 (1), 232–238.

Yoon, S.H., Rungraeng, N., Song, W., Jun, S., 2014. Superhydrophobic and superhydrophilic nanocomposite coatings for preventing *Escherichia coli* K-12 adhesion on food contact surface. Journal of Food Engineering 131, 135–141.

Zhao, Q., Wang, C., Liu, Y., Wang, S., 2007. Bacterial adhesion on the metal-polymer composite coatings. International Journal of Adhesion and Adhesives 27 (2), 85–91.

Zoutman, D., Shannon, M., Mandel, A., 2011. Effectiveness of a novel ozone-based system for the rapid high-level disinfection of health care spaces and surfaces. American Journal of Infection Control 39 (10), 873–879.

Part Three

Innovative Food Processing Technologies

Emerging Refrigeration and Freezing Technologies for Food Preservation

7

Judith Evans
London South Bank University, London, UK

7.1 Introduction

Refrigeration technologies for food chilling or freezing have changed little since the development of commercial refrigeration. Technologies have been based around the use of fluids such as air, water, or brine to transfer heat from the product. Although there have been developments in the techniques used, the efficiency of the processes has increased, and the refrigerant used has been updated, there has been little change in the fundamental principles of chilling and freezing food for over 100 years. There are a number of novel technologies that have undergone development since the mid-1980s, but to date none of these technologies has been able to achieve widespread commercial success. This has primarily been due to the scale of the equipment available, cost of the equipment, efficiency of the technology, and end user confidence in the technology when applied to critical processes in food manufacture.

This chapter covers several of these novel technologies and describes the level of development and commercial maturity of the technology. The chapter divides the technologies into those that are purely equipment-based (the equipment itself is novel) and those that are product-oriented (where the refrigeration equipment itself may not be novel, but it is used so that a novel product is created).

7.2 Refrigeration Based Technologies

7.2.1 Acoustic Refrigeration

Acoustic (sometimes termed thermoacoustic) refrigeration systems use sound waves to produce cooling. Cooling is achieved through compressing and expansion of an inert gas (usually helium, argon, air) or a mixture of gases in a resonator. Such devices usually are divided into "standing wave" or "traveling wave" systems (Figure 7.1).

In a closed, standing wave system, the main components are a closed cylinder, an acoustic driver, a porous media ("stack"), and two heat exchangers (a cold and a hot exchanger). In such a system the driver is typically a loudspeaker. Sound waves are used to create a resonant standing wave within the "stack." As the gas oscillates back and forth within the stack, it creates a temperature difference along the length of the

Innovation and Future Trends in Food Manufacturing and Supply Chain Technologies
http://dx.doi.org/10.1016/B978-1-78242-447-5.00007-1

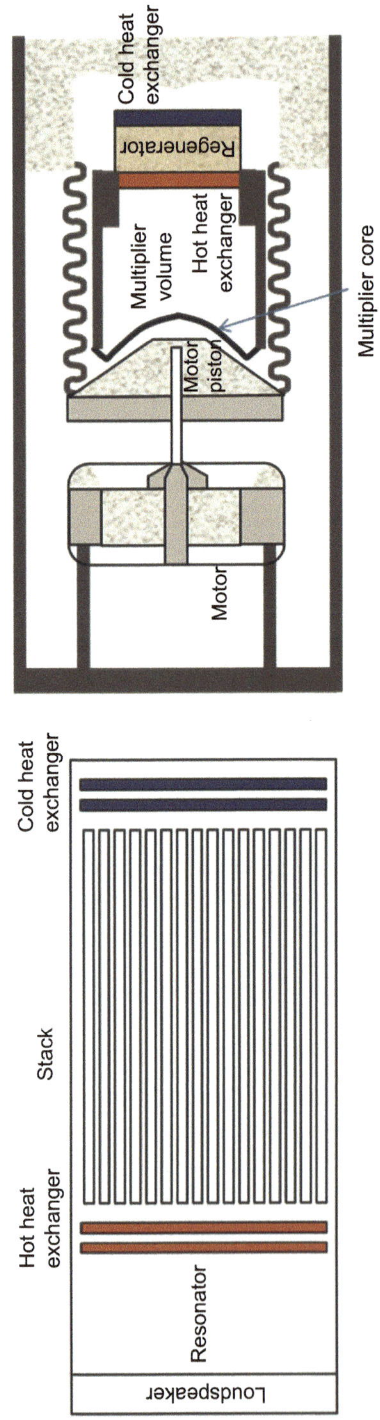

Figure 7.1 Diagram of a standing wave (left) and a traveling wave (right) acoustic refrigerator.

stack, due to expansion and compression by the sound wave. The temperature difference created removes heat from the cold side and rejects heat at the hot side.

In a closed, traveling wave system, the driver is a motor and piston. Unlike the standing wave system, the temperature difference for this system occurs in a regenerator rather than a stack. The regenerator consists of a matrix of small channels. The system is designed so that the air will oscillate between the hot and cold heat exchangers through the regenerator matrix as the pressure is increased and decreased. When the pressure is high, the gas moved toward the hot heat exchanger; and when the pressure is low, it moves toward the cold exchanger, thereby transferring heat.

Although thermoacoustic refrigerators have the potential to cover the whole spectrum of refrigeration down to cryogenic temperatures, the technology is most likely to be used for low capacity equipment initially (Tassou et al., 2010). The main benefit of the technology is that it uses environmentally safe, inert gasses such as air, argon, and helium. Systems can be open or closed; closed systems have shown the greatest potential to date. Flow-through systems (also referred to as open systems) would eliminate heat exchangers and reduce system complexity and cost, but more research is required into this configuration (Tassou et al., 2010).

Work at Penn State University has developed a demonstrator acoustic refrigerator for storage of ice cream, which is currently undergoing further development with the view of future commercialization (Poese et al., 2004). Prototype units have been developed, the most famous being the Ben and Jerry's ice cream freezer (2004).

Inefficiency in systems already built and tested is generally cited as being a result of inadequate tolerances in assembled apparatus, whereas heat exchangers are cited as the cause for high cost and complexity. Future developments could improve and optimize design and performance (Wetzel and Herman, 1997). Developments are needed in the design of stacks, resonators, and compact heat exchangers for oscillating flow. Almost 300 research papers have been published in the last 13 years dealing with numerical modeling or improving aspects of the technology. The large proportion of these papers originated from the USA and China.

Work is still needed to increase coefficient of performance (COP, which is the ratio of cooling provided to energy used for cooling) to the level of conventional vapor compression systems. Efficiency achieved so far is 0.1–0.2 of Carnot efficiency (the most efficient cycle allowed by physical laws); conventional refrigeration systems achieve 0.33 to 0.5 (Wetzel and Herman, 1997). Unless legislation prevents the use of more efficient vapor compression systems (for environmental or safety reasons), the efficiency of acoustic refrigerators will need to be improved to exceed that of vapor compression systems to enable uptake of this technology (Tassou et al., 2010).

7.2.2 Air Cycle

Air cycle is one of the oldest refrigeration technologies. Air cycle machinery was used onboard ships in the 1800s to maintain food temperature. However, the large reciprocating machinery was rapidly replaced at the beginning of the 1900s by smaller, lighter systems using other refrigerants as new technology developed. Today, high-speed

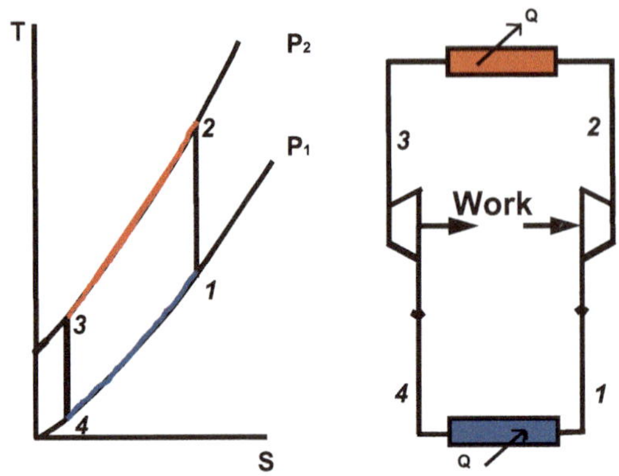

Figure 7.2 Air cycle on a T–S (Temperature–Entropy) diagram and shown diagrammatically.

turbo machinery is available that is compact and lightweight and therefore the use of air as a refrigerant is once again a commercial possibility.

The principle of the air cycle is that when air is compressed its temperature and pressure increases (1–2) (Figure 7.2). Heat is removed from the compressed air at constant pressure, and its temperature is reduced ideally while providing useful heat to high temperature processes (2–3). The air is then expanded and its temperature reduces as work is taken from it (3–4). The air then absorbs heat (gaining temperature) from low temperature processes at constant pressure (4–1), where it starts the cycle again.

The application of air cycle to food processing has many advantages such as safety, lack of flammability, and the lack of environmentally damaging gasses. Air cycle can be energy efficient at low temperatures or if heating as well as cooling is utilized.

A number of theoretical studies have indicated the potential for air cycle in food processing operations (Gigiel et al., 1992; Russell et al., 2000; Russell et al., 1998). Integrated heating and refrigeration is one of the applications with the highest theoretical potential. Theoretically, air temperatures up to 300 °C can also be obtained suitable for direct cooking or the production of steam. Alternatively, fast freezing of small products using the low temperatures available from air cycle systems will result in faster freezing, improving food quality, and either a smaller freezer, or a larger throughput through an existing freezer (Foster et al., 2011).

Air cycle has advantages for transport applications due to its low weight, use of a natural refrigerant, reliability, good performance at part load conditions, and robustness. Spence et al. (2004, 2005) developed an air cycle demonstrator using commercially available parts. Due to having to use less than optimal components, the fuel consumption was higher than for a vapor compression unit at full load (200%) and lower at part load (80%). Using a model to investigate the effect of using more optimal components, a coefficient of system performance (which is the ratio of cooling provided to total energy used) at full load of 0.53 at −20 °C and 7.8 kW was predicted that was 7% lower than an equivalent vapor compression system.

The main factor restricting take up of air cycle refrigeration is the lack of available components and their current high cost. Applications are most likely to be in low temperature applications, where air cycle can compete with cryogens and conventional direct expansion refrigeration technologies on efficiency (Brown et al., 2011).

7.2.3 Magnetic Refrigeration

Technologies such as magnetic cooling have potential advantages such as no harmful refrigerants and potentially high efficiencies above those of vapor compression technologies. Magnetic refrigeration takes advantage of the magnetocaloric effect, the ability of some metals to heat up when they are magnetized and cool when demagnetized. Much of the original work and most prototypes developed were based on the use of gadolinium magnets that are rather expensive. More recent work has looked for new materials that are cheap, have suitable transition temperatures, and exhibit a large magnetocaloric effect. Magnetic refrigeration has the prospect of efficient, environmentally friendly, and compact cooling for a wide field of applications.

Numerical models have been developed that show magnetic refrigeration has the potential for high efficiency. However, many of the prototypes and laboratory demonstrators have not been able to achieve the predicted efficacies yet, and the efficiencies are still well below commercially available vapor compression systems. To improve efficiency, energetic losses need to be reduced and system design aspects such as drive systems need to be better understood (Engelbrecht and Pryds, 2014).

Astronautics Corporation in America and Chubu Electric Power Co. Inc. in Japan have both produced rotary magnetic refrigerator systems. The highest COP reported for a near room temperature, permanent magnet system was 2.4. This was based on a 560 W cooling capacity at zero temperature span. For a 5 K temperature span and 20 °C sink temperature, the COP reduced to 0.6 and the cooling duty to 159 W (Lewis et al., 2007).

In 2008, Gschneidner and Pecharsky predicted that production of near room temperature, magnetic refrigeration systems would grow to 1000 units by 2015, by which time they would consider the technology to be commercialized. However, this level of development and growth has not occurred. Systems are still being developed for commercial use. Camfridge (backed by Cambridge University) began a project with Whirlpool in 2009 to develop a domestic refrigerator, but the planned demonstration unit is still under development (Whirlpool, 2009; Wilson et al., 2006). Cooltech has developed a magnetic refrigerator suitable for small commercial systems and have demonstrated the system on an Arneg refrigerator at the Euroshop exhibition in 2014. It would appear that in the next few years, further magnetic refrigerator demonstration units will be available with the possibility of commercialization in the domestic or small commercial refrigeration sectors.

7.2.4 Barocaloric Refrigeration

The barocaloric effect is analogous to the magnetocaloric effect but does not require strong permanent magnets (although it does require more energy to apply and remove the pressure). Much of the work to date appears to have been carried out in conjunction with, or alongside, research into the magnetocaloric effect and with the same materials. In the same way as a magnetic refrigerator produces a cooling effect by applying and removing a magnetic field, barocaloric refrigeration produces a cooling effect by applying and releasing pressure on solid materials (which exhibit barocaloric behavior). Some research has focused on using pressure to improve the magnetocaloric effect, while other research has seen the barocaloric effect as an alternative to magnetic cooling, which could overcome the need for powerful permanent magnets in magnetic refrigerators.

Applied pressure has been shown to change the entropy of a (manganite) solid in a significant way without the need for a magnetic field. The combined caloric effect of magnetic field and hydrostatic pressure on manganites offers some promise for magnetic refrigerators, but the practical realization is not likely to be simple (Szymczak et al., 2010). The pressures and materials currently being applied are producing a temperature change of up to 10 K/kBar (Hossain et al., 2004). Currently the energy required to achieve these pressures does not make the process attractive, but potentially efficiency improvements could possibly in future make the process more realistic (De Oliveira, 2011).

The technology is still at the experimental phase and materials and pressure cycles are being optimized. The main applications for barocaloric refrigeration appear to be in the domestic sector. Even when optimized the method is most likely to find application improving the magnetic refrigerators used in specialist applications. The pressures currently being applied (10 K/kBar) are likely to make systems prohibitively large and expensive for the very small cooling duties likely to be achieved.

7.2.5 Electrocaloric

Electrocaloric cooling is the electrical analogue of magnetocaloric cooling. Electrocaloric materials change their temperature when exposed to electric fields. The resulting changes in entropy and temperature of the material are known as the electrocaloric effect (Nesse et al., 2008). The technology does not use a conventional refrigerant, and so there are claimed environmental benefits as well as savings in energy use. The electrocaloric phenomena has been known since 1930 (Scott, 2011), but the technology remains at the experimental science stage with a small number of experimental prototypes and patents in existence.

Current research is divided into two areas: cryogenics and room temperature applications. Polymeric materials such as copolymers of (polyvinylidene fluoride) PVDF and trifluoroethylene are the most promising materials, but it is expected that crystals such as ammonium sulfate could give even better results if their ionic conductivity were greatly reduced (Scott, 2011).

Giant electrocaloric effects and large electrocaloric effects have been measured in thin film materials but not at temperatures useful for typical domestic refrigeration and not on a scale suitable to meet the cooling demands of domestic refrigeration. In 2006, researchers from Cambridge University reported in "Science" that thin films of perovskite lead zirconate titanate (PZT) showed a giant electrocaloric effect with the materials cooling down by up 7 °C in a field of just 25V (Mischenko et al., 2006).

At typical refrigeration temperatures, a 6.5 K temperature difference could be achieved; larger temperature differences (20–30 K) have only been measured at much warmer temperatures of 350–400 K (Scott, 2011). However, for commercially available $BaTiO_3$-based multilayer films, it has been claimed that an ideal General Electric Company plc (GEC) heat pump could deliver a cooling power of 22.5 W/kg, and it is hoped that through optimization of materials and design 2875 W/kg could be achieved (Maidment, 2012).

The technology still requires substantial development for practical use in the cold chain. Current challenges include fabricating multilayer films of the correct materials and then building a fridge and heat exchangers around them. In 2011, Scott identified the following challenges:

- Optimization of engineering design such as efficient and reliable thermal switching from source to sink under each cycle.
- Optimization of copolymers of PVDF and research of alternative materials such as ammonium sulfate crystals.
- Extension of the temperature range.

7.2.6 Hydraulic Refrigeration

Hydraulic refrigeration systems are vapor compression systems that achieve the compression and condensation of the refrigerant by entraining refrigerant vapor in a downflowing stream of water. They utilize the pressure head of the water to compress and condense the refrigerant.

The main benefits are that mechanical compressors or lubricants are not required, and the condenser is replaced by the water stream. The system is similar to vapor compression systems except that a water stream replaces the compressor and condenser. The refrigerant gas from the evaporator is entrained into the water stream at the top of the water column. As the water flows down the column, the refrigerant is compressed and cooled simultaneously. The condensed high-pressure liquid refrigerant is then separated from the water stream at the bottom of the column and returned to the evaporator via an expansion device. The compression ratio is therefore a function of the height of the water column, and the amount of cooling is a function of the water temperature.

The process is already in use on an industrial scale in the USA to compress air (where a free flow of water has been available) but has also been successfully applied to refrigeration systems. The major downside of the process is the large vertical height of the water column required. Separation of the refrigerant and the water stream is likely to be the major challenge practically (Rice and Hosterman, 1991).

Chau et al. (2001) designed and tested a multistage hydraulic refrigeration system using n-butane as the refrigerant. They concluded that hydraulic refrigeration had potential as an alternative to conventional vapor compression systems in applications where direct-contact heat exchange in the evaporator was desirable.

Systems are under development, but commercial availability of refrigeration systems using hydraulic compression is unlikely in the medium term. Even when commercialized, this technology will require more space and height than conventional systems. Its application will therefore be limited to industrial systems.

7.2.7 Optical Cooling

Optical cooling systems have no moving parts, and this could make them more reliable than conventional refrigeration systems. Claims are made that with development they could replace Stirling coolers, but the most likely application is on radiation balanced lasers where it will be used to compensate for internal heat generation.

Optical cooling works in a similar manner to magnetocaloric or barocaloric cooling; this technique also uses a solid-state material as the "refrigerant." Optical refrigeration is also referred to as "laser refrigeration" or "anti-Stokes fluorescent cooling." It is a technique for cooling a crystal (doped with ytterbium or thulium ions) by exciting the ions with a laser beam (Paschotta, 2011). The crystal is pulsed with light from a laser. The energy is absorbed and then emitted as fluorescence. When the energy of the fluorescence photons is larger than the laser light photons absorbed, there is a net energy loss and the material cools (Graydon, 2005). Most recent research work has concentrated on development of the crystals with future work required to integrate the technology into an operational refrigeration system.

The physical phenomenon of radiation cooling by anti-Stokes fluorescence was originally proposed in 1929 but only experimentally observed in 1995. ZBLAN glass has been cooled to 208 K from room temperature (Epstein et al., 1995), and 155 K has been achieved with $Yb:LiYF_4$ at a cooling power of 90 mW (Seletskiy et al., 2010). Theory suggests that 77 K could be achieved.

Although unlikely to be of practical benefit in the cold chain, optical cooling could be used to extend solid-state cooling achieved with thermoelectric systems into lower temperature regions; in cryogenic and spacecraft applications, optical cooling may bring size and weight advantages over thermoelectric or mechanical systems, where the load is less than 1 W and the temperature is between 20 and 200 K (Mills and Mord, 2006).

7.2.8 Stirling Cycle Variations

Stirling coolers can operate down to cryogenic temperatures. Stirling coolers are closed-cycle regenerative thermal machines that compress and expand a gas. Free piston machines use a moving magnet or linear machine unit to facilitate heat absorption and heat rejection respectively (but compression and expansion can also be performed by compressive waves as in an acoustic refrigerator). The gas is moved backward and forward between the hot and cold end spaces so that the gas is compressed and expanded (Figure 7.3).

Stirling coolers are available commercially, but there is still considerable ongoing research work to improve efficiency. Stirling units can have maximum cooling capacities up to 100 W with larger capacity units, up to 300 W, reported to be under development. Applications are in domestic and portable refrigerators and freezers as well as in beverage can vending machines. COPs of between 2 and 3 have been reported for cold head temperatures around 0 °C, and values around 1 for cold head temperatures approaching −40 °C. Although the rapid start-up and cool down of Stirling coolers is often cited as a major benefit, they are generally limited in their application by the

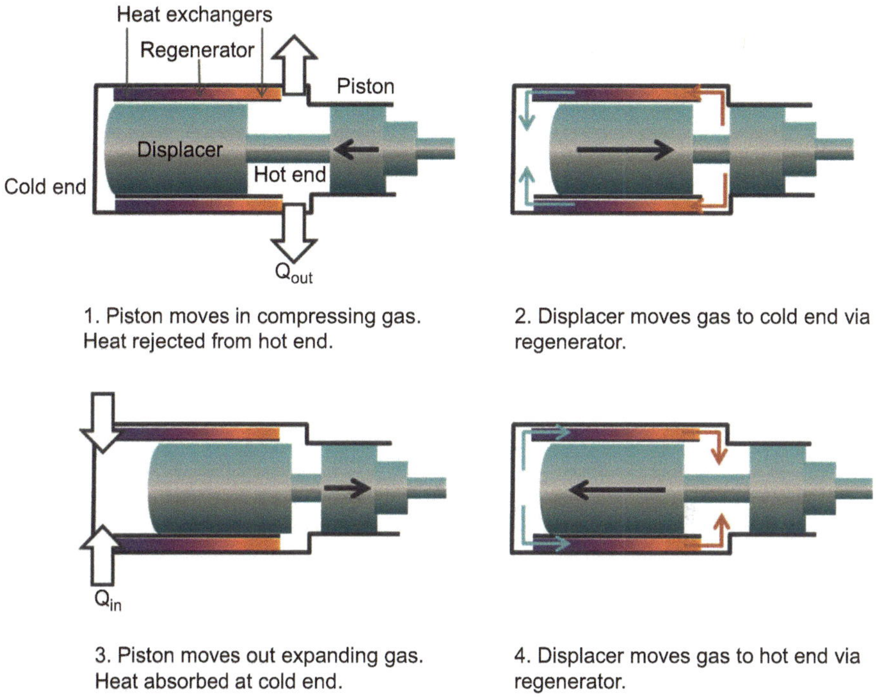

Figure 7.3 Piston and displacer movements in a Stirling cycle.

low heat transfer coefficient between the working fluid (gas) and the inside wall of the cylinder/heat exchanger; this generally makes them bulky and difficult to implement practically.

7.2.9 Thermionic Refrigeration

Thermionic cooling is based on the ability of a material to change temperature by applying an electric field under adiabatic conditions. The name thermionic arises from thermionic emission, which is the thermal excitation of hot electrons from a metal surface (Mahan, 2001). A thermionic device has two thin films separated by a vacuum layer. If a voltage is passed across the gap, the most energetic electrons on the negative side "jump" across to the positive side. As the electrons leave the negative side they get colder. Potentially such devices can be thermodynamically very efficient and could outperform classic direct expansion refrigeration systems.

Thermionic refrigeration is not currently applied to the cold chain, and most cited applications are for solid-state, on-chip cooling, or temperature regulation for sensors and other electronic devices. The major benefits would appear to be reduced direct emissions (i.e., no hydrochlorofluorine (HFC) refrigerants). Energy savings are expected by some researchers, but as the technology is still at the developmental stage there is limited practical evidence to support this. Based on a purely theoretical study, Mahan and Woods (1998) propose that efficiencies twice that of existing thermoelectric devices and comparable to existing vapor compression systems could be achieved but note that materials with a suitable work function (<0.3 eV) were not yet available. Brown et al. (2012), who compared a number of novel refrigeration technologies, were less positive about the potential of thermionic refrigeration stating that efficiencies did not appear as good as vapor compression systems and that the "food refrigeration applications are rated as poor."

7.2.10 Thermoelectric

Thermoelectric devices utilize the Peltier effect. This occurs when a direct electrical current is passed through the junction of two dissimilar conducing materials. This causes one junction to cool down (i.e., to absorb heat) and the other to heat up (i.e., to reject heat). The thermo elements are doped semiconductors, one is an n-type (with mainly negatively charged carriers) and the other a p-type (with mainly positively charged carriers) (Figure 7.4). Peltier coolers are commercially available and are mainly cut from bismuth telluride (Bi_2Te_3)-based materials. Peltier coolers have lower COPs than vapor compression refrigeration systems, and so their application in the food cold chain is limited. However, they can be cost-effective in small refrigeration systems (e.g., wine coolers, small beverage coolers, etc.) or where noise is an issue (e.g., hotel mini bars).

Thermoelectric power generation has also been suggested as a means to provide power on refrigerated transport vehicles. Their advantage is that they have no moving parts and no noise output. Heat to generate power can be sourced from the engine exhaust gas or solar radiation. A company called Hi-Z in Canada is developing the

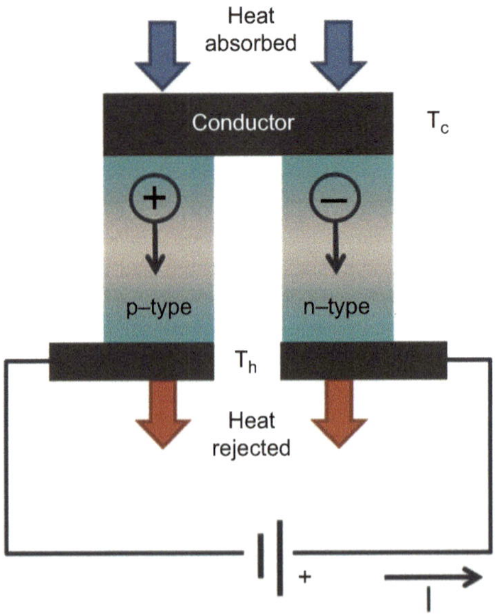

Figure 7.4 Thermoelectric cooler.

technology, and they aim to use 10–20 W modules to generate 5–10 kW to drive a vehicle refrigeration system (Hi-Z, 2008). Current modules appear to be generating up to 1 kW of power but have not yet been applied to large vehicle refrigeration systems (Hi-Z, 2014). Hi-Z also claims that although current energy conversion efficiencies are around 5%, they could rise to 25%.

7.2.11 Vortex Tube Cooling

Vortex tube cooling operates with compressed air as the working fluid and so is environmentally harmless with zero global warming potential (GWP) and zero ozone depletion potential (ODP). Compressed air is injected into a swirl chamber and exits via a longer tube as two air streams, one hot and one cold. As the compressed air enters the swirl chamber (or vortex generator), the air accelerates to a high rate of rotation (as much as 1,000,000 rpm). A small amount of the hottest gas is allowed to escape via the conical nozzle at the longer end of the vortex tube, and the remaining air returns down the center of the vortex tube to exit as cold air through the shorter end (Figure 7.5).

It is believed that a pressure difference occurs through the gas due to the centrifugal force. The resulting compression at the walls, expansion at the center, and heat transfer between the two streams within the vortex tube then result in the cold and hot air stream separation.

Vortex tubes are useful where inexpensive, nontoxic spot cooling is required and where a large amount of compressed air is available. Due to the energy cost of compressing the air initially and the inefficient conversion of that energy into useful cooling by vortex tubes, they do not present a low energy solution. Vortex tubes are available off-the-shelf with cooling duties up to 1700 W (with 43 °C temperature drop and an airflow rate of 2833 SLPM at 6.9 bar) (Newman Tools Inc. Ontario, Canada) (SLPM is standard liters per minute and is a mass flow rate based on air at 0 °C and 1.01 bar).

Unless a large amount of compressed air is available as a "waste" or "free" energy source, vortex tubes will only be suitable for spot cooling or where toxicity or flammability are a greater concern than energy.

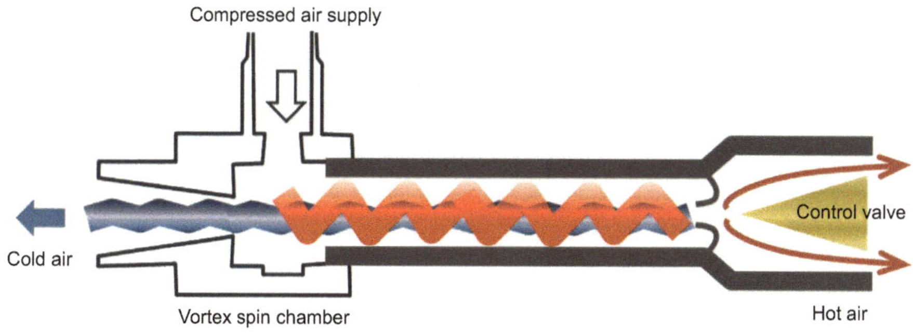

Figure 7.5 Vortex tube.

7.2.12 Water in Vapor Compression

The main advantages of water as a refrigerant are that it is nontoxic and nonflammable and has a GWP and ODP of zero. Small energy savings are possible over other refrigerants where warm evaporating temperatures (over 35 °C) or small (<10 K) temperature lifts are required.

The operation is quite similar to a conventional vapor compression system, but alternative compressor technology is required. The main disadvantage with water in vapor compression is that it works under a vacuum so the volumetric cooling capacity is very low and huge volumetric flow rates must be therefore be used. Water requires approximately 200 times larger volumetric flow rates than a conventional vapor compression system and about double the compression ratio (Müller, 2004). This means that plant becomes very large but on the plus side, pressure vessel testing is not required (Figure 7.6).

Water vapor chillers are already in use, but only offer energy savings (compared with conventional (HFC, hydrocarbon (HC), and ammonia) systems in a very limited range of applications and cost approximately twice the amount (Kühnl-Kinel, 2000). Ammonia is the closest challenger to water and offers better efficiency for most applications where its toxicity and flammability can be accommodated.

Centrifugal or radial compressors offer a pressure ratio of around 2.3 and so one to three stages are adequate for a water vapor chiller and have been used successfully. However, despite offering lower compression ratios, and so requiring up to eight stages, new axial flow compressors are now enabling the footprint to be reduced by up to one-third (Tepco, 2011), (Kühnl-Kinel, 2000). The Tepco system is currently undergoing reliability testing prior to release to market (Tepco, 2011).

Wave rotor technology (using compression waves to add energy to the working fluid, rather than pistons or turbine blades) is an old technology that has received attention as it promises improved efficiency and size reduction of the compressor required for a water chiller. Prototype refrigeration systems have been built and tested using three-port condensing wave rotors, and improvements in COP of up to 22% were measured (Kharazi et al., 2004). Most of the research was conducted by Michigan State University. The current problems with wave rotors are sealing and thermal expansion (Akbari et al., 2004).

Work by Efficient Energy GmbH used a high-speed direct driven centrifugal compressor to directly evaporate and condense water in a refrigerant cycle for air conditioning applications (Süß, 2015). The company claim that their system can halve the energy consumption of current state-of-the-art chiller technologies.

7.3 Technologies that are Product Oriented Technologies

7.3.1 Antifreeze Proteins

Antifreeze proteins lower the freezing point temperature. They are present in many fish and invertebrates as well as plants to prevent them freezing in cold weather or cold water. Further information on types of antifreeze proteins and their applications are contained in Hassas-Roudsari and Goff (2012).

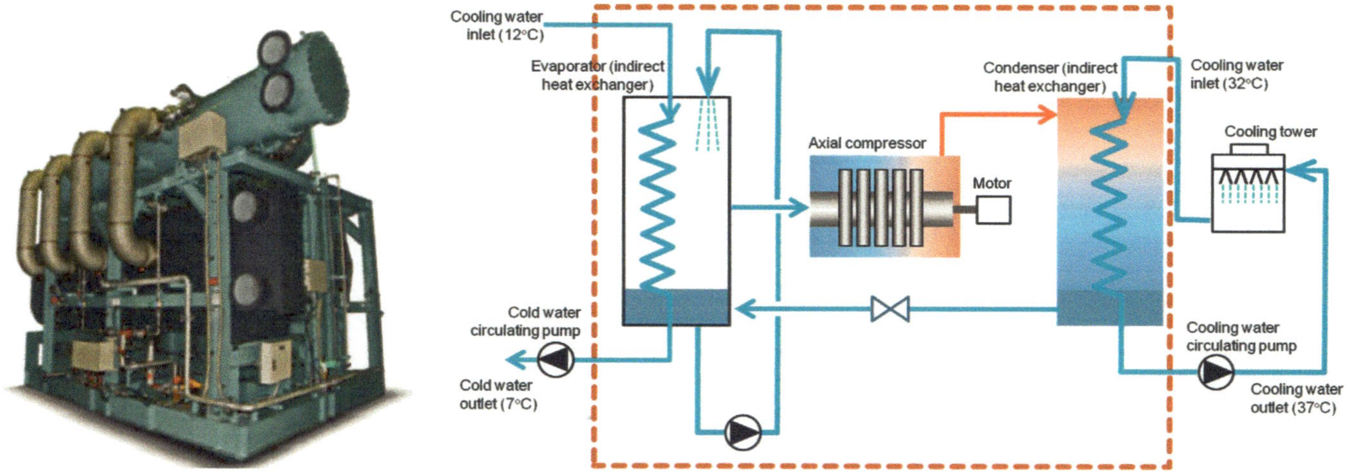

Figure 7.6 Photo and schematic of an indirect water vapor chiller. Taken from Tepco (2011).

By allowing a food to be stored at a low temperature while not being frozen the product latent heat does not need to be removed. This potentially can save energy as most of the heat extracted from products in freezing is from latent heat. Where freezing does occur the antifreeze proteins retard the growth of ice crystals.

Venketesh and Dayananda (2008) stated that antifreeze proteins have applications in food technology, but their uptake has been limited by high costs (due to low yields). To date, antifreeze proteins have been mainly used to prevent ice crystal growth in products such as ice cream. Commercial ice cream products with antifreeze proteins are now available in the market (Crilly et al., 2008). It has been suggested that the use of antifreeze proteins in meat may reduce drip loss (Payne et al., 1994; Payne and Young, 1995).

7.3.2 Dehydrofreezing

Dehydrofreezing has mainly been used for fruits and vegetables. A large part of the water (50–60%) in the product is removed prior to freezing using an osmotic solution and results in a freezing process that is faster and causes less ice crystal damage than a conventional process. Due to less water being contained in the product, the refrigeration heat load is also reduced. Texture and flavor are claimed to be superior to those of a conventionally frozen product.

Due to the product being dehydrated, there is less water in the product and therefore the latent heat is less. This is claimed to save energy (Mascheroni, 2007). Figure 7.7

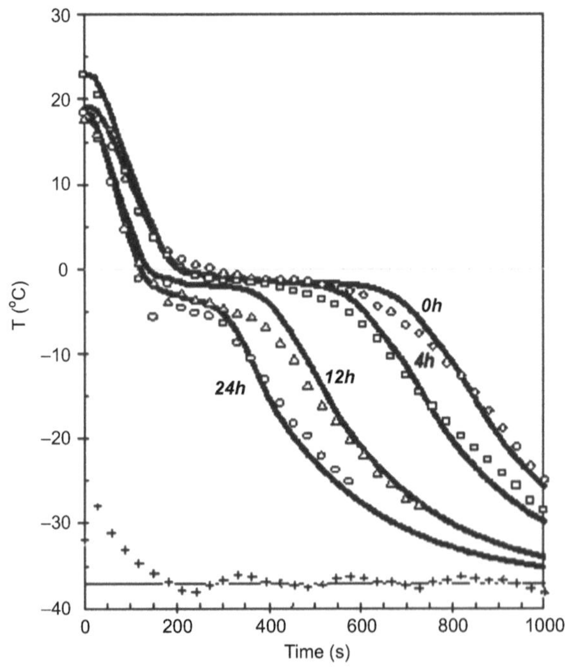

Figure 7.7 Product being frozen after dehydrofreezing for various times. From Mascheroni (2007).

shows the freezing times for a product after varied periods of dehydration. Greater periods of dehydration resulted in shorter freezing times. Further information about dehydrofreezing of fruits and vegetables can be found in James et al. (2014a).

7.3.3 Hydrofluidization

Hydrofluidization utilizes a pump that circulates refrigerant through orifices or nozzles to create agitating jets in a refrigeration vessel. This forms a fluidized bed of highly turbulent liquid that agitates product and generates extremely high heat transfer coefficients at the surface of the products (Fikiin, 1985, 1992, 2003). Suitable refrigerant media include brines, soluble carbohydrates (such as sucrose, invert sugar, glucose (dextrose), fructose, and other mono- and disaccharides), with additions of ethanol, salts, and glycerol. Freezing rates for fish and vegetables have been shown to exceed those for individual quick frozen products with heat transfer coefficients exceeding $900\,W/m^2K$ (Fikiin, 1992; Fikiin and Pham, 1985). This leads to low weight loss as the surface of the product is frozen extremely rapidly.

Detailed information is not available on energy consumed, but for small products that are frozen fast the energy use may be low. For some products, the heat transfer rates may be limited by the heat transfer rate through the product rather than that between the cooling fluid and the product; the increased pump energy required for hydrofluidization may therefore outweigh the benefits in some circumstances.

7.3.4 Ice Nucleation Proteins

Ice nucleation proteins reduce supercooling and are claimed to reduce freezing time. This is claimed to reduce energy cost and improve food quality mainly in liquid or semiliquid products (Watanabe et al., 1989; Li and Lee, 1998). Bacteria are the most widely recognized ice nucleation activators (usually of genera *Pseudomonas*, *Erwinia*, and *Xanthomonas*) and therefore care needs to be taken in their application to ensure that they are safe, nontoxic, and nonpathogenic (Li and Lee, 1995).

7.3.5 Magnetic Field Freezing/Cells Alive System (CAS)

Magnetic resonance freezing (MRF) utilizes continuous magnetic wave vibrations which impede ice crystallization. When cooling is applied this allows water to be supercooled below its freezing point. When the magnetic field is removed the product can freeze rapidly. This has claimed benefits of small ice crystals, little cellular damage, and low water loss during the freezing process. Published information on MRF technologies is rather limited with much retained within commercial companies.

The CAS was developed by the ABI Corporation, Ltd., of Chiba, Japan. Applying a magnetic field during freezing is claimed to avoid cell wall damage during the freezing process. The CAS vibrates water in foods with magnetic fields. This is claimed to prevent freezing, even at temperatures as low as −10°C. When the magnetic field is turned off, the water in the food instantly freezes. This creates small ice crystals and less damage to cell walls. It is not clear how ice crystal structure is maintained as ice crystals over time will agglomerate (Ostwald ripening).

There is very little published work on magnetic field freezing. Work by Suzuki et al. (2007) created a copy of the commercially available CAS and assessed freezing with and without a magnetic field on quality of sweet potato, spinach, fish, agar gel, and water. In assessment of color, oiliness, smell, umami, fatty taste, strength, and texture they found no significant difference between samples frozen with and without a magnetic field. In addition, work by James et al. (2014b) found little beneficial effect of the CAS system of freezing of garlic.

7.3.6 Perfusion

Meat carcasses must be chilled to below 7 °C before leaving the slaughterhouse. Typically this is done by passing refrigerated air over the surface of an eviscerated and dehided carcass. Because the cooling medium is only acting on the outer surface, it can take many hours for the temperature at the center of the carcass to drop below 7 °C. Often fan energy is as high as 30% of the overall energy consumption. In vascular perfusion chilling, a cold fluid is circulated through the intact vascular system theoretically offering significant reductions in cooling time and negating the need for cooling fans. Reducing the time required to chill carcasses would have substantial benefits to the meat industry in terms of both quality and energy usage. Such a treatment is still in the development stage, but systems not aimed to provide full chilling have been used in the USA and Australia to remove blood from carcasses and claim improved hygiene (Dikeman et al., 2003; Wang et al., 1995).

7.3.7 Pressure Shift Freezing

Two main processes exist: high-pressure assisted freezing and high-pressure shift freezing (HPSF). High-pressure assisted freezing occurs under constant high pressure, while temperature is lowered to the corresponding freezing point. Freezing occurs from the outside of the product with ice nucleation in the outer area of the food which grows radially into the center. There is little evidence that structure or texture is any different from product frozen at atmospheric pressure. Potentially, freezing times can be less than at atmospheric pressure.

HPSF is potentially a method to produce frozen food with small ice crystals and consequently little tissue damage and high quality. At atmospheric pressure ice has a freezing point of 0 °C. With increasing pressure the freezing point of water is lowered until at 207,500 kPa, the freezing point is −22 °C. HPSF involves increasing the pressure of a sample and reducing the temperature to create supercooled water. At this point the pressure is released to create ice instantaneously. Figure 7.8 shows a phase diagram for water showing the process. The pressure can be released slowly over several minutes or rapidly over a few seconds (Figure 7.9). This produces small ice crystals of granular shape almost instantaneously throughout the sample. When the pressure is released, there is a large release of heat of fusion and a consequent rise in sample temperature. Improvements in texture and histological damage of samples

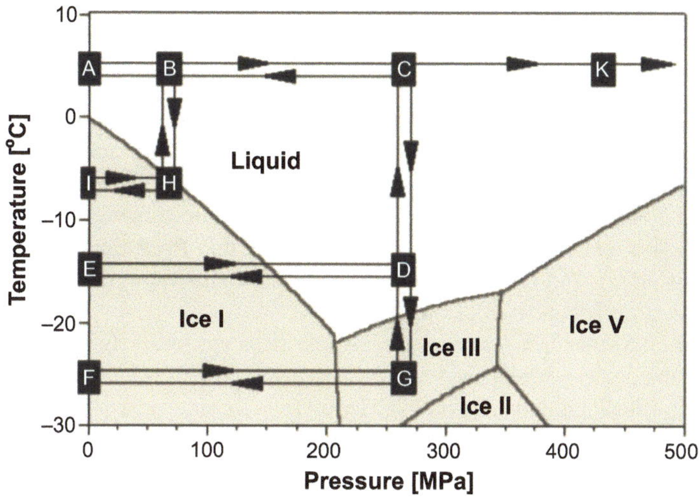

A – B – C – D – C – B – A	Subzero storage without freezing
A – B – H – I	Pressure-assisted[1] freezing
I – H – B – A	Pressure-assisted[1] thawing
A – B – C – D – E	Pressure-shift[2] freezing
E – D – C – B – A	Pressure-induced[3] thawing
A – B – C – D – G – F	Freezing to ice III
F – G – D – C – B – A	Thawing of ice III
A – B – C – K – ice VI	Freezing above 0 °C

[1]Assisted: Phase transition at constant pressure
[2]Shift: Phase transition due to pressure change
[3]Induced: Phase transition initiated with pressure change and continued at constant pressure

Figure 7.8 Water phase diagram and high pressure effects on the phase diagram.
From Fikiin (2003).

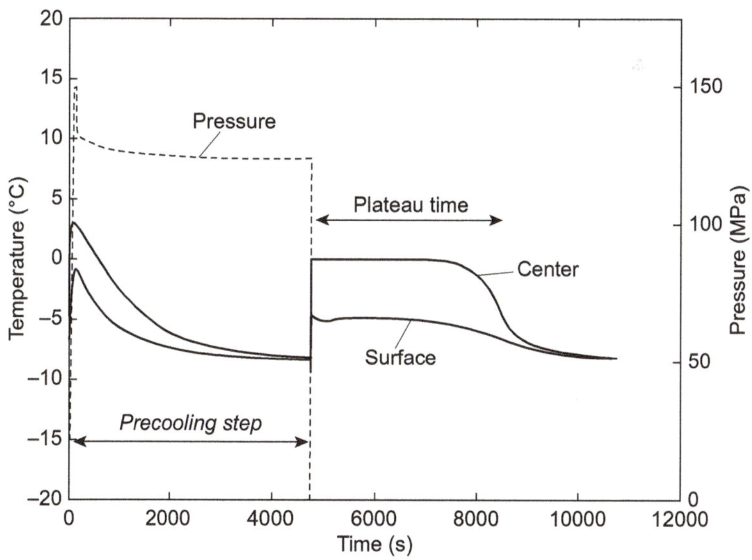

Figure 7.9 High pressure-shift freezing of an agar GEP sample from 120 MPa/-8 °C to 0.1 MPa. Temperature of the pressure/cooling medium: 8.5 °C.
From Mascheroni (2007).

have been reported, but deteriorations in color, water holding capacity, and texture of meat products have also been reported and appear to be related to applied pressure (Fernandez-Martın et al., 2000; Massaux et al., 1998). HPSF has the advantage of deactivating some microbes, with the level of inactivation ranging from about 1 to 8 log cycles. For example, Picart et al. (2004) found that *Pseudomonas fluorescens, Micrococcus luteus*, and *Listeria innocua* in salmon mince were reduced by between 1.2 and 1.3 log cycles under HPSF treatments.

Although HPSF has many advantages, sample sizes are relatively small, the vessels required for the process are expensive, and the freezing process is a relatively slow batch operation with small throughputs. Although ice crystals are small and uniform, the maintenance of such structures needs uniform temperature control in frozen storage, and this may be difficult to practically achieve. Some claims have been made that pressure-shift freezing reduces the latent heat of a product, but there are few practical data to substantiate this.

7.3.8 Superchilling

Superchilling or partial freezing, subcooling or deep chilling is the process of taking a product just below its freezing point and then holding it at this high freezing temperature under strict temperature control. The aim is to limit the degree of ice formation to between 10% and 50% of the product's water fraction. This lower level of ice formation (as opposed to conventional frozen storage) reduces the amount of damage caused by the formation of larger ice crystals associated with standard freezing. For most foodstuffs, the temperature range for successful superchilling should be −0.5 to −4.0 °C (Aune, 2003; Olafsdotti et al., 2006; Gallart-Jornet et al., 2007; Medina et al., 2009; Stevik et al., 2010a), and strict temperature is needed as even small fluctuations in temperature (e.g., ±0.5 °C) have large effects on the degree of ice in the product.

Superchilling also increases the product shelf life compared with fresh storage as temperature is the most critical factor controlling the growth of microorganisms in foods, especially fish and meat (Zeng et al., 2006; Dunn and Rustad, 2007, 2008). At superchilled temperatures, the growth of many microorganisms is either reduced or inhibited though this technique but cannot match the length of acceptable storage achievable in standard frozen storage (−18 to −20 °C). To date, superchilling has mainly been applied to fish and meat as these are high value products, and the benefits of extended shelf life have the greatest benefits. Superchilling has also been applied to several other products such as cheese and shellfish.

There appears to be some potential problems with using superchilling to limit physical deterioration in foods. This is because it has been reported by a number of authors (Sheard et al., 2001; Bao et al., 2007; Magnussen et al., 2008) that superchilling increases drip loss from high moisture foods such as fish and meat. This increase in drip loss compared with conventional fresh storage is due to surface freezing of the food, leading to the formation of ice crystals, which can disrupt the cell membranes and extracellular spaces, allowing liquid to escape upon thawing. Several authors have also reported the formation of white spots on the surface of superchilled muscle foods, with Small et al. (2011) showing it in beef, Dunn and Rustad (2007) in cod fillets, and Dunn and Rustad (2008) in salmon fillets. Loss of product color was reported in shrimp by Huidobro et al. (2002), and deleterious changes in product textural qualities have been shown in fish by Dunn and Rustad (2007, 2008) and Bahuaud et al. (2008). Medina et al. (2009) reported that superchilling by ice slurry, when applied to whole fish can lead to dulling of the fish color and cloudiness developing in the fish eyes. This is a problem, as eye appearance is often used by graders as a sign of freshness, with cloudiness developing in less fresh fish. They concluded that these quality alterations were brought on by chilling the fish too quickly, with the temperature of the slurry being reduced too low before the surface of the fish had begun to freeze.

Superchilling of food can be achieved using a number of methods which include: liquid ice/ice slurry (a mixture of small ice crystals in suspension), air blast chilling, superchilled brine, and cryogenic freezing using mediums such as dry ice, liquid CO_2, and liquid N_2 (Watt and Herring, 1974; Lan and Farid, 2004; Bao et al., 2007; Magnussen et al., 2008; Bedecarrats et al., 2009; Medina et al., 2009; Erikson et al., 2011). Aubourg et al. (2007) and Medina et al. (2009) reported that the use of an ice slurry gave approximately a fourfold increase in the rate of heat exchange compared to flake ice, which would allow a product to be cooled much more efficiently and inhibits evaporative losses from the product.

To maintain food in a superchilled state a separate cold chain and distribution system would be required, and so superchilling is primarily a technique that is applied during processing and storage. Magnussen et al. (2008) stated that for the incorporation of widespread superchilling to be adopted into food processing and distribution, then it would not only require upgrading of refrigeration systems with better temperature control to avoid fluctuations, but would also require the further educating of distribution/retail staff along with the consumer on the identification and handling of this new type of product. When these requirements are accomplished, then superchilling would lead to extended shelf life and quality of foods. An alternative approach is called "thawing during distribution" and involves superchilling the product at the production facility with distribution at 5 °C and relying on the large thermal buffer of the latent heating capacity to allow the product to thaw slowly. Alder-Nissen et al. (2013) used frozen cod as an example for this thawing technique and estimated that the shelf life of the cut fish could be extended by as much as two days over a conventionally chilled product.

7.3.9 Supercooling

Supercooling (or subcooling) is analogous to superchilling, but where water remains unfrozen. Potentially, this is an energy efficient technology as energy is not used to freeze water. There is limited published information on supercooling and no information has been identified on the energy or carbon implications. Published work covers supercooling of garlic (James et al., 2009), strawberries (Martins and Lopes, 2007), tomatoes (Cox and Moore, 1997), cauliflower (Fuller and Wisniewski, 1998), and a range of vegetables and fish (James et al., 2011). Earlier work from the 1920s (Diehl and Wright, 1924) has reported that apples could sometimes be subcooled. Grapes, lemons, and naval oranges have also been reported by Lucas (1954) to supercool. A summary of products that have been shown to subcool and the level of subcooling are shown in Table 7.1.

Table 7.1 **Supercooling of Products from Published Literature**

	Degree of Supercooling (°C)	Supercooling Point (°C)	References
Apples	4		Diehl and Wright (1924)
Broccoli	2.3	−4.4	James et al. (2011)
Carrot	1.1	−2.7	James et al. (2011)
Cauliflower		−6.5 to −9.5	Fuller & Wisniewski (1998)
Cauliflower	3.7	−5.2	James et al. (2011)
Cod	3.9	−5.3	James et al (2011)
Garlic	10.3	−13.0	James et al. (2009)
Grapes	3		Lucas (1954)
Herring	5.6	−9.2	James et al. (2011)
Leek	1.4	−3.3	James et al. (2011)
Lemons		−6.1	Lucas (1954)
Oranges	1.5		Lucas (1954)
Parsnip	0.6	−2.8	James et al. (2011)
Prawns (previously frozen), "jumbo"	3.8	−5.9	James et al. (2011)
Prawns (previously frozen), "large"	4.6	−6.5	James et al. (2011)
Shallot	3.8	−5.4	James et al. (2011)
Squid	6.6	−8.6	James et al. (2011)
Strawberries		−0.3 to −4.6	Martins and Lopes (2007)
Tomatoes		−4.3 to −4.5	Cox and Moore (1997)

Generally supercooling is accentuated when products are slowly cooled. It is thought that vibration to movement can cause sudden freezing. A detailed review of supercooling technologies and their application is available in Stonehouse and Evans (2015).

7.3.10 Ultrasound-Assisted Freezing

Power ultrasound uses sound energy to accelerate freezing. Ultrasound can create cavitation in cells, which promotes ice nucleation and accelerates heat and mass transfer. Usually low frequency high-energy power ultrasound is used to accentuate food freezing. Ultrasound can potentially fracture ice crystals leading to small crystal sizes and in some cases better product quality. Although the technology has considerable promise for high value products, considerable research is still required for industrial application (Zheng and Sun, 2006). Freezing times are claimed to be less than in a conventional freezing process indicating that energy could be saved in the refrigeration process (Figure 7.10). Longer ultrasound treatments have been found to reduce freezing times (Figure 7.11) as has applying ultrasound during the phase change period (Mascheroni, 2007).

A detailed review of the technology is available in Zheng and Sun (2006).

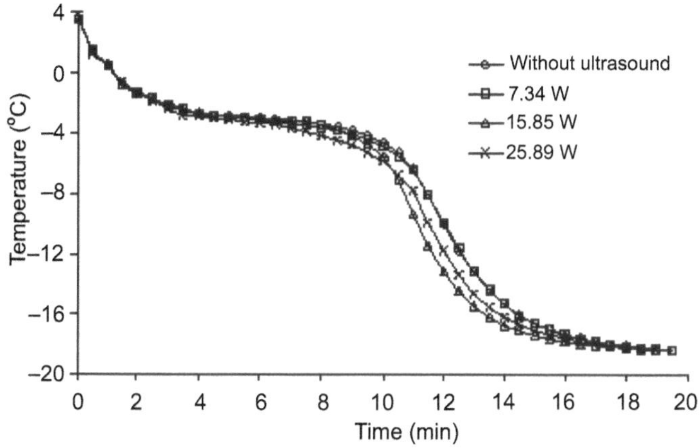

Figure 7.10 Temperature versus time curves for potato samples under different ultrasonic power levels applied totally for 2 min at the coolant temperature of −18 °C.
From Mascheroni (2007).

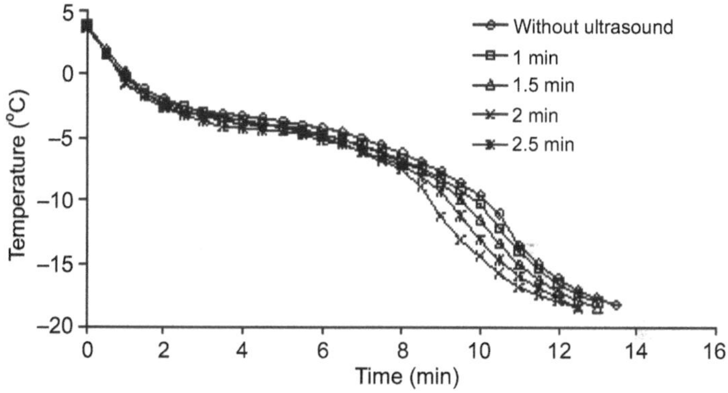

Figure 7.11 Comparison of freezing curves for potato samples under different exposure times to 15.85 W power ultrasound at the coolant temperature of -20 °C.
From Mascheroni (2007).

7.4 Future Trends

A large number of technologies exist that are competing to replace conventional direct expansion refrigeration systems. These systems may either be totally novel or may be older technologies that now are more attractive due to improved components. Many of the systems have been in development for long periods. This has often been due to high manufacturing costs and reliability issues that often take long periods to overcome. Many systems have definite advantages as they operate using benign refrigerants. Ongoing regulatory initiatives to reduce the use of high global warming refrigerants provide a huge opportunity for these systems to flourish if they can be proven to be efficient, reliable, and produced for reasonable cost.

Refrigeration technologies that are food-related tend to be applied by food producers and to be successful need to provide some substantial benefit above current technologies. The technologies that possibly have the greatest potential are those that provide some consumer benefit such as improved food quality where the consumer may be happy to pay more for a high quality product. Technologies that do not have direct consumer benefits need to provide the food manufacture with a benefit such as reduced costs, more flexible storage, or enhanced shelf life.

Sources of Further Information and Advice

Journals and books:

International Journal of Refrigeration (http://www.journals.elsevier.com/international-journal-of-refrigeration).

Operations in Food Refrigeration. Rodolfo H. Mascheroni. CRC Press.

Refrigeration Systems and Applications (2nd Revised edition). Mehmet Kanoglu and Ibrahim Dincer, Wiley.

Professional institutions:

IIR http://www.iifiir.org/

ASHRAE http://www.ashrae.org/

References

Akbari, P., Nalim, R., Müller, N., 2004. A review of wave rotor technology and its applications. In: Proceedings of IMECE04. ASME International Mechanical Engineering Congress, Anaheim, California, USA.

Alder-Nissen, J., Akkerman, R., Frosch, S., Grunow, M., Loje, H., Risum, J., Wang, Y., Ornholt-Johansson, G., 2013. Improving the supply chain and food quality of professionally prepared meals. Trends in Food Science & Technology 29, 74–79.

Aubourg, S.P., Losada, V., Prada, M., Miranda, J.M., Velazquez, J.B., 2007. Improvement of the commercial quality of chilled Norway lobster (*Nephrpos norvegicus*) stored in slurry ice: effects of a preliminary treatment with an antimelanosic agent on enzymatic browning. Food Chemistry 103, 741–748.

Aune, E.,J., 2003. Superchilling of foodstuff, a review. In: Proceedings of the International Congress of Refrigeration 2003. International Institute of Refrigeration, Washington, DC, USA.

Bahuaud, D., Morkore, T., Langsrud, O., Sinnes, K., Veiseth, E., Ofstad, R., Thomassen, M.S., 2008. Effects of −1.5 °C super-chilling on quality of Atlantic salmon (*Salmo salar*) pre-rigor fillets: cathepsin activity, muscle histology, texture and liquid leakage. Food Chemistry 111, 329–339.

Bao, H.N.D., Ararson, S., Lorarinsdottir, K.A., 2007. Effects of dry ice and superchilling on quality and shelf life of Arctic charr (*Salvelinus alpines*) fillets. International Journal of Food Engineering 3, 1–27.

Bedecarrats, J.P., David, T., Castaing-Lasvignottes, J., 2009. Ice slurry production using supercooling phenomenon. International journal of Refrigeration 33, 196–204.

Ben and Jerry's, 2004. http://www.acs.psu.edu/thermoacoustics/refrigeration/benandjerrys.htm.

Brown, D., Stout, T., Dirks, J., Fernandez, N., October 1, 2012. The prospects of alternatives to vapor compression technology for space cooling and food refrigeration applications. Energy Engineering: Journal of the Association of Energy Engineering 109 (6), 7–20.

Brown, T., Foster, A.M., Evans, J.A., 2011. Mathematical modelling of air cycle systems for combined heating and cooling. In: The 23rd IIR International Congress of Refrigeration, Refrigeration for Sustainable Development August 21–26, Prague.

Chau, D.S., Rice, W., Phelan, P.E., Whitfield, K.E., Wood, B.D., June 2001. Experimental results for a hydraulic refrigeration system using *n*-butane. International Journal of Refrigeration 24 (4), 325–337.

Cox, D.R.G., Moore, S.R., 1997. A Process for Supercooling, Patent WO 97/18879.

Crilly, J.F., Russell, A.B., Cox, A.R., Cebula, D.J., 2008. Designing multiscale structures for desired properties of ice cream. Industrial & Engineering Chemistry Research 47, 6362–6367.

De Oliveira, N.A., March 1, 2011. Barocaloric effect and the pressure induced solid state refrigerator. Journal of Applied Physics 109 (5) Article number 053515.

Diehl, H.C., Wright, R.C., 1924. Freezing injury of apples. Journal of Agricultural Research 29, 0099–0127.

Dikeman, M.E., Hunt, M.C., Addis, P.B., Schoenbeck, H.J., Pullen, M., Katsanidis, E., Yancey, E.J., 2003. Effects of post exsangination vascular infusion of cattle with a solution of saccharides, sodium chloride, and phosphates or with calcium chloride on quality and sensory traits of steaks and ground beef. Journal of Animal Science 81, 156–166.

Dunn, A.S., Rustad, T., 2007. Quality changes during superchilled storage of cod (*Gadus morhua*) fillets. Food Chemistry 105, 1067–1075.

Dunn, A.S., Rustad, T., 2008. Quality of superchilled vacuum packed Atlantic salmon (*Salmo salar*) fillets stored at −1.4 °Cand −3.6 °C. Food Chemistry 106, 122–131.

Engelbrecht, K., Pryds, N., 2014. Progress in magnetic refrigeration and future challenges. 6th International Conference on Magnetic Refrigeration at Room Temperature (Thermag VI). Proceedings: Victoria, Canada, September 7–10.

Epstein, R.I., Buchwald, M.I., Edwards, B.C., Gosnell, T.R., Mungan, C.E., 1995. Observation of laser-induced fluorescent cooling of a solid. Nature 377, 500.

Erikson, U., Misimi, E., Gallart-Jornet, L., 2011. Superchilling of rested Atlantic salmon: different chilling strategies and effects on fish and fillet quality. Food Chemistry 127, 1427–1437.

Fernandez-Martın, F., Otero, L., Solas, M.T., Sanz, P.D., 2000. Protein denaturation and structural damage during high pressure shift freezing of porcine and bovine muscle. Journal of Food Science 65 (6), 1002–1008.

Fikiin, A.G., 1985. Method and system for immersion cooling and freezing of foodstuffs by hydrofluidization. Invention Certificate No. 40164, Bulgarian Patent Agency INRA.

Fikiin, A.G., 1992. New method and fluidized water system for intensive chilling and freezing of fish. Food Control (Oxford) 3 (3), 153–160.

Fikiin, A.G., Pham, V.H., 1985. System for examination of heat transfer regimes during hydrorefrigeration of foodstuffs. Invention Certificate No. 39749, Bulgarian Patent Agency INRA.

Fikiin, K., 2003. Novelties of Food Freezing Research in Europe and Beyond. Flair-Flow Europe.

Foster, A.M., Brown, T., Gigiel, A.J., Alford, A., Evans, J.A., 2011. Air cycle combined heating and cooling for the food industry. International Journal of Refrigeration 34 (2011), 1296–1304.

Fuller, M.P., Wisniewski, M., 1998. The use of infrared thermal imaging in the study of ice nucleation and freezing of plants. Journal of Thermal Biology 23, 81–89.

Gallart-Jornet, L., Rustad, t, Barat, J.M., Fito, P., Escriche, I., 2007. Effect of superchilled storage on the freshness and salting behaviour of Atlantic salmon (*Salmo salar*) fillets. Food Chemistry 103, 1268–1281.

Gigiel, A.J., Chauveron, S., Fitt, P., 1992. Air as a replacement for CFC refrigerants. In: IIR Congress 'Cold '92', Buenos Aires, Argentina, 7–9 September.

Graydon, O., April 15, 2005. Optical Refrigeration Breaks Record. Opto & Laser Europe (Magazine). http://optics.org/article/22004.

Hassas-Roudsari, M., Goff, H.D., April 2012. Ice structuring proteins from plants: mechanism of action and food application (review). Food Research International 46 (1), 425–436.

Hi-Z, 2008. Available from: http://www.hi-z.com.

Hi-Z, 2014. http://www.hi-z.com/hz-20.html.

Hossain, Z., Strässle, Geibel, C., Furrer, A., 2004. First-order calence transition and barocaloric effect in $EuNi_2(Si_{1-x}Ge_x)_2$. Journal of Magnetism and Magnetic Materials 272–276, 2352–2354.

Huidobro, A., Lopez-Caballero, M.E., Mendes, R., 2002. Onboard processing of deepewater pink shrimp (*Parapenaeus longirostris*) with liquid ice: effect on quality. European Food Research & Technology 214, 469–475.

James, C., Hanser, P., James, S.J., 2011. Super-cooling phenomena in fruits, vegetables and seafoods. In: 11th International Congress on Engineering and Food (ICEF 2011), Athens, Greece, May 22–26, 2011.

James, C., Seignemartin, V., James, S.J., 2009. The freezing and supercooling of garlic (*Allium sativum* L.). International Journal of Refrigeration 32 (2009), 253–260.

James, C., Purnell, G., James, S.J., 2014a. A critical review of dehydrofreezing of fruits and vegetables. Food and Bioprocess Technology 7 (5), 1219–1234.

James, C., Reitz, B., James, S.J., 2014b. The freezing characteristics of garlic bulbs (*Allium sativum* L.) frozen conventionally or with the assistance of an oscillating weak magnetic field. Food and Bioprocess Technology 8 (3), 702–708.

Kharazi, A.A., Akbari, P., Müller, N., 2004. An application of wave rotor technology for performance enhancement of R718 refrigeration cycles. In: American Institute of Aeronautics and Astronautics, 2nd International Energy Conversion Engineering Conference, Rhode Island.

Kühnl-Kinel, J., 2000. New Age Water Chillers with Water as Refrigerant. CERN, Geneva, Switzerland.

Lan, S.H., Farid, M.M., 2004. Experimental analysis of cryogenic freezing of food. Journal of Chemical Engineering of Japan, Special issue for the 9th congress of Asian Pacific Confederation of Chemical Engineering 37, 304–309.

Lewis, J.S., Chaer, I., Tassou, S.A., 2007. Fostering the development of technologies and practices to reduce the energy inputs into the refrigeration of food. Reviews of Alternative Refrigeration Technologies Centre for Energy and Built Environment Research, School of Engineering and Design, Brunel University. (DEFRA Energy project).

Li, J., Lee, T.C., 1995. Bacterial ice nucleation and its potential application in the food industry. Trends in Food Science and Technology 6, 259–265.

Li, J., Lee, T.C., 1998. Bacterial extracellular ice nucleator effects on freezing of foods. Journal of Food Science 63 (3), 375–381.

Lucas, J.W., 1954. Subcooling and ice nucleation in lemons. Journal of Plant Physiology 29, 245–251.

Magnussen, O.M., Haugland, A., Hemmingsen, A.K.T., Johansen, S., Nordtvedt, T.S., 2008. Advances in superchilling of food – process characteristics and product quality. Trends in Food Science & Technology 19, 418–424.

Mahan, G.D., 2001. Chapter 3 Thermionic refrigeration. Semiconductors and Semimetals, vol. 71. Elsevier. ISBN: 9780127521800. pp.157–174.

Mahan, G.D., Woods, L.M., May 4, 1998. Multilayer thermionic refrigeration. Physics Review Letters 80 (18).

Maidment, G., 2012. SIRAC, Sustainable Innovation – A Technology Review. London South Bank University. http://www.sirach.org.uk/P496FT327131.

Martins, R.C., Lopes, V.V., May 2007. Modelling supercooling in frozen strawberries: experimental analysis, cellular automation and inverse problem methodology. Journal of Food Engineering 80 (1), 126–141. http://dx.doi.org/10.1016/j.jfoodeng.2006.05.009.

Mascheroni, R.H., 2007. Overview on new freezing technologies. In: Refrigeration, Energy and Environment: Challenges and Perspectives 5th Alcuefood Seminar (São Paulo) – March 27–29, 2007.

Massaux, C., Bera, F., Steyer, B., Sindic, M., Deroanne, C., 1998. High hydrostatic pressure freezing and thawing of pork meat: quality preservation, processing times and highpressure treatments advantages. In: Abstracts of High-Pressure Bioscience & Biotechnology; IV Joint Meeting of Japanese and European Seminars on High-Pressure Bioscience and Biotechnology; Heidelberg, Germany, 1998, p. 86.

Medina, I., Gallardo, J.M., Aubourg, S.P., 2009. Quality preservation in chilled and frozen fish products by employment of slurry ice and natural antioxidants. International Journal of Food Science and Technology 44, 1467–1479.

Mills, G., Mord, A., February–March 2006. Performance modeling of optical refrigerators. Cryogenics 46 (2–3), 176–182. http://dx.doi.org/10.1016/j.cryogenics.2005.11.015.

Mischenko, A.S., Zhang, Q., Scott, J.F., Whatmore, R.W., Mathur, N.D., 2006. Giant electrocaloric effect in thin-film $PbZr_{0.95}Ti_{0.05}O_3$. Science 311 (5765), 1270–1271.

Müller, N., 2004. Turbo Chillers Using Water as a Refrigerant. Michigan State University.

Nesse, B., Chu, B., Lu, S., Wang, Y., Furman, E., Zhang, Q.M., August 8, 2008. Large electrocaloric effect in ferroelectric polymers near room temperature. Science (Magazine) 321 (5890), 821–823.

Olafsdottir, G., Lauzon, H.L., Martinsdottir, E., Oehlenschlager, J., Kristbergsson, K., 2006. Evaluation of shelf life of superchilled cod (*Gadus morhua*) fillets and the influence of temperature fluctuations during storage on microbial and chemical quality indicators. Journal of Food Science 71, S97–S109.

Paschotta, R., 2011. Optical Refrigeration. Encyclopedia of Laser Physics and Technology. www.rp-photonics.com/optical_refrigeration.html.

Payne, S.R., Young, O.A., 1995. Effect of pre-slaughter administration of antifreeze proteins on frozen meat quality. Meat Science 41, 147–155.

Payne, S.R., Sandford, D., Harris, A., Young, O.A., 1994. The effects of antifreeze proteins on chilled and frozen meat. Meat Science 37, 429–438.

Picart, L., Dumay, E., Guiraud, J.-P., Cheftel, J.C., 2004. Microbial inactivation by pressure-shift freezing: effects on smoked salmon mince inoculated with *Pseudomonas fluorescens*, *Micrococcus luteus* and *Listeria innocua*. LWT – Food Science and Technology 37 (2), 227–238.

Poese, M.E., Smith, R.W.M., Garrett, S.L., van Gerwen, R., Gosselin, P., 2004. Thermoacoustic refrigeration for ice cream sales. In: 6th IIR Gustav Lorentzen Natural Working Fluids Conference, Current applications and opportunities, 29th August to 1st September, Glasgow, UK.

Rice, W., Hosterman, C., 1991. Hydraulic Refrigeration System and Method. US patent US5056323 A.

Russell, S.L., Gigiel, A.J., James, S.J., 1998. Progress in the use of air cycle technology in food refrigeration and especially retail display. AIRAH Journal 55 (11), 20–25.

Russell, S.L., Gigiel, A.J., James, S.J., 2000. Development of a fluidised bed food freezing system that can use air cycle technology. Proceedings of IChemE Food & Drink ISBN:0 85295 438.

Scott, J.F., 2011. Electrocaloric materials. Annual Review of Materials Research, vol. 41, 229–240.

Seletskiy, D.V., Melgaard, S.D., Bigotta, S., Di Lieto, A., Tonelli, M., Sheik-Bahae, M., 2010. Laser cooling of solids to cryogenic temperatures. Nature Photonics 4 (3), 161.

Sheard, P.R., Taylor, A.A., Savage, A.W.J., Robinson, A.M., Richardson, R.I., Nute, G.R., 2001. Factors affecting the composition and amount of 'white exudate' from cooked bacon. Meat Science 59, 423–435.

Small, A., Sikes, A., O'Callaghan, D., 2011. Development of a Deep Chilling Process for Beef and Sheepmeat. Published by Meat & Livestock Australia Limited, Sydney, Australia.

Spence, S.W.T., Doran, W.J., Artt, D.W., McCullough, G., 2005. Performance analysis of a feasible air-cycle refrigeration system for road transport. International Journal of Refrigeration 27, 381–388.

Spence, S.W.T., Doran, W.J., Artt, D.W., 2004. Design, construction and testing of an air cycle refrigeration system for road transport. International Journal of Refrigeration 27, 503–510.

Stevik, A.M., Dunn, A.S., Rustad, T., O'Farrell, M., Schulerud, H., Ottestad, S., 2010a. Ice fraction assessment by near-infrared spectroscopy enhancing automated superchilling process lines. Journal of Food Engineering 100, 169–177.

Stonehouse, G.G., Evans, J.A., March 2015. The use of supercooling for fresh foods: a review. Journal of Food Engineering 148, 74–79.

Süß, J., 2014–2015. New ways to use water as a refrigerant. Proceedings of the Institute of Refrigeration 7–2.

Suzuki, T., Takeuchi, Y., Masuda, K., Watanabe, M., Shirakashi, R., Fukuda, Y., Tsuruta, T., Yamamoto, K., Koga, N., Hiruma, N., Ichioka, J., Taki, K., 2007. Test report on effectiveness of magnetic field for food freezing process. JSRAE Japanese Society for Refrigeration and Air Conditioning Engineering 10, 22–24.

Szymczak, R., Kolano, R., Kolano-Burian, A., Pietosa, J., Szymczak, H., 2010. Cooling by adiabatic pressure application in $La_{0.7}Ca_{0.3}MnO_3$ magnetocaloric effect material. Journal of Magnetism and Magnetic Materials 332, 1589–1591.

Tassou, S.A., Lewis, J.S., Ge, Y.T., Hadawey, A., Chaer, I., 2010. A review of emerging technologies for food refrigeration application. Applied Thermal Engineering 30, 263–276.

Tepco, 2011. Development of the World's First Water Vapour Chiller with Axial Compressor. Press release January 19, 2011.

Venketesh, S., Dayananda, C., 2008. Properties, potentials, and prospects of antifreeze proteins. Critical Reviews in Biotechnology 28, 57–82.

Wang, Y., McGinnis, D.S., Segada, R.R., Jones, S.D.M., 1995. Vascular infusion of beef carcasses effects on chilling efficiency and weight change. Food Research International, 28 (4), 425–430.

Watanabe, M., Watanabe, J., Kumeno, K., Nakahama, N., Arai, S., 1989. Freeze concentration of some foodstuffs using ice nucleation-active bacterial cells entrapped in calcium alginate gel. Journal of Biological Chemistry 53 (10), 2731–2735.

Watt, D.B., Herring, H.K., 1974. Rapid chilling of beef carcasses utilising ammonia and cryogenic systems: effects on shrink and tenderness. Journal of Animal Science 38, 928–934.

Wetzel, M., Herman, C., 1997. Design optimization of thermoacoustic refrigerators. International Journal of Refrigeration 20 (1), 3–21.

Whirlpool, 2009. Whirlpool to Explore Revolutionary Magnetic Refrigeration Concept. Press release October 10, 2009, Whirlpool Communications and Corporate Relations, Whirlpool Europe.

Wilson, N., Ozcan, S., Sandeman, K., Burdett, P., 2006. Overview of magnetic refrigeration. Proceedings of the Institute of Refrigeration 2006–07 4–1.

Zeng, Q.Z., Thorarinsdottir, K.A., Olafsdottir, G., 2006. Quality changes of shrimp (*Pandalus borealis*) stored under different cooling conditions. Journal of Food Science 70, 459–466.

Zheng, L., Sun, D.-W., 2006. Innovative applications of power ultrasound during food freezing processes: a review. Trends in Food Science and Technology 17 (1), 16–23.

Advanced Heating Technologies for Food Processing

8

Norman Maloney, Michael Harrison
C-Tech Innovation Ltd, Capenhurst Technology Park, Chester, UK

8.1 Introduction

Volumetric heating (or full-body heating) is a term used for technologies that cause heating to occur throughout the whole volume of a material. This characteristic can result in significant process and quality benefits when applied to food processing. This article introduces these techniques—ohmic, radio frequency (RF), and microwave (MW) heating—so that a comparison of their use for a given application can be made.

Traditional thermal processing techniques are typically surface heating techniques, where the heat is transferred to the rest of the material being processed from a heated surface by conduction in solids, and conduction/convection in liquids. Such techniques require the surface of the product to become significantly hotter than the bulk of the material which can result in thermal damage to the surface if increased processing rates are sought. Thus, where these techniques are used, heating rates can be limited by this potential for surface damage.

In contrast, with volumetric heating techniques, heat is generated throughout the whole volume of the material, the heating process being independent of thermal conduction/convection mechanisms. These volumetric heating techniques result from the use of electromagnetic fields at operating frequencies located at various positions on the electromagnetic spectrum (see Figure 8.1).

The mechanisms by which heat is generated in the product varies with the operating frequency of the electric field: at low frequencies (less than 10^4 Hz) and radio frequencies (10^4–10^8 Hz) the effects are primarily due to resistive heating caused by the movement of ions/electrons in the electric field. At higher frequencies (e.g., microwave heating (10^8–10^{11} Hz) molecular dipole rotation is the main heating mechanism, whereas at higher frequencies still (e.g., infrared (IR)) the dominant mechanism is one of molecular vibration, and at visible frequencies is by the elevation of electrons to higher energy levels. As the frequency of the electromagnetic field increases the energy penetrates less deeply into the product. At IR and visible frequencies, for most food products, the energy will be absorbed very rapidly as it enters the product resulting in essentially surface heating; though for other materials, e.g., some polymers, IR heating can provide volumetric heating.

Typical volumetric heating technologies of use in the food processing industries include microwave, radio frequency, and ohmic heating. Common to these volumetric heating methods is the fact that they do not require hot surfaces, or rely on conductive or convective heat transfer through a product.

Innovation and Future Trends in Food Manufacturing and Supply Chain Technologies
http://dx.doi.org/10.1016/B978-1-78242-447-5.00008-3

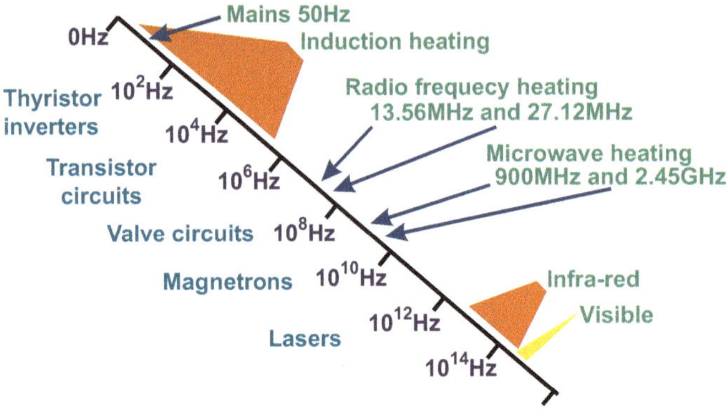

Figure 8.1 The electromagnetic spectrum.

Ohmic heating is a proven technology for processing high-quality fluid food products. It can be used to pasteurize, sterilize, or cook foodstuffs ranging from juice and dairy products to soups and stews. With ohmic heating, as for all volumetric heating techniques, the absence of hot surfaces in contact with the product leads to reduced thermal degradation and thus improved product quality. It is also capable of extremely rapid heating and is a very efficient heating technique. This makes ohmic heaters a good choice in areas where traditional heat exchangers can cause problems, such as fouling and product quality reduction from overheating, and where products are difficult to heat because of a large solid content or the product contains delicate particulates. Ohmic heating involves passing an alternating current (AC) (typically 50 Hz to 100 kHz) through the product being processed. Due to the electrical resistance of the product, this current flow then causes "ohmic" heating throughout, giving a uniform temperature profile. Any conductive and pumpable product can be successfully heated including most food products. The rapid and uniform heating made possible with the technology lends itself to high temperature short time processes with potential benefits in terms of product taste, color, and texture.

Radio frequency and microwave heating (collectively known as dielectric heating techniques) are widely used in the food industry, for both fluid and solid products, because of the volumetric heating that they provide, which often results in improved product quality and efficiency improvements due to reduced processing times and improved heating uniformity. Dielectric heating is applicable to materials which are very poor electrical conductors and so-called "lossy" dielectrics. The presence of dissolved salts and the dipolar nature of water molecules mean that water is well suited to dielectric heating. The presence of water in most foodstuffs enables them to be readily heated. Consequently, direct volumetric heating at large power densities is possible which enables rapid preheating, cooking, puffing, and selective drying to be achieved.

Radio frequency and microwave heating use different frequencies on the electromagnetic spectrum, in the MHz region. Frequencies below 300 MHz are normally referred to as RF and above 300 MHz as microwave.

The lower frequency and longer wavelength of RF means that RF energy penetrates deeper into materials, heating throughout the product. By comparison, in most high moisture content foods, microwaves typically only penetrate a few centimetres or so. In this way radio frequency gives true "volumetric" heating, whereas microwave heating is often considered quasivolumetric. Additionally, the lower frequency and longer wavelength of radio frequency compared with microwave heating generally makes it easier to manage electromagnetic emissions and to engineer product infeed and outfeed. Microwave heating, however, is generally cheaper and uses more general purpose applicators. Radio frequency and microwave heating are currently used widely in industry sometimes as the sole source of heat but often in combination with other heat sources, such as IR or hot air or steam.

The three techniques are discussed in more detail below, some basic theory and characteristics explained, and common methods of application described. Together this gives some insight into where they can be applied.

8.2 Ohmic Heating

8.2.1 Overview of Ohmic and Conventional Heating

Ohmic heating (also known as a Joule heating, and electric resistance heating) is an electrical heating technology in which heat is internally generated within the material being processed due to its natural electrical resistance. Heat is produced directly within the fluid as an electric current passes through it (see Figure 8.2), rather than being transferred into the fluid from a hot surface. This has a number of benefits in comparison to conventional heat exchangers for "difficult" fluids such as slurries, highly viscous materials, liquids containing large or delicate particles, and materials susceptible to damage from hot surfaces.

The most commonly used heating techniques for liquids and slurries rely on conductive and convective heat transfer from a hot surface. This heat can be provided by a hot medium (typically steam) via a heat exchanger (e.g., shell and tube or plate type) or a vessel with a heated jacket. This heating process requires a temperature gradient to transfer heat to the process liquid so the heat transfer surfaces need to be at

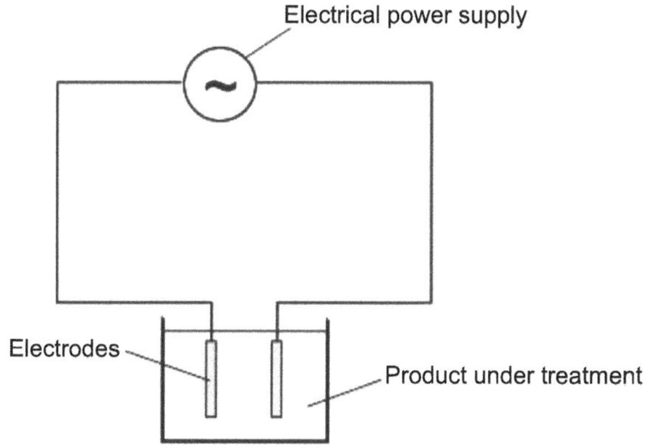

Figure 8.2 Basic ohmic heating.

a significantly higher temperature than the product. This can cause certain products to become thermally degraded or burnt onto the hot surfaces adversely affecting product quality. This fouling of the surfaces for certain products leads to reduced heat transfer rates, poor heating efficiency, and a requirement for frequent cleaning. An additional limitation of conductive heat transfer processes when heating very viscous fluids and fluids with particulates is that heat transfer can be very limited, and thus the heating process may be slow and can result in nonuniform temperature profiles. In these cases a significant level of mechanical agitation can be required to obtain the required level of heat transfer and uniformity. This agitation can often badly damage the physical structure of the product.

As ohmic heating causes heat to be generated in the product directly, with no heated surfaces, the problems highlighted above are mostly avoided. Thermal degradation on hot surfaces does not occur leading to improved product quality (color, flavor, nutrient retention), and as ohmic heating is not dependent on turbulence effects, there is no loss in performance for low flow rates or high viscosity fluids. Particulates in the product stream will also be heated by the electric field resulting in much improved uniformity of heating requiring no mechanical shear which can cause product damage.

Typically, any product that can be pumped and which has a suitable electrical conductivity (of between 0.01 and 10 S/m) can be successfully processed using ohmic heating. As most pumpable foodstuffs will have a significant moisture content and usually therefore a suitable electrical conductivity, a wide range of foods can be processed by ohmic heating. Typical electrical conductivities of a variety of foods are discussed in Section 8.2.4. Ohmic heaters are very efficient with over 95% of the applied electrical power converted to useful heat in the product (in 50 Hz systems). The wholly electrical nature of the technology also means that the process is highly controllable with the capacity for rapid start-up. However, as the technology is an electrical heating technique the cost of electrical energy must be considered when contemplating ohmic heating.

Ohmic heating is not a new technology (Jones, 1897); it was developed in the early years of the twentieth century for the pasteurization of milk (Anderson and Finkelstein, 1919). However, the "electropure process" as it was then known was later discontinued, largely because of a lack of suitable electrode materials. Work on ohmic heating recommenced in the 1980s when work at the United Kingdom Electricity Council Research Center led to the development of a viable ohmic heating process targeted initially at the sterilization of foods containing large particulates (Simpson, 1983; Simpson, 1994; Stirling and Coombes, 1990). Since its industrial development in the 1980s, ohmic heating has been installed in many sites worldwide and has been shown to offer improved food quality, in terms of color, vitamin retention, particulate identity of foods, and to provide a longer shelf-life for pasteurized products compared with hot surface heaters (Castro et al., 2004; Leizerson and Shimoni, 2005b; Anderson, 2008).

8.2.2 Food Quality Using Ohmic Heating

A major driver for the use of ohmic heating in food processing is the potential quality benefits that can arise from its use. Typical areas of food quality improvement quoted in the literature include improved color, texture/structure of particulates, flavors, and nutrient retention. Rahman (1999) reported that ohmic heating produced a high quality product with minimal structural, nutritional, or organoleptic changes, manufactured in a short operating time. Similarly, Avasoo and Johansson (2011) showed that the rapid, uniform heating and lack of hot surfaces in ohmic heating enabled the production of a strawberry jam with better preserved color, strawberry flavor, and vitamin C content compared with a microwave preservation technique and equal to the reference jam quality.

Work by Leizerson and Shimoni (2005b), in tests where orange juice was heated to pasteurization temperatures in very short times (of the order of about 1 s), examined the effect of ohmic heating on the shelf-life of orange juice compared with conventional pasteurization. They examined levels of ascorbic acid (vitamin C) concentration, pectin esterase (PE) activity (cloudiness), color, and five representative flavor compounds (decanal, octanal, limonene, pinene, and myrcene) and related these to the quality of the orange juice. They found no significant difference in the levels of vitamin C concentration or PE activity between the two heating methods, though the ohmically treated juice appeared less cloudy. Both ohmic heating and conventional heat treatment had the same effect on microbial load for 105 days, but sensory testing indicated that ohmically heated samples retained their flavor for twice as long as the conventionally heated juice at >100 days, testing showing that the levels of the flavor compounds examined were significantly higher in the ohmic treated samples after storage than in the pasteurized samples (Leizerson and Shimoni, 2005a). Panellists could not distinguish between fresh and ohmically heated samples, since they exhibited similar levels of flavor compounds.

Other work by Vikram et al. (2005) looked at the kinetics of vitamin degradation and changes in color as measure of carotenoids. The degradation kinetics of vitamin C and color are discussed and of the four methods studied, ohmic heating gave the best result and better vitamin retention at all temperatures. Anderson (2008) also reports improved flavor retention when using ohmic heating, suggesting that this is because of the shorter residence times which can be achieved in ohmic heating systems, as well as the lack of hot surfaces. The effect is explained by the fact that "the rate of increase in heat deactivation of microorganisms with increase in temperature is greater than the rate of increase of cooking reactions and vitamin destruction. To minimize cooking reactions it is therefore beneficial to heat the food at the highest possible rate to the highest possible sterilization temperature" (Skudder, 1993).

In other work, as reviewed by Christian and Leadley (2006) products such as fruit juices and concentrates, shelf-stable milk, puddings, soups, and liquid egg products were studied. Heating was shown to be rapid, uniform, and was seen to have a reduced impact on the organoleptic properties of the products. Similarly, other researchers have examined six "stew" recipes, which were sterilized using ohmic heating and analyzed after 3 years of storage. The color, appearance, flavor, texture, and overall food quality were described as excellent, "indicating that ohmic heating technology has the potential to provide shelf-stable foods closely equivalent to those prepared from scratch" (Yang et al., 1997). For another product containing beef and carrots in gravy, the experimentation "clearly illustrated that the particulates were heated directly by

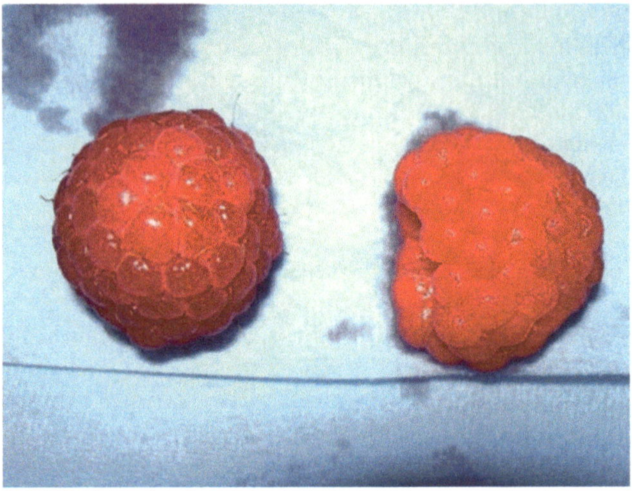

Figure 8.3 Ohmic processed raspberry (right) compared with unprocessed raspberry, using a batch process.
C-Tech Innovation.

electrical resistance, to a level similar to that of the liquid phase." It could be seen that "there was not a large difference in the residence-time distribution of the particulates throughout the ohmic system" (Parrott, 1992).

Another major quality benefit that is commonly reported is improved product and particulate identity due to the rapid volumetric and the low shear nature of the process. Products with berries can be continuously processed with excellent berry appearance and identity (see Figure 8.3). When used in the processing of meat products several researchers Ozkan et al. (2004) have shown that the quality and mechanical properties of hamburgers cooked with a combination of conventional and ohmic heating were not different from hamburgers cooked with conventional heating.

8.2.3 Microbial Death Kinetics for Ohmic Heating

A major concern when considering the use of ohmic heating is the effectiveness of the technique in achieving the required levels of microbial kill in all parts of the food product, particularly as ohmic heating is currently primarily used as a pasteurization/sterilization technique. Ohmic heating is fundamentally a thermal processing technique, and work has been carried out to confirm that ohmic heating does indeed perform at least as well as conventional heating processes. Palaniappan et al. (1992) studied the effect of ohmic heating on microbial death kinetics and compared the death kinetics of yeast cells under time–temperature histories as identical as possible, and found no difference between conventional and ohmic heating. The Food and Drug Administration (FDA) published a comprehensive review of microbial death kinetics data regarding ohmic heating (USFDA, 2001).

Another main area of work on microbial death kinetics in ohmic heating has been in identifying whether microbial kill is affected by electric field effects or solely by thermal treatment. Ohmic heating has generally been considered to be a thermal process, but nonthermal effects might also operate, due to electric field causing additionally effects such as electroporation or similar damage to cell wall membranes. Work in this area has indicated that the log reduction times of *Bacillus subtilis* spores were significantly reduced when using ohmic heating compared with other thermal processing at identical temperatures (Cho et al., 1999). These researchers used a two-step treatment process involving ohmic heating followed by holding and heat treatment, which, they reported, accelerated microbial death kinetics; they suggested that electroporation may positively influence microbial death kinetics. Work by other researchers also suggests that a mild electroporation type mechanism may operate during ohmic heating (Ruan et al., 2001). These athermal effects are cited as being more pronounced at low frequencies.

So, at the present time, it appears that considering microbial death as only a function of temperature may result in a conservative design assumption.

8.2.4 Ohmic Heater Design

Ohmic heating is normally implemented as a continuous process, although batch processing is a possibility. In designing a continuous flow ohmic heater it is necessary to consider both fluid flow and electric field characteristics. A uniform electric field is desirable to give a uniform heating profile in the fluid; however, temperature uniformity will only be achieved if the flow characteristics of the fluid are also appropriate (plug flow). Where plug flow is not attainable, the ohmic heating effect will still lead to a good level of temperature uniformity compared with surface heating techniques, however, some mixing may be required to ensure a wholly uniform temperature profile or it may sometimes be appropriate to fit a short pipe section after the ohmic heater to allow temperatures in the product to equilibrate.

The power density "p" (Watt/m^3) generated in an ohmic heater is dependent upon the electrical conductivity of the fluid and the electric field strength applied across it.

$$p = E^2\sigma \tag{8.1}$$

p, power density (W/m^3)
E, electric field strength (V/m)
σ, electrical conductivity (S/m)

Critical parameters in the design of an ohmic heater are the electrical conductivity of the fluid, the surface area and separation of the electrodes, the cross-sectional area and the applied voltage. The heating power in Watts in an ohmic heater is

$$P = V^2A\sigma/s \tag{8.2}$$

P, power (W)
V, applied voltage (V)
A, area through which current flows (m^2)
s, electrode separation (m)

The mean conductivity will be determined by the product and the product temperature rise because conductivity increases with temperature, although it may be possible to adjust the conductivity by formulation or ingredient manipulation. Voltage will be limited by electrical breakdown considerations of insulation materials and within the heater and product and would generally not exceed 5000V. Another critical factor in the design of ohmic heaters is the electrical current density A/m^2 at the surface of the electrodes. An excessive current density can lead to the occurrence of electrochemical effects, which are undesirable in the context of food heating and must therefore be avoided. The threshold value for current density depends upon the fluid being processed and the material of construction of the electrodes. Additionally the frequency of the applied electric field has a significant effect on the current density that can be employed with a given electrode material. To avoid electrochemical effects the electrical supply should be an AC supply, with the current transmitted to the product using electrodes made from a suitable material selected to avoid unwanted electrochemical effects. Two main approaches are typically used: the use of standard mains frequency (50/60Hz) with electrodes constructed from dimensionally stable anode materials such as titanium with a noble metal or inert metal oxide coating, or carbon-based materials such as graphite or diamond (Stirling, 1992; Stirling, 1993). An alternative approach is to use a higher frequency electricity supply; this can permit the use of a wider range of materials, including the most common choice of stainless steel. The use of a high frequency AC causes any electrochemical reactions that may occur at the surface of the electrode material to be rapidly reversed before the species formed diffuse away, thus the formation of any electrochemical reaction products is minimized.

Another material selection issue in the design of ohmic heaters is that of the material of construction of the main body of the ohmic heater. The current passing to the fluid from the electrodes must pass through the fluid itself and not through the body of the ohmic heater, thus the material of construction of the main body must be chosen to be electrically nonconducting. The body must also be designed to be capable of withstanding the process temperature and pressure conditions. Typical materials of construction for ohmic heater bodies include polymers (such as polypropylene) or sometimes glass, or alternatively metal pipe sections that are internally coated or lined with nonconducting polymer materials, e.g., polytetrafluoroethylene, perfluoroalkoxy, tetrafluoroethylene-perfluoropropylene are used.

8.2.4.1 Product Properties—Electrical Conductivity

As discussed previously, a critical parameter in the design and operation of ohmic heaters is the electrical conductivity of the food being processed. The electrical conductivity of a food is made up of a number of components, some of which increase electrical conductivity (such as dissolved salts, acids, and moisture content and mobility), and some which tend to decrease conductivity (such as fats and sugars). Electrical conductivity increases with temperature, thus the variation of this product parameter throughput the whole process temperature range needs to be understood. In the absence of actual data an estimate of conductivity throughout the temperature range can be made using a linear relationship, if the conductivity at room temperature is known. Figure 8.4 presents such data for skimmed milk. Typically, the electrical conductivity

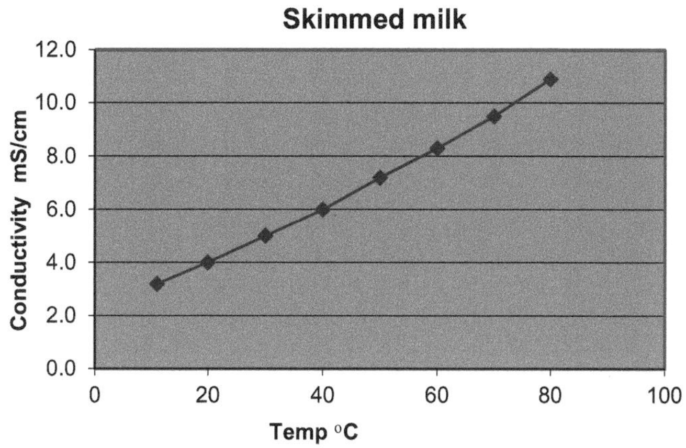

Figure 8.4 Typical conductivity variation with temperature.

Table 8.1 Conductivity Increase Cooking Vegetables with Ohmic Heating

	Conductivity Increase in Cooking (S/m)	
	Raw (S/m)	**Cooked (S/m)**
Carrot	0.033	0.763
Potato	0.08	0.623
Yam	0.034	0.393

of an aqueous food product will rise by 2–3% per °C, or expressed another way, will double or treble over a temperature rise of 50–70 °C.

Some processes, e.g., cooking, which involves cellular breakdown, can result, however, in higher conductivity rises. When using ohmic heating to cook vegetables, the conductivity rises by an order of magnitude during the cooking process (see Table 8.1).

Some data for the electrical conductivity of foodstuffs are given in Table 8.2.

In multiphase mixtures the electrical conductivity of both the solid and liquid phases are important. If liquid and solid phases possess similar electrical conductivities they will generally heat at approximately the same rate. When there are differences in the electrical conductivity of the particulates and the carrier fluid, heating will be less uniform but heating of the particulates will occur to some extent, leading generally to better temperature uniformity than would have been achieved in conventional heat transfer equipment. It is possible for the particulates to be heated at either a faster or slower rate depending upon the relative conductivities of the two phases and also upon the packing density and shape of the particles. Indeed for some mixtures, the relative conductivity of the particulates and carrier fluid are such that the particulates are heated more rapidly than the fluid leading to the possibility of reduced processing times/rapid cooking.

Table 8.2 **Electrical Conductivity of Foods**

Food Products Electrical Conductivity	Electrical Conductivity (S/m) at 25 °C		Electrical Conductivity (S/m) at 25 °C
Vegetables		**Juices**	
Vegetable pieces	0.1–0.06	Tomato juice	0.863–1.5
Fruit pieces	0.05–0.15	Orange juice	0.567
Potato	0.04	Carrot juice	1.1
Carrot	0.025–0.042	Apple juice	0.25
Onion	0.022	Cranberry juice	0.1
Mushrooms	0.022	Orange juice	0.35–0.5
Cucumber	0.027	Pineapple juice	0.4
Turnip	0.026		
Parsnip	0.025		
Brussels sprouts	0.009	**Others**	
Tomato	0.045	Pickles and chutney	2.0–3.0
Leeks	0.007–0.032	Savoury sauces	1.6–1.8
Celery	0.026	Soups	1.4–1.8
Courgettes	0.017–0.032	Pet food	0.7–1.2
Pepper	0.048	Full-cream milk	0.52
Pear	0.041	Dairy desserts (vanilla/custard)	0.38–0.50
Apple	0.023	Beaten egg	0.4
Pea	0.17	Margarine	0.027
Tomato, chopped	0.4–0.5	Sugar syrup	0.001
Fruit in syrup	0.2	Milk	0.41
		Potable water	0.01–0.1
Meat		Beer	0.15
Chicken	0.08–0.37	Liquid egg	0.09
Pork	0.087–0.76	Jams and compotes	0.0005–0.05
Beef	0.42–0.44	Custard	0.25
Minced beef, lamb	0.8–1.2	Honey	0.02
Heterogeneous meat products	0.6–3.0		

Adapted from various sources including Mitchell and de Alwis (1989) and Palaniappan and Sastry (1991).

Much modeling work has been carried out assessing heating of multicomponent products and tailoring/modifying the conductivity of different components (Zhu et al., 2010; Benabderrahmane and Pain, 2000). Ohmic heating is generally seen to be advantageous for such multicomponent products.

8.2.4.2 Continuous Flow Ohmic Heating

Most commonly, in industrial applications ohmic heaters are employed in continuous flow mode. Simple open flow channel geometries that prevent damage to products, reduce the effects of fouling, and make cleaning easier are most common (Simpson, 1994). Heating can be very fast with large and uniform temperature rise occurring in a short heating zone. In a flow through unit a number of electrode arrangements are possible, chosen to suit the required operation (see Figure 8.5).

Alternative electrode arrangements can also be configured exemplified by the schematics in Figure 8.6. Each will have their own heating characteristics that need to be studied to assess suitability for a given process.

From Eqn (8.2) it can be seen that heating power is determined by the geometry of the heater, the conductivity of the product and the applied voltage. Limits are imposed for current density, which further dictates the form and design of heaters. Furthermore, residence/heating time and volume within a heater are also geometry and power related. As the electrical conductivity of the fluid increases with temperature it is often practised that different applied voltage or increased electrode separations are implemented along the ohmic heater to balance power or current density. For colinear and staggered rod designs, separation can easily be altered by using longer tube lengths between each successive electrode module. In parallel plate designs the same effect can be achieved by using several parallel plate modules in series increasing the spacing of the plates or applying different voltages on different plates.

Ohmic units are commonly mounted vertically or on a vertical incline, with product flow from bottom to top to exclude gas from the system as the presence of large gas bubbles can lead to electrical arcing. Heat generated in the product will be passed from the product to the electrodes by thermal conduction; in some products that are very heat sensitive this can cause the electrodes to reach a temperature where thermal degradation effects may occur (e.g., liquid egg). In such cases the electrodes can have a cooling system built into them to ensure they remain below the temperature of the product (Simpson, 1983).

Ohmic heating is an accepted process which lends itself to combination with aseptic filling. A typical ohmic heater aseptic filling process (see Figure 8.7) will involve a high quality pump (most commonly of positive displacement type), capable of providing a consistent flow rate without any significant flow pulsations and without causing any unwanted product damage. Pump types in use are twin-piston or double acting piston or Bornemann twin screw, or rotary piston/lobe. A suitable flow meter can facilitate improved control and feed forward control methods. After the ohmic heater, a "holding pipe" of suitable length will generally be required to ensure the product is held at the appropriate temperature for the appropriate time. It will usually be preferable to cool the product as rapidly as possible in order to minimize any ongoing temperature related damage to the food and thus maximize any benefits due to the rapid heating. This cooling duty will typically be provided by a tube in-tube heat exchanger unit. Following cooling an aseptic receiver tank is usually employed before the product is passed to the aseptic filler or other suitable packaging unit.

A limiting factor to be considered in the design/installation of ohmic heaters is the availability of a suitable electricity supply. Nevertheless, large installations sterilizing 7 t/h of diced fruit for aseptic filling operate at powers of 500 kW. A large fruit juice line operating at 380 l/min might have a pasteurizing thermal requirement of 2 MW, half of which might come from ohmic heating. Heat recovery and energy integration is important to minimize the running costs associated with the electrical energy requirements.

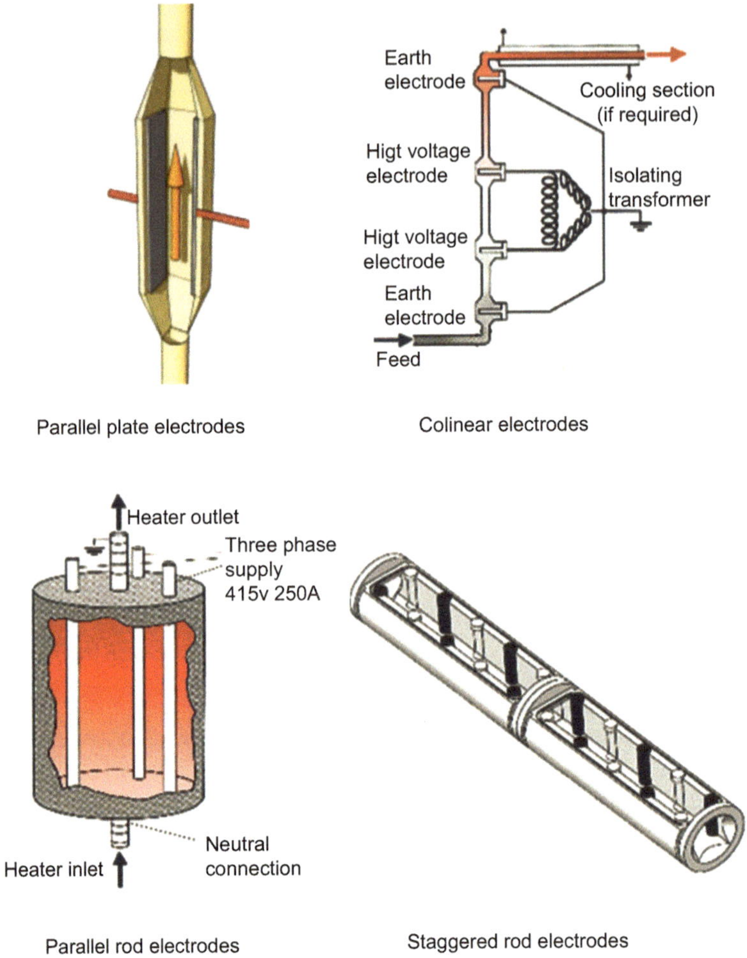

Figure 8.5 Some basic electrode arrangements in ohmic heaters.

- The parallel plate arrangement is most suitable for low conductivity fluids (<0.25 S/m) and also offers a benefit where there are large solids due to the completely unrestricted flow channel. The electric field uniformity is optimized in this geometry, giving very even heating. The design can usually operate at standard voltages, e.g., 240 or 480 V (Simpson, 1983).
- The colinear design is a better option for medium to high conductivity fluids as wider electrode separations are possible. The electrodes can be positioned in the fluid stream as collars/rings around a pipe (which provide a fully unrestricted flow channel) or bullet-type insertions or can utilize pipe sections or pipe bends themselves as electrodes. For most applications this design requires a higher voltage than the parallel plate (1–5 kV). Current distribution is important: if the electrodes are not designed correctly, areas of high current density can occur at the leading edges of the electrodes which can produce localized boiling and arcing if not designed correctly (Stirling and Coombes, 1990).
- The parallel rod design, often three rods on a three phase supply, has been used as a low cost economic heater arrangement. The design is less expensive to construct than parallel plates or colinear designs, but provides less even heating of the medium. As a result the fluid must often be mixed after heating to even out the temperature, reducing its suitability for heating solids without causing damage to them and making this arrangement less favored where precise temperatures are required (Simpson and Stirling, 1995).
- The staggered rod arrangement is an alternative low cost option, but can provide more even heating than the parallel rod design.

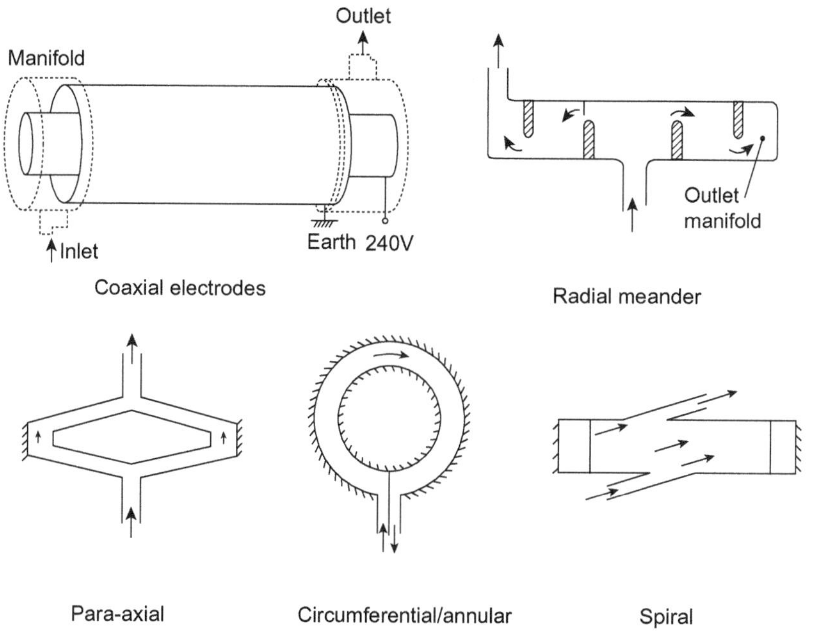

Figure 8.6 Varied ohmic heater electrode arrangements are possible.

Often preheating will form part of the process as in blanching with diced fruit so the product may only require ohmic heating from, e.g., 50 °C to 90 °C for sterilizing such products.

8.2.4.3 Batch Ohmic Heaters

Although continuous flow systems are the most common ohmic heater arrangement, batch ohmic heater designs are possible, with a number of electrodes mounted within a suitable vessel. Batch systems may be preferred for applications requiring long holding times such as evaporation systems or where cooking of a product is occurring. Another common use of batch ohmic heaters is in experimental units for use in the laboratory (see Figure 8.8). Generally batch ohmic heaters will require some level of mixing to avoid any thermal stratification that may occur.

8.2.5 Control of Ohmic Heating

The delivered power is the principal aspect to be controlled in ohmic heating as this determines the product heating rate and the product temperature reached (Power = Volts × Amps or Volts^2/Resistance). Typically, an accurate and tightly controlled final product temperature is required to achieve pasteurizing or sterilizing while maintaining the benefit of minimal processing and avoiding overheating. Ohmic heating can do this very well. Power control is achieved mostly by voltage control. Some systems use flow control as a means to control product temperature; however, this does not make the best use of the inherent controllability of the heating method.

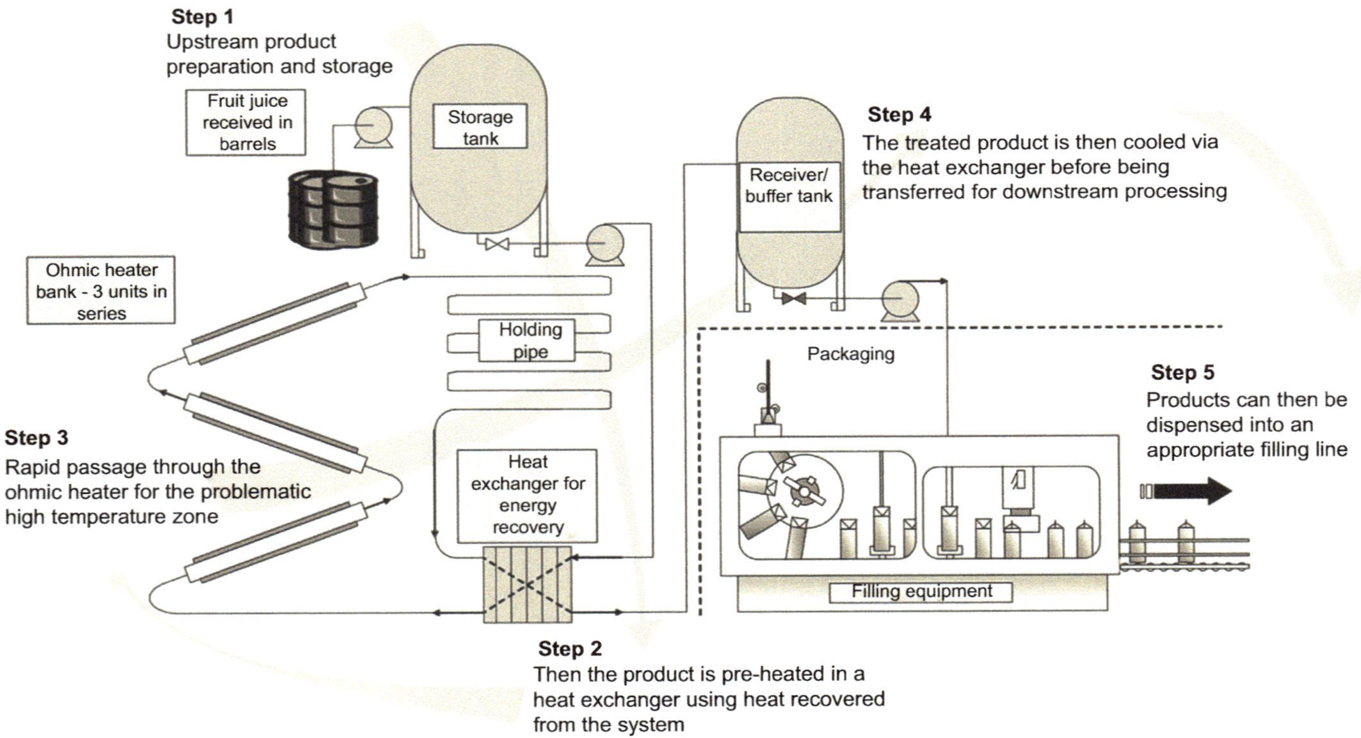

Figure 8.7 Example configuration for an ohmic heater aseptic filling process.

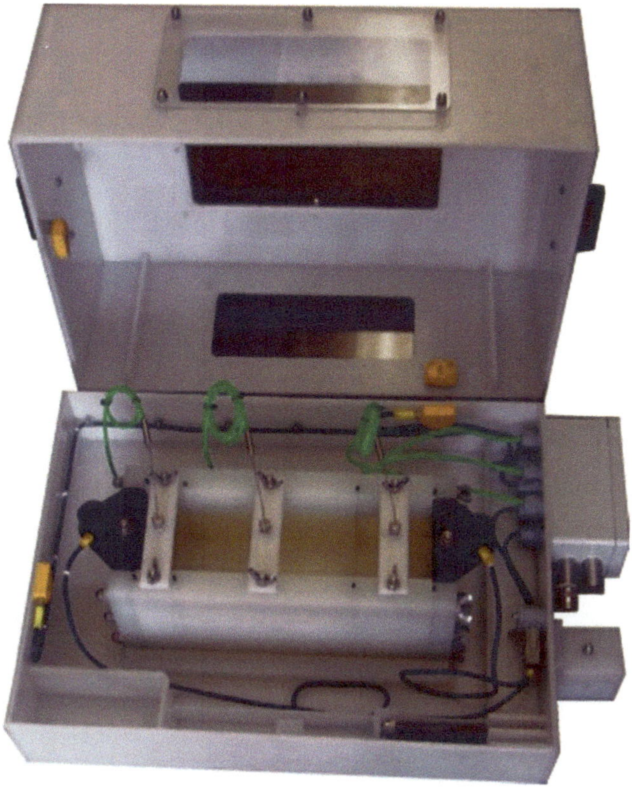

Figure 8.8 Laboratory-scale batch ohmic heater unit.
C-Tech Innovation.

It is important that the electric current drawn by the system is controlled at below the rated current of the electricity supply and that the heater operates within design limits for voltage and current density. The current drawn by any given design of ohmic heater depends upon the applied voltage and the resistance of the fluid (Ohm's law, I=V/R). As fluid conductivity with product temperature, the ohmic heater must be designed to compensate for this variation otherwise an excessive current may be drawn.

Control of the ohmic heating for systems at normal mains frequency (50/60 Hz) is usually by thyristor voltage control. Control of the power drawn can be performed by the thyristor using either burst fire or phase angle control with phase angle control preferred to give good control of the current since phase angle gives more effective control of the voltage. The thyristor output voltage can input to the primary of a step-up transformer where higher electrode voltages are needed for colinear heater arrangements. Multitapped transformer secondaries with different winding taps can be specified to cater for a wide voltage range to cover a wide product conductivity range. Auto tap changers can be specified. For high frequency ohmic designs, control can be undertaken in a similar way by the use of inverter technology to generate the high frequency supply. These rectify the mains voltage to a high direct current (DC) voltage,

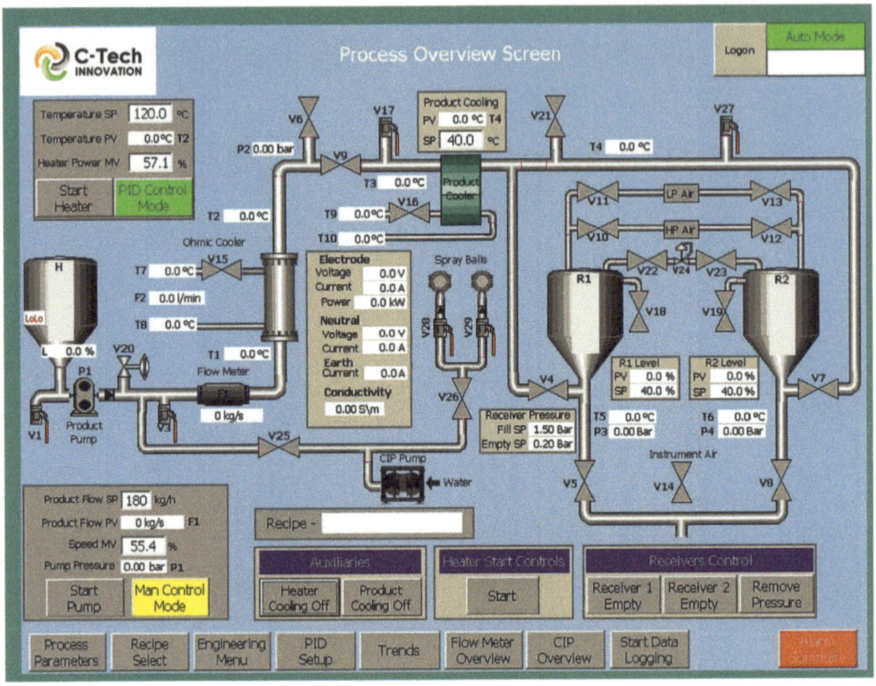

Figure 8.9 PLC/human machine interface (HMI) control schematic for an ohmic heating process. C-Tech Innovation.

which is then switched by a bridge of four transistors in a stepped AC waveform. The width of the on pulse is used to alter the on/off ratio; this is used to control the power to the heater, similar to phase angle control with a thyristor package.

The product temperature from the ohmic unit can be controlled by the thyristor unit and a proportional integral derivative control loop measuring the output temperature of the fluid from the heater. Alternatively, the power required for a given process can be calculated (using $Mc_p\Delta T$ where M is the mass flow rate, c_p the specific heat capacity, and ΔT the temperature increase required for the product being processed), and this power can then be reliably input by a feed forward control system. Because the voltage and current are easily measured and precisely known, the power is precisely known (power = V × I, volts × current) and therefore can be accurately controlled. This feed forward control is very effective at start-up and can result in no temperature overshoot and allows very accurate outfeed temperature control based on accurate infeed temperature and flow measurement.

Ohmic heating lends itself to fully automated programmable logic controller (PLC) control. Figure 8.9 shows a PLC controlled ohmic heating process designed for the continuous sterilizing and cooling of particulate foods with controlled pressurized receivers to maintain pressure for a wide range of products and viscosities.

8.2.6 Advantages and Disadvantages of Ohmic Heating

Ohmic heating offers a number of process advantages dependent upon the application. These advantages are typically due to the lack of hot surfaces (reduction in thermal degradation and associated fouling, reduced discolourization, and caramelization), as well as the rapid, uniform heating caused by volumetric heating. The fast response/controllability of the technology allows rapid start-up, and good turndown ratios.

Initially developed in the 1980 for the aseptic processing of fluids containing particulates, ohmic heating is particularly well suited to the heating of such fluids, allowing the particulates to heat at a similar rate to the carrier fluid without the requirement for high levels of agitation/shear which can cause product damage. To produce shelf-stable foods it is necessary to heat all parts of the food to the required temperature for the appropriate time. Where thermal processing is undertaken in containers (such as tins) the quality of the food is reduced by the need for the food near the container wall to be excessively heated in order to ensure the food at the center of the container is heated sufficiently in an appropriate time. Where large particulates are required to be processed, conductive heat transfer through the particulates can prove to be a limiting factor, necessitating the overprocessing of the liquid phase in order to ensure that the center of each particle is sterilized. Ohmic heating overcomes these heat transfer problems as the particulates and carrier liquid are all heated by the electric field.

As the heat is generated directly in the fluid being treated and the heating is purely resistive heating the efficiency of the process is very high (typically>95% for mains frequency equipment) the only losses in the system occurring in the cables and control system. For higher frequency systems the efficiency is slightly lower due to losses in the generation of the high frequency supply; however, heating efficiency is still high.

Another processing advantage exists for applications requiring high process temperatures, where the use of ohmic heating can give high efficiency and high quality heating, in a situation where the use of steam as the heat transfer medium would be prohibitively expensive (due to the high pressures required to handle high temperature steam). For steam at 200 °C, which would typically be required to heat a product stream to 180 °C (in applications such as the production of caramel flavourings (Ng et al., 2014), the steam supply pressure required would be in excess of 15 bara; "the saturation pressures of steam at temperatures above 200 °C mark this as the limiting value for general use" (Dove Thermal Engineering). High temperature processes designed to sterilize food waste prior to anaerobic digestion and to partially hydrolyze waste have been realized using ohmic heating at operating temperatures from 160 °C to 300 °C. Such processes prior to anerobic digestion plant can also operate with flash depressurization which explosively disrupts the particulate matter yielding increased gas production due to partial hydrolysis and the increased availability of nutrients for the digestion process. There is no inherent limit to the operating temperature for a given fluid provided the temperatures, pressures, and mechanical issues such as thermal expansion, sealing, material compatibility, pressure containment are properly designed and dealt with.

The advantages of ohmic heating can be summarized as:

- Rapid heating of liquid and particulates with even temperature distribution. Required processing temperature can be rapidly achieved with short residence times in the heater. This leads to increased flavor and vitamins compared with traditional heat exchangers.
- No hot surfaces, resulting in a low risk of product damage from burning or overprocessing.
- Handles high solids/particulate content. Low shear maintains particulate shape and appearance, e.g., berries. Particulate foods up to 30 mm are suitable for ohmic heating.
- High efficiency—95% of input energy transferred to the product.

- Highly controllable—good start-up, shut-down performance. Can be easily integrated with existing control systems.
- Robust/high resistance to fouling/low maintenance—automatic cleaning using standard clean in place operations effective due to minimal fouling of surfaces. Heaters can incorporate cooled surfaces for further mitigation against fouling with difficult products.
- Good for high viscosity products.
- Good for high temperature applications, e.g., working temperatures > 150 °C.
- Low shear and low pressure drop through heaters.
- Can avoid the infrastructure and inefficiencies associated with steam heating.

Ohmic heating is subject to a number of limitations which can be summarized as:

- It is only feasible for products with a suitable electrical conductivity.
- Predominantly used for those products which are pumpable and can thus ensure sufficiently good current transfer to and from electrodes.
- Units are designed to suit a specific product conductivity and design becomes more difficult for a range of products and process conditions.
- The electric nature of the heating effect means that sites considering ohmic heating will need a suitable electrical supply, and the cost of electrical energy compared with other fuels must be taken into account when considering ohmic heating.

8.2.7 Applications of Ohmic Heating

Ohmic heating can be utilized in a range of applications and industries, from chemical processing to waste treatment, however, it was initially developed for the pasteurization/sterilization of food products, and this remains the major area of application for the technology. Ohmic heating can be applied to a wide variety of foods, including liquids, solids, and fluid–solid mixtures/slurries. Common foods which use ohmic heaters are tomato products, fruit preparations, fruit juice, liquid egg, jams, compotes, soups, casseroles, and ragouts. As long as the food product can be pumped and has an electrical conductivity within a suitable range it will generally be suitable for ohmic heating. Fluids with a conductivity of between 0.1 and 2.0 S/m can be easily processed by ohmic heating; this range covers most food products. Ohmic heaters can be designed for fluids with conductivities outside this range, 0.01 to 10 S/m, but certain products such as honey and those with an electrical conductivity lower than 0.01 S/m would not be suitable.

Ohmic heating is often selected for applications which prove problematic for normal heating methods. Typically, fluids with high solids content, or which have a tendency to burn on to hot surfaces, require the use of specialist equipment such as scraped surface heat exchangers which can result in a lower quality product through mechanical damage. By direct heating of the fluid with no hot surfaces and minimal obstacles to solids movements ohmic heaters can be more effective.

Ohmic heating has been used commercially to produce liquid egg product in the United States (Mans and Swientek, 1993). The production of liquid egg poses severe processing problems due to the very temperature sensitive nature of the product and the changes in its rheology that occur as it is thermally processed. Ohmic heating has proved to be an effective technique for processing this difficult product. Papetti's Hygrade Egg Products of Elizabeth, New Jersey, USA, installed a 200 kHz system to process between 5 and 10 t/h of liquid egg heating them to 60–64 °C and holding for 3.5 min. After processing, the eggs were shown to have a 12-week shelf-life, though some cartons regularly surpassed this shelf-life.

Newly Wed Foods, a company in the United Kingdom, produces Japanese style crumb using an ohmic baking process. The dough is baked using only ohmic heating giving a crustless loaf.

Ohmic heating has also been applied to the cooking of meat products experimentally both in-batch and continuous form. This has shown very effective in achieving rapid cooking compared with conventional retrort or oven or smokehouse cooking with potential energy requirements of 211–252 kJ/kg for ohmic compared with 1200–8100 kJ/kg for conventional processes with a cook time reduction of 82–97% (De Halleux et al., 2005). Lyng and McKenna (2007) shows a reduction in cook time for meat emulsions (12 °C to ~80 °C) from 85 min with steam cooking to less than 5 min with ohmic heating. Maintaining electrical contact and conduction is a challenge for this ohmic application but methods are being trialled and cooking methods combining an ohmic section with a dielectric heating RF section are in development for continuous cooking of meats.

Ohmic heating has potential in batch or continuous designs for fast cooking of vegetables and pulses. Demonstrated rapid cooking of a range of food products are shown in Figure 8.10. These fast cook times were achieved by heating products in a saline solution by ohmic heating with some ohmic current then maintained during a simmer

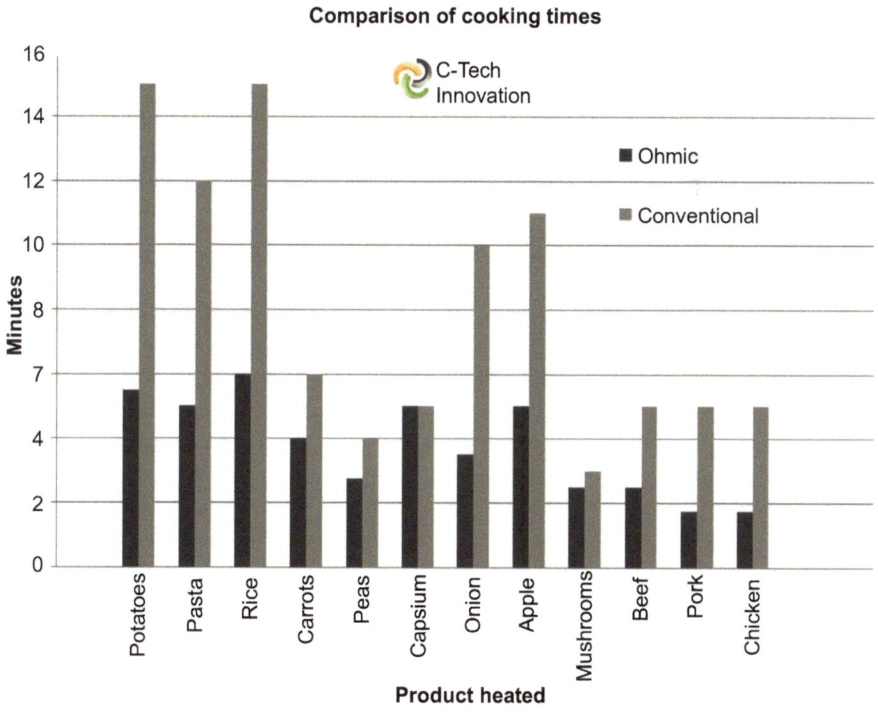

Figure 8.10 Faster ohmic cooking/simmer compared with conventional boil/simmer.

period. Pinto beans were also cooked in 20 min ohmically, compared with 45 min with conventional boiling. Ohmic blanching has also been investigated by a number of workers, and has been found to offer considerable benefits compared to conventional processing, particularly where large particulates are being processed (Eliot-Godereaux et al., 2001; Icier et al., 2006); however, its use at industrial scale is limited.

Ohmic heating has also been shown to enhance drying rates (Lima and Sastry, 1999; Zhong and Lima, 2003; Wang and Sastry, 2000) and extraction yields (Lima and Sastry, 1999; Zhong and Lima, 2003; Halden et al., 1990) in certain fruits and vegetables. Ohmic pretreatment gave increased juice yield and reduction in input work. Compared with the raw material the juices extracted from pretreated apple, samples were visually similar in quality and the yields were significantly greater (595 ml/kg from untreated compared to 653 ml/kg from ohmically processed).

High temperature processes that operate at temperatures above 160 °C are an emerging area of development using ohmic heating with application areas such as production of caramel color, waste food, and animal by-products sterilizing hydrolysis and pretreatment processes prior to anaerobic digestion.

Significant installers of ohmic heating and aseptic filling processes in recent years with over 30 installations are Asepsystems and Emmepiemme for the processing of tomato sauces, diced tomato and tomato pastes, fruit preparations and purees, diced fruit and vegetables, and whole strawberries, at powers ranging from 50 to 500 kW, with throughputs ranging from 400 to 7000 kg/h. Ohmic heating has shown itself to be a versatile and advantageous heating technique. Its application has not been as widespread as might be expected due, it is suggested, to the relatively few suppliers and the somewhat bespoke nature of the process design. With appropriate knowledge and good design and new players entering the supply chain ohmic heating applications are destined to grow significantly producing very high quality products particularly where particulates and enhanced taste and texture are desired.

8.3 Radio Frequency and Microwave Heating

Both RF and microwave heating are used in the food industry because of the "volumetric" heating that they provide, with heat being generated within the food itself. This often results in a better quality product with no overheating of the surface. The use of these technologies allows foods to be raised to the required processing temperatures in shorter times compared to conventional heating (which relies primarily on thermal conduction for heat transfer to the core of a material) very often with energy efficiency improvements.

8.3.1 Overview of the Physical Principles of RF and Microwave Heating

RF and microwave heating are collectively known as dielectric heating. There are two broad bands of frequencies used for dielectric heating. Those below 300 MHz tend to be called RF, while those above are known as microwaves (in essence they are all radio waves, microwaves merely being radio waves of very short wavelength).

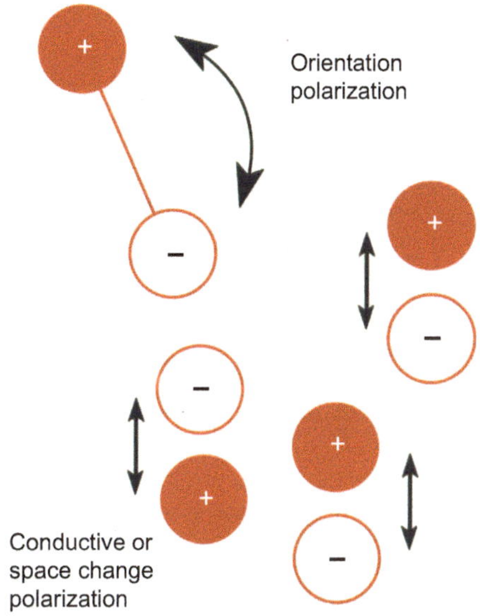

Figure 8.11 Microwave and RF dipolar and ionic/charge loss mechanisms.

Specific frequencies have been allocated for industrial, scientific, and medical use (ISM bands). The commonly used ISM bands for radio frequency are 13.56, 27.12, and 40.68 MHz and for microwave 896 MHz (UK)/915 MHz and 2.45 GHz and 5.8 GHz.

When a nonconducting (dielectric) material is placed in an alternating electric field at high frequency, energy will be absorbed and heat generated throughout the whole of the material due to a number of effects. Heating may be due to more than one of these mechanisms (see Figure 8.11).

- Rapid reversal—or polarization—of polarized components causes volumetric heating as this realignment of dipoles is resisted by frictional and elastic properties of the material.
- Movement of charge
  - **a.** Joule heating where electrical charges move from their mean position, with this heating mechanism dominated by electrical conductivity.
  - **b.** Interfacial space charge effects where again electrical charge migrate and moves within the material typically to and from discontinuities in the material.

Dielectric materials exhibit polarization because their molecular structure has strongly bound electrons, whereas conductive materials have free or loosely bound electrons. Microwave heating is more dominated by the polarization mechanism, whereas RF heating draws on the conductive charge movement mechanism as well. A more detailed explanation of these effects can be found in the literature (Metaxas and Meredith, 1993).

8.3.2 *Loss Factor and Loss Tangent*

A fundamental property of dielectric materials is the so-called complex relative permittivity, given by $\varepsilon_r = \varepsilon' - j\varepsilon''$, where ε'' is the imaginary component of permittivity attributed to charge and dipole relaxation phenomena and charge conduction phenomena, and ε' is the dielectric constant. The ε'' component of the complex permittivity is

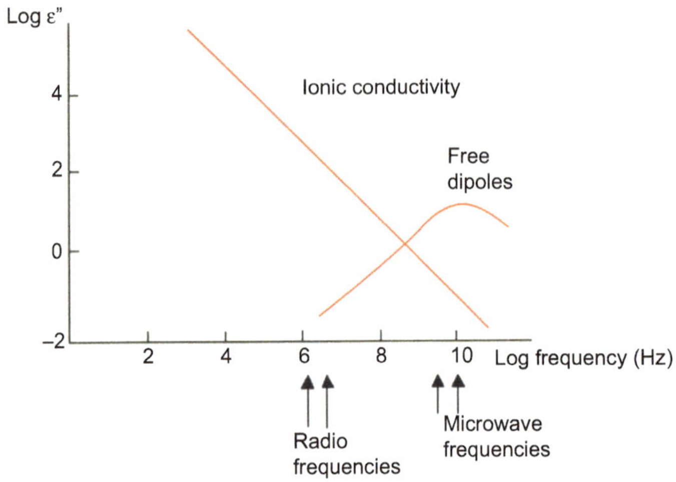

Figure 8.12 Dipole and ionic conduction loss versus frequency for water.

called the loss factor and describes the ease of heating a material: it is a measure of the ability of a material to absorb electromagnetic energy due to the heating mechanisms described above within the material.

Another commonly used term to describe the heatability of a material is its loss tangent. The loss tangent, tan δ, is measured as the tangent of the phase angle between the electric field generated in the material and the applied field. At the molecular level, it can be considered as an indication of the average "friction" effect contributed by each polarized component. Loss tangent and loss factor are related by:

$$\text{Loss factor } \varepsilon'' = \varepsilon' \tan\delta \tag{8.3}$$

The two components of loss factor vary with both the frequency of the applied electric field and the temperature of the material. Foodstuffs often consist of more than one substance, which means that heat will be generated in each substance according to its own loss factor. Water is a very common and high loss constituent within foods, and is a component that is often a dominant factor in heating. Figure 8.12 shows dipolar and conduction loss vs frequency for water. Some typical dielectric data for foods is given in Table 8.3.

8.3.3 Power Dissipation

The power dissipated in a material by dielectric heating is given by:

$$p = 2\pi f E_i^2 \varepsilon_0 \varepsilon'' \tag{8.4}$$

p, power (W/m^3)
f, frequency (Hz)
E_i, Electric field strength or voltage gradient within the material (V/m)
ε_0, the permittivity of free space (F/m)
ε'', product loss factor

Table 8.3 Dielectric Properties of Foods

	Dielectric Properties ε′/ε″ for Food Products		
	30 MHz	**1 GHz**	**2.5 GHz**
Fruit/vegetables			
Apple		61/9	57/12
Banana		64/20	57/19
Carrot		68/20	65/15
Onion		66/14	63/15
Orange		73/15	70/15
Peach		72/14	68/14
Pear		68/13	64/14
Potato		66/30	61/19
Strawberry		73/15	71/13
Green pea puree		70/14	
Carrot puree		77/19	
Salsa		58/34	
Meat/fish			
Beef (8.6% fat)		56/22	52/17
Beef 8.6% fat frozen -20C			5.9/1.0
Beefsteak -10C frozen	3.7/0.9	5.3/0.2	5.3/0.2
Fish, cod frozen	5.5/1.4	4.0/0.7	
Fish, cod	83/250	66/40	
Ham frozen			10/5
Ham cooked			45/24
Lard +82C		2.6/0.14	2.5/0.15
Others			
Bread dough		24/12	20/10
Bread		8/2	4.5/1.2
Butter frozen			3.9/3.9
Butter			4/4
Honey	37/3	14/7	
Oil (corn)		2.7/0.17	2.6/0.14
Wheat	3.7/3.7	2.8/0.28	2.6/0.26
Cottage cheese		60/28	58/18
Ice	3.8/0.7	3/0.004	3.2/0.003

Von Hippel (1954), Meredith (1998), Datta (2005).

Thus, the amount of power that can be absorbed by a foodstuff is dependent upon the frequency, the dielectric loss factor of the food, and the electric field strength in the material which in turn may depend on the dielectric constant. Higher frequencies generate more heat within the food stuff for a given field strength in the material.

8.3.4 Penetration Depth

Another factor important in the absorption of RF and microwave energy in foods is the penetration depth. The penetration depth in a material is defined as the depth at which the heating power density has fallen to 1/e (i.e., to 37%) of its surface value. The penetration depth D_p for most foods where $\varepsilon'' \leq \varepsilon'$ is approximated by:

$$D_p \sim \frac{\lambda_0 \sqrt{\varepsilon'}}{2\pi \ \varepsilon''} \tag{8.5}$$

Thus, the depth to which the dielectric heating effect penetrates a given foodstuff is determined largely by the material's loss factor which in turn varies with frequency (lower dielectric loss results in larger penetration), the wavelength of the energy in free space ($\lambda_0 = c/f$), and also the relative dielectric constant ε' which does not vary significantly with frequency. In general, materials with loss factors greater than 0.02 can be regarded as candidates for dielectric heating. RF heating of materials with loss factors below this value is likely to be precluded.

8.3.5 Electrical Breakdown and its Avoidance

All materials have a finite dielectric strength after which they are subject to electrical breakdown and damage. Electrical breakdown results in arcing which can damage product and equipment result in burning and can even initiate fire. For clean dry air the electrical breakdown strength is 3000V/mm. For normal processing conditions this drops to around 100–300V/mm. In practice, therefore, voltage gradients (electric field) should be limited to these lower values.

Product shape and form significantly affect the electric field in a dielectric applicator. Product fill, for example, between RF electrodes and electrode configuration (e.g., whether there is air above or around the product) affect developed electric fields. In general electric fields are higher external to the product by a factor of the ratio of dielectric constant of the material to the dielectric constant of the surrounding dielectric medium. As such it should be remembered that air gaps above a product will cause a voltage gradient which may result in electrical breakdown. This can become particularly important in RF applicators where, for a given power density in the material, the electric field required in the product is higher than for microwave frequencies, see Eqn (8.4).

There are many other factors, however, that can contribute to electrical breakdown. Field concentrations occur at the edges of product and structures within applicators. This is affected by product shape, product spacing, and distance to adjacent structures. For example, the situation presented by a frozen rectangular block of chicken breast is different to that of chicken legs which can present projecting sharper bones and bones in varying proximity to each other. A conveyor belt which becomes contaminated by product that is not subsequently cleaned off can lead to drying, heating, carbonization of the contaminant, and ultimately arcing.

Avoidance of breakdown must be minimized by good design, such as avoiding sharp edges, maintaining suitable clearances, appropriate power control with regard to product load and type. Excessive build up of dirt, dust, moisture, or drips should be avoided or minimized. Purging the area around the product with air to avoid build up of ionized gas is also beneficial. Experience and good engineering design is required particularly for high power applications or difficult applications where small and low loss loads are present.

Arcing can therefore occur in dielectric applicators due to poor power control, bad design, poor product loading control, contamination build up, over drying of product, or a combination of these and other factors. An initiated arc will usually persist so arc detection systems may be fitted which result in the power being switched off thus extinguishing the arc. Belt cleaning systems are often implemented on continuous machines. Fire detection systems and fire suppression systems are not uncommon on industrial machines.

8.3.6 Generation and Application of RF and Microwave Power

RF and microwave techniques are distinctly different in how they are generated and applied. RF in general is applied between electrodes or between electrodes and a ground plane with the applicator dimensions being generally less than a wavelength in free space (22.1 m at 13.56 MHz). These electrodes can be configured to tailor the electric field applied to the product. Microwaves are applied as guided or radiated energy with applicators of size comparable to the wavelength or many times the wavelength. On account of this, electric field maxima and minima will generally occur in microwave applicators leading to nonuniform thermal profiles, and methods are needed to distribute the microwaves, or disturb standing wave patterns that arise, to give uniform heating. In both instances product movement is also often used to promote uniformity.

8.3.6.1 RF Power Generators

Older technology, still much used, is based on a so-called class C self-excited oscillator that has a high efficiency of operation. RF power units may vary from a few 100 W to several 100 kW. This system for industrial powers generally uses a thermionic vacuum triode valve to generate useful power in a resonant tank circuit arrangement which is close coupled to the RF applicator (see Figure 8.13).

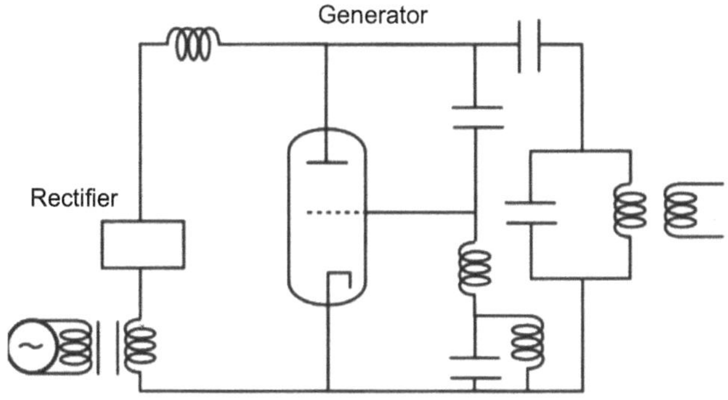

Figure 8.13 Class C self-excited RF generator.

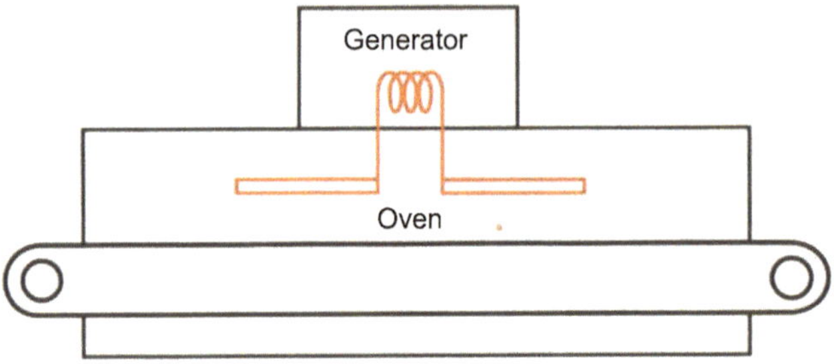

Figure 8.14 RF conveyor oven, close coupled class C generator.

The operating frequency is generally chosen to coincide with allocated ISM frequencies (industrial, scientific, and medical allocated frequency bands) where radiated emissions limits from equipment and electromagnetic compatibility (EMC) requirements with telecommunication applications are more relaxed. Typically the transmission line between amplifier and load needs to be short and the generator is usually integrally mounted on the product applicator with power being drawn from the generator by an output transformer stage (see Figure 8.14). Power control is often by physical movement of the applicator load capacitor which can cause the frequency to shift. This makes EMC compliance more difficult for these applicators. Class C advantages are simplicity and reduced cost compared with 50 Ω systems discussed next.

Modern 50 Ω RF generators typically comprise a low power signal source followed by a number of amplifier stages to deliver usable powers for process heating. Powers can range from a few 100 W to 10s of kW using solid state generators and to 100s of kW using triode valves amplifiers. The RF source is normally a solid state crystal frequency oscillator operating at a precise frequency followed by solid state amplifier stages or triode amplifier. The output stage from the generator is configured to have a characteristic impedance, most commonly 50 Ω, which allows power to be transmitted using standard 50 Ω coaxial components to a tuned applicator with negligible losses. The applicator and load to be heated must also be made to appear to be 50 Ω for low loss power transfer to the load. An impedance matching network, consisting of variable capacitors and inductor components, is normally integrated with the applicator to achieve this. Advantages of this 50 Ω technology are the ability to guarantee frequency control and to more easily meet EMC emissions regulations. The generator can be located many 10s of meters from the applicator and the technology allows sophisticated control to be implemented. With this technology forward to and reflected power from the applicator can be very accurately measured allowing very accurate and fast PLC or microchip control.

RF generators are available at the ISM frequencies 13.56, 27.12, and 40.68 MHz corresponding to wavelengths of 22.12, 11.06, and 7.37 m respectively. Powers are available over a wide range from a few watts, to 30 kW from solid state equipment and up to 200 kW from triode valve equipment. RF generation efficiencies are relatively high at around 75%, however, additional losses occur in RF coupling into the product in an applicator leading to overall efficiencies ranging from 50% to 70%.

8.3.6.2 RF Applicators and Control

Radio frequency energy may be applied to the product to be heated in a number of ways but essentially the product forms the "dielectric" of a capacitor which in its simplest form may be two parallel plates across which the RF voltage is applied.

RF applicators are almost invariably inductive, capacitive, and resistive tuned circuits which draw power from a generator which can be local or remote. The two most common applicator arrangements for RF processing are class C applicators where the generator is close coupled to the applicator and 50 Ω RF applicators where the generator can be remote and power is delivered via a coaxial transmission line or coaxial cable and introduced onto an electrode array via an impedance matching network. With the class C applicator, the method for power control often involves physical movement of the applicator/load capacitive and inductive elements, e.g., movement of an electrode. This method of power control causes the frequency to shift making it more difficult to maintain frequency within the allocated ISM bands. Where 50 Ω RF technology is used, very accurate power delivery to the product is possible, this power can be measured and controlled, for example, to match a measured product mass flow. Dielectric properties information about the product can also be inferred from the matching network which can give useful information about the product condition, e.g., moisture level or degree of defrosting. Figure 8.15 shows class C and 50 Ω RF arrangements.

For RF heating, electrode geometry and separation, product shape and size, and position relative to the electrodes, dielectric media surrounding the product (air or other) strongly affect the product interaction and heating. The very long wavelength of RF means that the electric field between parallel plate electrodes (product shape and media considerations excepted) can be very uniform; together with the deep penetration of RF this can lead to very uniform volumetric heating.

For safety, the electrodes will be housed within a metal enclosure which may have inlet and outlet ports or apertures or tunnels for transport of the product. As with microwave heating, this metal enclosure serves to contain the electromagnetic energy and isolate high voltages within them from the operator. On account of its long wavelength, containment of RF energy is much easier than for microwave energy and very large openings, several meters across can be used without leakage when properly engineered. Product infeed and outfeed arrangements without doors are therefore common and relatively simple with RF applicators.

There are a number of possible RF applicator/electrode arrangements which are described below.

8.3.6.3 RF Throughfield

The simplest RF applicator is the "throughfield" which consists of a pair of parallel electrodes (see Figure 8.16), normally made of aluminum or nonmagnetic stainless steel, between which the product is placed. The food acts as the dielectric component of a capacitor. The electrodes can be in contact with the material but for continuous food applications, an air gap is usually required. One electrode is usually a high voltage electrode operating at up to 10 kV with the other electrode a ground potential electrode. The lower electrode can be a conveyor or have a conveyor running across it.

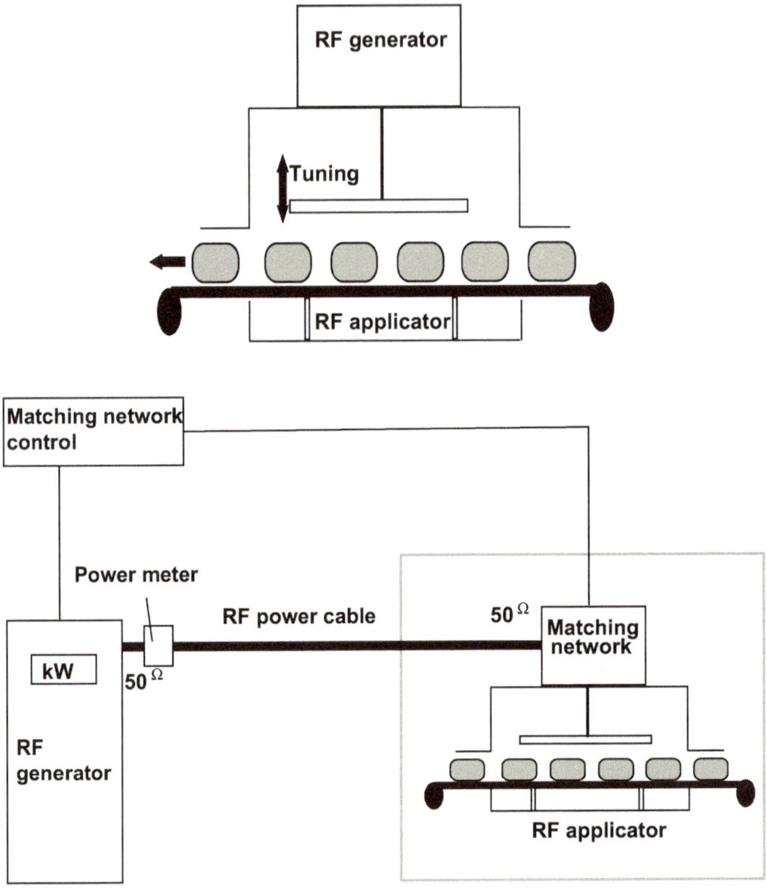

Figure 8.15 RF control arrangements, above class C electrode movement, below 50 Ω RF with automatic impedance matching network.

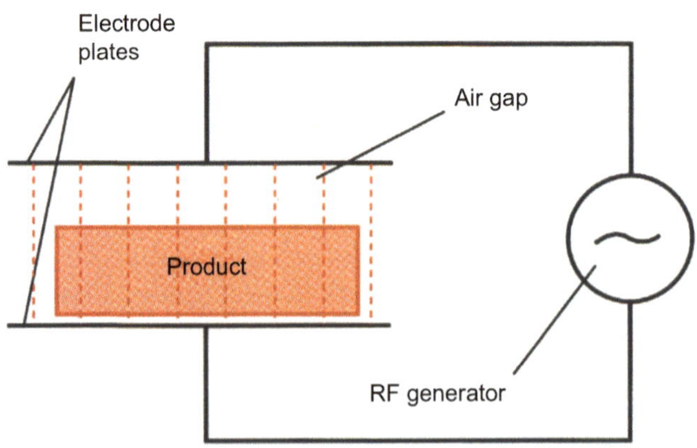

Figure 8.16 RF throughfield applicator.

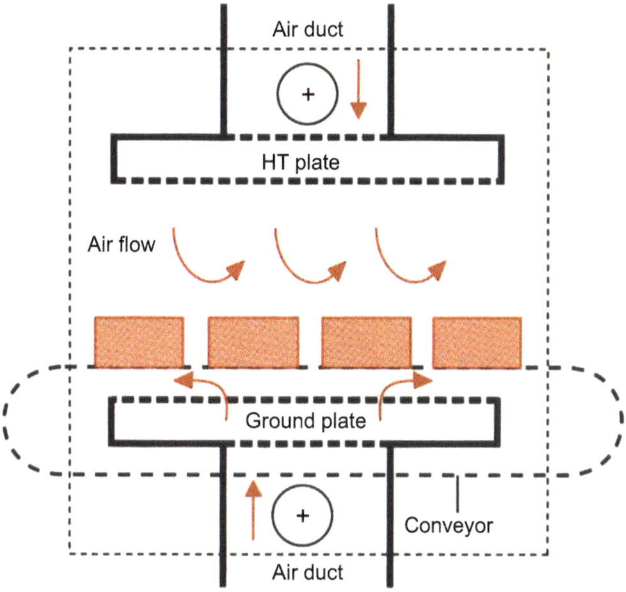

Figure 8.17 RF throughfield applicator with ARFA (Air Radio Frequency Assisted) air delivery plenums.

A common feature of RF electrodes, as they are metallic structures, is that they can double up as conveyors or hot air delivery plenums or can be used to shape or apply pressure to a component. A common applicator is the Air Radio Frequency Assisted (ARFA) oven which uses such plenum air delivery electrodes (see Figure 8.17).

8.3.6.4 RF Strayfield and Staggered Throughfield

This strayfield system consists of an array of electrodes, in the form of rods or bars located in the same plane as the material to be heated (see Figure 8.18). This arrangement produces a fringing electric field more along the material being processed. Though this field is not as uniform as that in the throughfield system is more suited to the continuous processing of thin substrates (up to ~10mm thick).

The staggered throughfield electrode system is a hybrid of throughfield and strayfield electrode types, in which electrode rods or flat bars are used but arranged above and below the material to be heated. This type of field enables thicker materials to be processed without substantial differential heating occurring.

8.3.6.5 Microwave Power Generation

Microwave energy can be produced by a number of devices: Klystrons, Gyrotrons, traveling wave tubes, and magnetrons, as well as solid state devices. The first four devices are vacuum tube devices and can deliver high power from a few 100W to 100s of kW. Solid state microwave generators are currently available at up to 500W. The most common and economically available microwave generator is the

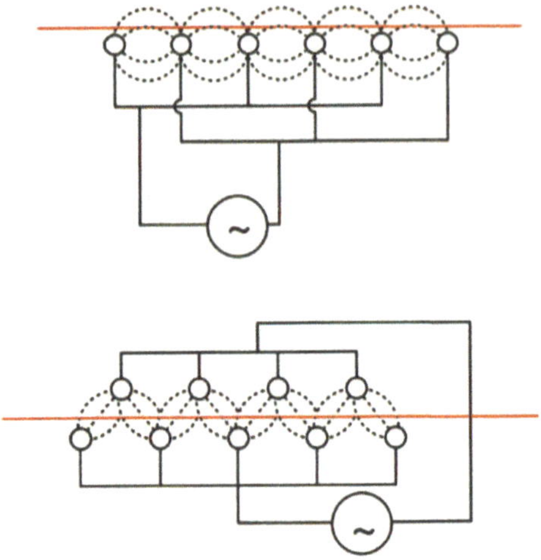

Figure 8.18 RF fringfield (strayfield) and staggered throughfield applicators.

magnetron oscillator. This is a vacuum tube device with a cathode and surrounding anode chambers. It operates by controlled release of microwave energy from anode resonant chambers. The microwave energy is derived from an accelerated electron cloud with the electrons generated by thermionic emission from the cathode. The anode cavities attract the electron cloud under the influence of a high voltage DC potential between anode and cathode and an antenna in one of these cavities is typically used to radiate energy from the magnetron (see Figure 8.19). A magnetic field is commonly used to control the power output. Increasing the magnetic field stops the electron cloud reaching the cavity chambers and shuts off the power.

Microwave generators for industrial food processing applications have microwaves generated almost exclusively by magnetrons. Of the three common ISM frequencies 896/915 MHz, 2.45 GHz, and 5.8 GHz the 900 and 2450 MHz frequencies predominate. The wavelength in free space at 900 MHz is 333 mm and for 2450 MHz it is 122 mm. Microwave generators are available in powers of 1–30 kW at 2450 MHz and commonly 15–100 kW at 900 MHz. Microwave generation efficiencies are relatively high and around 80–85% can be achieved for higher power units. Typical applicators can then achieve coupling into a good lossy product with efficiencies ranging from 60% to 90% giving overall efficiencies of 50–75%.

The magnetron within a generator has a finite operating life and would normally carry a modest operating hours manufacturers' guarantee, but its life will vary with manufacturer, duty cycle, environment, and type, typically between 5000 and 15,000 h. Industrial higher power magnetrons can be rebuilt as a cheaper alternative to replacement by new.

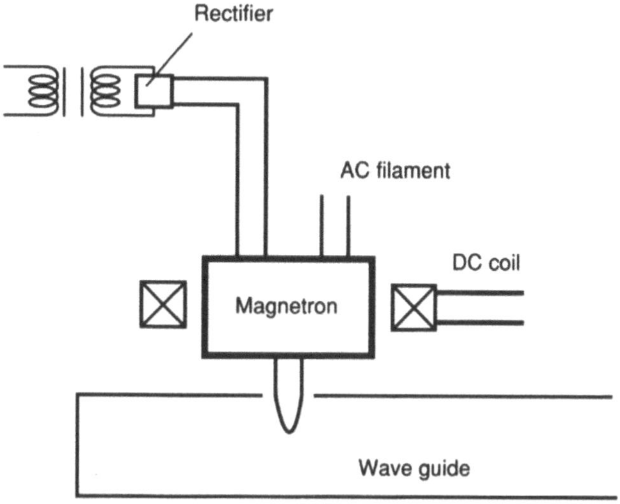

Figure 8.19 Schematic of microwave magnetron and waveguide launch.

8.3.6.6 Microwave Cavities for Heating the Product

The physical principles, heating mechanism, power dissipation equation, and penetration depth have been described above. There are distinct engineering and applicator features applicable to microwave processes which are described now in more detail. In essence the microwave applicator (microwave cavity) is a metal enclosure/box into which the microwave energy is launched by various means depending on the physical shape and size of the product, for example, block or sheet, and the heating requirements and the shape and size of the microwave cavity.

Transmission of microwaves from generator to applicator is by a waveguide or coaxial cable. Waveguide is by far the most common and is capable of transmitting higher powers. Waveguide comprises hollow metallic duct usually rectangular (or less commonly circular) in cross section with internal cross-sectional dimension greater than a minimum cut-off dimension ~$0.5\lambda_0$ and less than λ_0. Waveguide is normally constructed from highly conductive materials to reduce transmission losses. The most common waveguide material is aluminum and with this material transmission loss is low (see Table 8.4). Less conductive materials, e.g., stainless steel with higher resistivities are less desirable and should be restricted in length if possible. Waveguide can be used to transmit energy considerable distances and 10s of meters is quite common. Generators thus can be located at a distance from an applicator if desired.

The applicator or microwave cavity is invariably a metallic enclosure of a high integrity of construction which acts as a Faraday cage to prevent emission and radiation of microwaves. Microwave energy is reflected internally at the metallic walls. Applicator types can be multimode, mono-mode, traveling wave, depending on their size relative to the operating frequency wavelength and the configuration of the cavity. Applicators can be batch or continuous, tunnel or in-pipe/tube arrangements.

Table 8.4 Waveguide Sizes, Continuous Powers, Transmission Losses

Microwave Waveguide	Frequency, MHz	Dimensions, mm	Dimensions, inches	Power Capability, kW	Power Loss, Watts/10 m Aluminum Waveguide 25 kW 2450 MHz and at 75 kW 900 MHz
WG9A = WR340	2450	86 × 43	3.40 × 1.7	25	Approx. 1 kW or 1.4%
WG4 = WR975	900	248 × 124	9.75 × 4.875	75 to 150	Approx. 1.8 kW or 7%

The most common microwave cavity is the multimode cavity. Here the cavity dimensions are large relative to the free space wavelength λ_0 (see Figure 8.20). Waveguide is terminated at the cavity skin as a rectangular or circular opening usually with an environmental sealing microwave transparent window to segregate the waveguide from the cavity and to prevent gases or atmospheres from the cavity making their way into the waveguide. A cavity can be a closed batch type applicator with a door/other opening method or a tunnel structure with infeed and outfeed doors or apertures.

Repeated wave patterns (standing waves) exist within such a cavity with reinforcement of the electric field in periodic locations which can lead to higher heating at these locations. To achieve more uniform heating a whole series of methods are employed:

- Some asymmetry in cavity dimensions is usually incorporated to reduce standing waves. For example, a cuboid would not be a preferred geometry.
- In industrial applicators multiple microwave entry locations or ports/windows are commonly used which give rise to more uniform electric field through a product load. Often launch points are located on multiple sides with different planes of electric field polarization.
- An alternative launch method can involve antennae which can be rotating.
- Rotating metallic surfaces called mode-stirrers are often employed within the cavity which scatters the energy within the cavity. Mode-stirrers disturb the electric field patterns within the cavity giving more uniform electric field on average within the product.
- Product movement helps significantly toward uniform heating, rotation being common in batch cavities and longitudinal movement (conveyorized or indexing) in a tunnel.
- Other specific and specialized launch arrangements, e.g., horns, slotted waveguide, circular polarized feeds (Ferrite Inc.) are also sometimes applied at or inside the cavity skin depending of the process and product being targeted.

Additional considerations of significance are product placement and proximity to the microwave entry points and the cavity wall, variability of dielectric properties of the medium or multicomponent media being heated, the form and size and thermal properties of the media being heated. The influence of electric field concentration that can arise around product edges, electric field concentration or screening due to

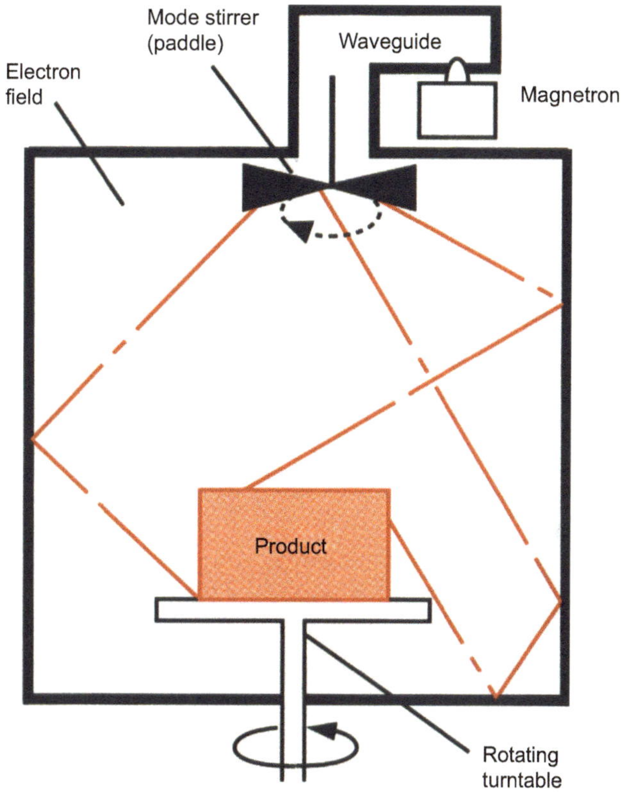

Figure 8.20 Multimode microwave cavity.

structures within the microwave cavity, influence of other dielectric materials on electric field in product adjacent to them also affect heating uniformity.

Multimode microwave cavities and ovens several meters wide, by a few meters high, by 10s of meters long are quite common, with microwave powers up to several 100s of kW being used. Microwave applicators are more general purpose and flexible in use than RF systems and thus are used for a wider range of applications.

Another common applicator type is the single mode cavity (or cavities of restricted/limited modes) as in Figure 8.21. This arrangement comprises a smaller cavity with dimensions capable of supporting microwave propagation of a single (or of restricted and limited) mode only. The single mode applicator results in a very high electric field in a precise location within the cavity. This can be useful for applications which require a high power density into well-defined and small product shapes and for low loss materials.

Fibers and filaments and grains can be easily positioned in the location of highest electric field for processing. Continuous in-pipe heating/cooking can be relatively easily performed, for liquids and pumpable products, by conveying them in a microwave transparent tube through a microwave applicator of these sorts. Such applicators with similar

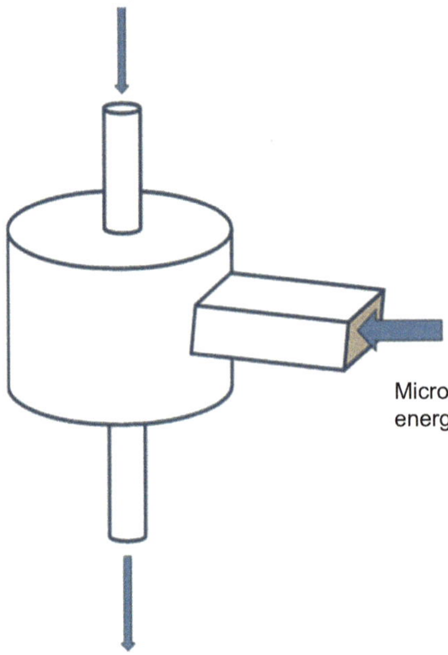

Figure 8.21 Circular single mode microwave cavity.

restricted mode of propagation have been used with good effect for continuous microwave heating of liquids and chemical reactants (Wharton, 2014; Perkin et al., 2012).

Combined processes which utilize hot air and steam together with the microwave, are used and may give rise to improved, more economic solutions compared with microwave only. There are well known techniques and methods for the introduction and distribution of heated and gaseous atmospheres within microwave processes, both batch and continuous. Tube bundles, for example, can pass gases but attenuate microwave transmission. Continuous systems with wide in and outfeed ports are possible using pin chokes to attenuate microwave emissions as an alternative to semi continuous systems using door or closing mechanisms. With such pin chokes apertures of ~ $\lambda_0/4$ in height can be attenuated. Specialist knowledge is required for effective design of microwave cavities and their safe use. Microwave technology can also be used in reduced pressure applications such as vacuum drying more readily than RF since the electric fields used are generally lower and so less likely to cause arcing.

8.3.7 Overview of the Benefits and Applications of Dielectric Heating

It is convenient to consider the application of RF and microwave heating in a generic way—as dielectric heating—because often the choice of RF or microwave may not be clear cut until the specific application requirements are considered in detail: technically both approaches may often be possible in principle. The main differences between radio frequencies and microwaves, as far as an industrial user is concerned, are in the type of applicator used and how the dielectric heating is applied to the product, the frequency of the energy applied and its effect on heating and penetration depth. The design of the applicator unit including containment of the electromagnetic energy and the method of controlling the power are quite different for microwave and RF processes. The presence of water in most foodstuffs, which readily absorbs microwave and RF energy, enables them to be readily heated volumetrically. Direct, volumetric heating at large power densities may be utilized to give rapid preheating, cooking, puffing, and drying.

As has been said previously, conventional heating can be a slow process and overheating of the surface may occur before the center has reached the required temperature, especially if attempts are made to accelerate the heating rate. Uniform heating is a common aim for dielectric heating. The volumetric nature of the heating helps a great deal in achieving this, however, many other factors affect uniformity, such as size and form of the product, whether it is homogeneous or contains components with different dielectric properties and how these properties change with temperature or with any phase change (e.g., ice melting to become water, or water evaporating to become steam). The type and design of the applicator used, and the chosen frequency of the microwave or RF power used also affects uniformity, as do any thermal losses that may occur from the surface of the product.

Hybrid heating technologies combining RF and microwave energy with, e.g., hot air, steam, and IR can realize enhanced processes which depending on the application can often give better results than conventional or dielectric only processes. With combined heating technologies, benefits from the different heating sources can be engineered, such as surface color and flavor development from external hot gases or infrared heating. The dielectric process component is often only a modest percentage of the total process energy requirement (10–40%); however, its effect delivered directly into the product can bring significant process improvements in speed, quality or productivity. Hybrid processes utilizing a proportion of dielectric heating also reduces the cost associated with the generally more expensive dielectric component (equipment capital and running cost and the engineering requirements) when compared to dielectric RF and microwave only processes. RF-assisted baking of biscuits can reduce the conventional later oven zone length needed to achieve low moisture contents by up to 50%.

Process benefits include

- Reduced processing times with fast start-up.
- Improved product quality.
- Reduced footprint.
- Increased energy efficiency.
- Faster response to orders.
- Increased opportunity for new product development, e.g., with snack foods and healthy alternatives to frying.

The main disadvantages of these techniques are the perceived high capital cost of the equipment and the fact that each product to be heated must have suitable dielectric properties to enable absorption of electromagnetic energy—a possible limitation if processing of a wide range of products is required. As a rule of thumb, dielectric heating techniques favor applications where a high quality, value added product is sought, with characteristics which cannot easily be obtained with conventional heating techniques and where the throughput is low to medium. In such situations, the capital cost of the equipment when viewed against the benefits obtained can be acceptable. A good example is the cooking of bacon slices for subsequent use as food topping; the high power density achieved with microwave energy leads to compact equipment with large throughputs—installations of over 500kW of microwave energy are typical.

Established RF and microwave processes include

- Continuous cooking of bacon and toppings, several t/h.
- RF defrosting of meat, chicken, fruit, and fish defrost at t/h, tempering up to several t/h.
- Microwave defrosting and tempering meats and butter, defrost at t/h, tempering up to several t/h.
- Microwave pasteurizing of ready meals and packaged products.
- Preheating boost heating and reheating of foods.

Typically, these involve installed microwave powers of 10–75kW common for RF and microwave, 100s of kW for microwave in processes with throughputs ranging up to several t/h depending on the application. Common belt widths are up to 1.2m, however, bespoke applicators up to 3m wide are feasible. Machine lengths range from a few meters to 10s of meters. Equipment suppliers exist worldwide. Some well-known suppliers are Sairem SAS, BCH Ltd, DanTech Ltd, Selo Bollans Ltd, Strayfield Ltd, Stalam S.p.A, AMTek Microwaves, Ferrite Microwave Technologies, Microdry Inc, and Thermex Thermatron.

Some applications are specifically suited to microwaves or more commonly use microwaves.

- Cooking of bacon.
- Microwave puffing of snack foods.
- In pack sterilization, "Micvac" process.
- Cooking of extruded potato products.
- Tempering of frozen foods, particularly meat products.
- Microwave vacuum and microwave-assisted vacuum drying.

8.3.7.1 Dielectric Drying

Products can be dried particularly effectively with RF and microwave with either the dielectric heating providing all the energy for evaporation or, when the material has suitable internal moisture transport properties, assisting a conventional convective drying method. In the latter case the volumetric nature of the dielectric heating can increase moisture mobility within the product which speeds up drying. In these circumstances convective heating is generally still required to evaporate the moisture and carry away the water vapor. Often drying can proceed many times faster than conventional drying, sometimes at lower temperatures and with better product quality, all by the introduction of a modest level of dielectric heating into the process (25–50%).

Figure 8.22 (a) Board drying. Top with RF uniform moisture, flat. Below no RF, case hardened, less flat. (b) Biscuit baking. Top with RF uniform moisture, flat. Below no RF, case hardened, less flat.
C-Tech Innovation.

Convective surface heating can be very efficient when product is wet but it becomes inefficient when residual moisture has to be removed from the center of an almost dry product. In these circumstances the direct volumetric heat transfer that dielectric heating provides can be used to great advantage. Furthermore, since the electromagnetic energy may couple with wetter material in preference to the drier material, it is able to provide a moisture leveling effect. Figure 8.22(a) and (b) shows an example of how a case hardened surface can be avoided using RF-assisted drying/baking.

Volumetric heating can lead to a number of benefits. For example, in the drying of pasta, color was enhanced and the product had a better "bite" (Smith, 1979). The equipment for pasta drying, a combined hot air/microwave system, was more compact than a conventional dryer; the overall length of the equipment being reduced from 120–160 to 27 feet. It has also been noted that improved rehydration can result following dielectric-assisted drying (Khraisheh, 2004) owing to lack of shrinkage. A further example of improved rehydration following RF-assisted drying is shown in Figure 8.23.

For temperature-sensitive materials, drying in vacuum conditions permits the drying to occur at reduced temperatures which can result in improved product quality as well as increased drying rates for dehydration of fruits, vegetables, and herbs. Microwave heating provides an easier method for transferring energy into the vacuum chamber, compared with conventional heating, as the vacuum vessel can also form the microwave cavity by using suitable microwave windows to allow transmission of the microwaves. Microwave frequencies are preferred to RF frequencies as they operate with lower electric fields which are less prone to dielectric breakdown/arcing in the low pressure atmosphere (Perkin, 1983).

8.3.7.2 Postbaking

Up to 30% of the length of a conventional oven is used for final drying. A process which utilizes the targeted moisture removal by RF is postbaking of biscuits (Jones, 1981). Dielectric heating seeks out water in preference to the other ingredients of

Figure 8.23 Improved rehydration (left) following RF-assisted drying.
C-Tech Innovation.

the biscuit mix and is able to provide a more efficient means of moisture removal at the later stage of baking. There are some specific RF electrode array designs, e.g., staggered throughfield and strayfield which can particularly take advantage of this moisture profiling in thin products (Rowley, 2001). RF heating gives effective control of moisture content and uniformity of this through the product. Benefits are up to 40% increase in productivity, virtual elimination of checking, better control of product color and longer shelf life. While RF heating can be applied as a standalone postbaking application, it can also be combined with convective heating in the latter stages of a conventional oven (see Figures 8.17 and 8.24 showing ARFA applicators).

8.3.7.3 Defrosting and Tempering of Frozen Products

Tempering (raising the temperature of frozen products to just below freezing point) is an ideal application for microwave and RF heating. The deep penetration of RF and microwave energy (generated at the 900MHz band) raises the temperature of the core and the surface at a similar rate and without the surface being exposed to an external higher temperature medium. Tempering equipment can bring the temperature of a 27kg block of meat up to just below freezing point from deep frozen in 2min, whereas conventionally this would typically take 24h. Other common products are fruit, butter, and fish.

RF defrosting can reduce process times from 24h for conventional or higher ambient temperature defrosting to 40min for RF defrosting with dramatically reduced microbiological bacteria count (see Figure 8.25). Drip loss with chicken defrosting has been shown to reduce from 7% to 0.5%. Continuous defrosting can permit production of meat or fruit finished products on demand directly from deep frozen stored

Figure 8.24 ARFA baking applicator.
Courtesy of Petrie Technologies.

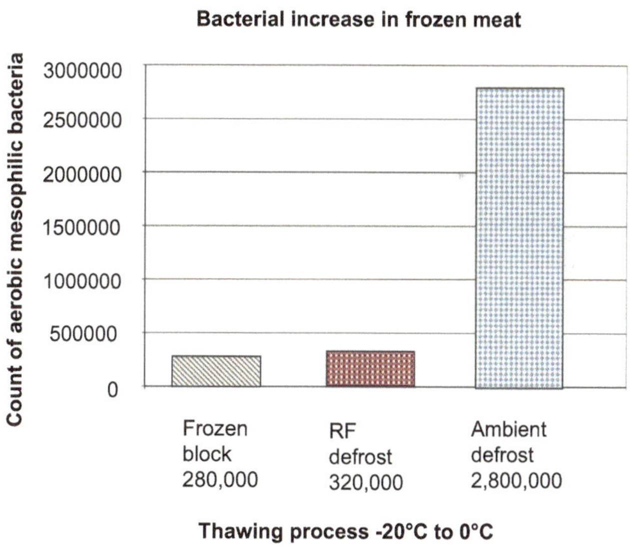

Figure 8.25 Microbiological comparison of RF defrosting with ambient defrosting.
Data Petrie Technologies.

ingredients giving improved product uniformity, quality, and yield, reduced microbiological loading, and ability to respond very quickly to customer demands.

Defrosting where the product is required to be brought up to 0 °C or above is more difficult than tempering and does introduce a challenge for dielectric heating. The dielectric loss of pure ice is low while that of water is much higher. This relative

Table 8.5 Dielectric Properties of Ice and Water

Dielectric Properties of Water and Ice	Temperature °C	ε′/ε″		
		30 MHz	1 GHz	2.5 GHz
Water ice	−12	3.8/0.7	3/0.004	3.2/0.003
Water distilled	25	78/0.4	77/5.2	77/13
Water distilled	85	58/0.3	56/1	56/3
Water NaCl solution 0.1 mol/kg	25	76/480	76/30	76/20
Water NaCl solution 0.5 mol/kg	25	76/2400	68/140	68/54

Adapted from Meredith, 1998.

difference varies with frequency, and also with impurities and ionic content (e.g., salt content). Table 8.5 presents some dielectric properties of ice and water.

During tempering and defrosting, water is progressively melted within the frozen material in increasing proportions as the temperature approaches 0 °C. The melted ice in the form of water is more susceptible to dielectric heating and thus heats more than the ice, which may result in thermal runaway. This can lead to nonuniformity and localized overheating in regions of the product, and in some cases localized cooking where parts of the product reach quite high temperatures while other parts or components are still partially frozen. To reduce the nonuniformity power should be reduced as the temperature increases.

RF and microwave are proven and very attractive methods for fast tempering and defrosting. The variation in dielectric loss is less at RF than at microwave frequencies; this in combination with the greater penetration and more uniform electric field obtained at RF may be the reason that it has an advantage over microwaves for defrosting, particularly for sensitive products such as frozen fruit or fish. Nevertheless, the majority of existing applications are tempering using microwave energy which is somewhat easier and cheaper to implement with applicators which are more general purpose. Figure 8.26 shows a 1 t/h industrial RF defroster processing frozen chicken.

8.3.7.4 Cooking and Baking

A range of modern domestic and commercial microwaves exist where microwave and conventional heating technologies are combined giving equipment capable of rapid cooking and baking of high quality foods. At industrial scale dielectric heating technology has been used in applications including microwave cooking of bacon, cooking of extruded vegetable products and baking of bread. Combination systems of RF or microwave heating together with hot air or steam are usual for these applications in order to give the product quality required; the hot air giving color and flavor development while the dielectric heating gives high quality/controlled internal product

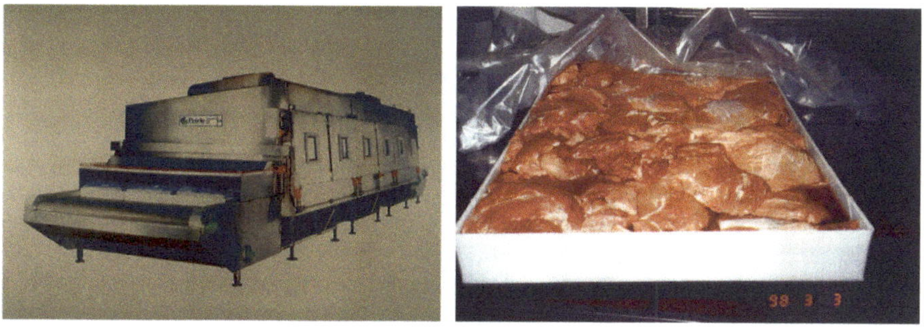

Figure 8.26 1 t/h 0 °C RF defroster.
Courtesy of Petrie Technologies.

Figure 8.27 Industrial microwave/steam cooking process.
Courtesy of Petrie Technologies.

structure and enhanced processing rates. Figure 8.27 shows a large microwave plus steam cooker operating at > t/h scale. Figure 8.28 shows a microwave tunnel for heating Chinese spring rolls from 10 °C to 70 °C, capacity 5000 spring rolls per hour, using 40 kW 2450 MHz microwaves.

8.3.7.5 Melting of Fats

Melting of blocks of fats and related products has been undertaken successfully with RF and microwave heating even though the dielectric loss factor is not particularly large. Figure 8.29 shows an RF example using a strayfield electrode array where the RF field fringes into the fat with limited penetration. The fat melts and passes through the electrode array and out the heating field preventing further heating.

Figure 8.28 Microwave heating of spring rolls.
Courtesy of Sairem SAS.

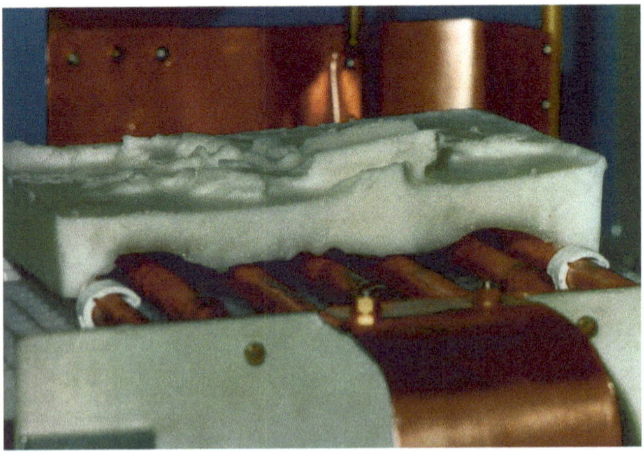

Figure 8.29 RF fat melting.
C-Tech Innovation.

8.3.7.6 Novel Processes

Some of the more novel processes which are possible due to the specific nature of the volumetric heating provided by dielectric heating include

- Puffing grains and filaments for products such as cereal and snack food.
- Pasteurizing solid food products without drying; for products such as fresh pasta.
- Color/texture control in baking.
- Marinading and tenderizing of meats.

- Proving of dough.
- Microwave heating for pasteurizing and sterilizing is now being combined with aseptic filling, "Aseptia" process.
- Microwave and RF heating supplying heat in freeze drying.

8.3.7.6.1 Puffing

Cereal products such as rice and grains can be puffed and expanded in a continuous applicator (much like popcorn in a domestic microwave). A configuration has been developed using a single mode microwave cavity resulting in high electrical field in a defined physical location. When combined with an appropriate through flow of air the residence time of the grains in the microwave cavity becomes self-controlling. When puffed the grains are entrained and carried away by the air flow. Such single mode type configurations are inherently suited to highly defined product forms and location such as might be obtained with grains, fibers, filaments, and extruded items.

8.3.7.6.2 Baking with Controlled Surface Color

Microwave and RF-assisted baking has resulted in production of bread products with control of color from conventional darkened crust to complete absence of color (ohmic heating has also successfully been applied for baking bread, for Japanese crumb, without color or crust). A number of large industrial RF and microwave process lines have produced bread at a 2000 to 2500 loaves/hour scale where the crust is not colored or hard as conventionally results from a high temperature hot air convective baking oven (Wiggins and Cauvin, 2015; Anon. The Grocer, 2006; Anon. The Grocer, 2007; Awuah et al., 2014). Conventionally bread baking results in a significant mass of the ingredients contained in the crust, sometimes up to 40% and some manufacturers cut-off this crust to achieve a crustless bread. So-called crustless, or invisible crust, products produced using RF or microwave-assisted baking achieves this desired absence of color without cutting the crusts off. This has a significant cost saving in terms of mass and value of ingredients to produce the end product. These processes use polymer tins which are transparent to RF and microwaves in conjunction with low oven temperatures where such white or par-baked product is required. Where color is required RF and microwave processes of different design have also been applied with metal tins in combination with high oven temperatures and at times with impingement hot air for added high heat transfer to develop color quickly.

8.3.7.6.3 Alternative Cooking Methods

Microwave and microwave-assisted steam/hot-air cooking has been applied for cooking starch and extruded potato and vegetable formed products/shapes as an alternative to conventional deep fat frying. The processes are desirable for producing finished product which contains reduced levels of fat or reduced levels of acrylamide. High-intensity electric fields can rival the energy input of conventional deep fat frying to produce healthier foods. These processes have been applied both cooking without oil or using microwave heating as a precooking process with reduced final acrylamide or using microwave heating as well as deep fat frying in oil at the same time and exploiting the microwave transparency of the oil to lead to reduced fat content and acrylamide (Erdogdu et al., 2007; Sahin and Sumnu, 2008). With the processes which eliminate the oil as a cooking medium, the fat uptake in such products is dramatically reduced and a controlled fat and flavor level can then be achieved through formulation or postcook application.

8.3.7.6.4 Advanced Meat Cooking Processes

RF cooking (Zhang et al., 2006; McKenna, 2006) has been demonstrated with dramatically improved productivity for meat cooking and better deep tissue and bone marrow cooking to give a faster cooking operation than conventional processes. The deep penetration possible with RF heating has shown significant reductions in cooking times for leg and shoulder hams. Preheating and postcook heating using microwave or RF have also been applied to improve cook yields and quality, often with perceived better appearance, e.g., less uncooked red blood in joints and within bones.

For pumpable meat products continuous in-pipe, in-tube cook processes are available using microwave or RF processes where a microwave/RF transparent polymer tube is configured within a microwave or RF applicator. Figure 8.30 shows continuous cooking of meat with RF. These processes have the potential to revolutionize the cooking of sandwich meat and other extruded meat formulations. Conventional ovens and retorts cooking in molds take several hours, whereas continuous RF cooking takes minutes. Novel shaped products can also be continuously cooked in pipe with the dielectric methods. The economics of these processes using dielectric heating, particularly the capital cost of the equipment, needs careful assessment, however.

Marinading with RF-assistance is also reported for tenderizing meat (McKenna, 2006; Kichlu, 1988).

8.3.7.6.5 Proving for Dough Products

RF, microwave, and ohmic heating are good candidate technologies for improving the dough proving process. An RF or ohmic dough prover in particular could achieve very accurate energy delivery and temperature control into the proving dough. Successful application of these processes is limited (Wiggins and Cauvin, 2015). The deep

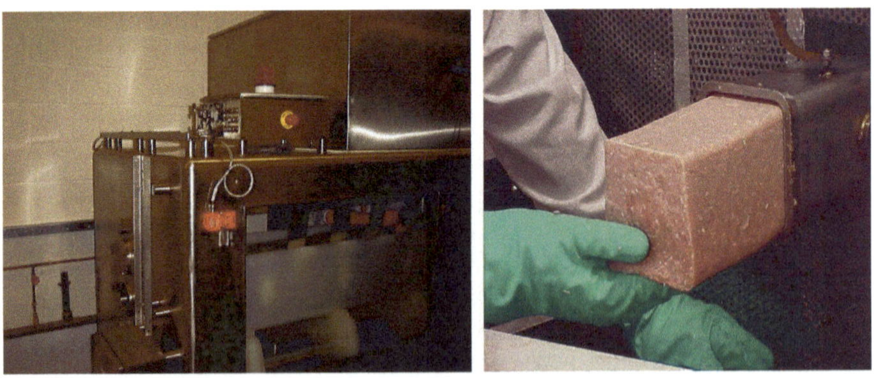

Figure 8.30 Continuous cooking of pumped meats with RF.
Courtesy of Petrie Technologies.

penetrating volumetric nature of these methods, along with precise energy delivery, and, if needed, active temperature control, should enable these processes to deliver tangible benefits.

8.4 Safety

8.4.1 Safety Aspects of Ohmic Heating

As the ohmic heating process, by its very nature, involves the conduction of large electric currents at potentially lethal voltages, the process has the potential for electrocution, electric shock, and burning if not properly designed and maintained. Safe designs for the technology have been developed and proven, and the use of ohmic heating in industry is well established. Safe operation is achieved by ensuring that the electrical energy is contained within the body of the ohmic heater. In order for ohmic heating to function, the walls of the ohmic heater unit should be made from electrically insulating material (as described in Section 8.2.4), thus the outer pipe walls are not exposed to the electric field. If a lined metal pipe is used, then damage to the lining could potentially allow current to pass to the outer pipe wall and thus all such outer metal piping must be earthed, and typically the earth current monitored so that any fault that might arise due to damage to a pipe lining can be detected immediately and power switched off.

Similarly any electrical current flowing out from the ohmic heater body through the product stream could lead to an electrical potential in upstream of downstream piping. To prevent this voltage reaching a hazardous level, the upstream and downstream piping must be securely earthed. To ensure the level of electrical current leaving the ohmic heater body through the product stream in this way does not exceed safe and legally defined levels, the electrical resistance of this current path must be made to be sufficiently high. This can be done by employing an electrically insulated length of pipe upstream and downstream of the final ohmic power electrodes, chosen to be of sufficient length to cause the electrical resistance of the food product being processed to be large enough so as to present a significant electrical resistance, thus ensuring the current flow in the region is below the required level. In order to minimize the length of pipe required to restrict the "leakage" current, the voltage of the final electrode should be maintained as at close to earth voltage as possible. To do this, where possible, the ohmic electrodes at the inlet or outlet should be wired to be at earth or neutral voltage. For designs where this is not appropriate it is an option to employ additional electrodes wired at earth or neutral voltage specifically for safety purposes.

Other safety considerations when using ohmic heating processing include ensuring no overpressurization occurs within the system. In ohmic heating processes there is no limiting temperature at which the system will settle; if electrical power is available the product will continue to heat and can reach high temperatures and pressures if the power input is not properly controlled. For this reason it is important that the system is properly monitored and that the electrical power input is controlled at a level appropriate for the flow rate through the system. A particular hazard can arise if fluid flow through the unit stops as heating will continue if the heater is not switched off leading to rapid temperature rise and potential pressure rise. For this reason it is important to monitor the flow rate and to switch off power to the system if flow stops. Additionally the system should also be fitted with suitable pressure relief valves, typically positioned between the pump and the ohmic heater, as is usual, but also close to the outlet of the heater.

8.4.2 Safety and Regulatory Issues for Microwave and RF Heating

In common with all electrical installations, the volumetric heating technologies—ohmic, microwave, and RF—present a variety of risks and hazards and so correct installation and running procedures must be undertaken. It should be noted that these volumetric techniques are highly developed and operated successfully and safely in industrial environments with good safety histories.

Generator equipment and applicators often involve high voltages which can be up to 15 kV. RF and microwave equipment have the potential for electric shock and burning. Processing chambers (applicators) must be properly designed and constructed so that there is negligible radiation of energy which might otherwise cause harmful heating of personnel. All RF and microwave equipment and applicators must comply with strict EMC requirements covering interference with other equipment and electrical infrastructure. Strict regulations for RF and microwave equipment apply regarding radiated emissions (electric field, magnetic field, and power) for personnel safety in addition to the EMC requirements. All equipment must in addition comply with directives for safe mechanical and electrical construction and operation. These regulations are not presented in detail here rather some key facts and practices are highlighted and sources for key regulatory requirements are identified. Reference sources for information regarding EMC regulations, exposure limits, and ISM bands are International Commission on Non-Ionising Radiation Protection (ICNIRP) Guidelines and BS EN 55011.

Safe operation to avoid electric shock is achieved by insulation or isolation of live conductors in the context of RF and ohmic often by the enclosure of live conductors in interlocked enclosures or pipework. Microwave and RF applicators are also designed so that high levels of attenuation of energy transmission are designed in at all aperturesincluding product infeed and outfeeds. Human tissue is highly receptive to heating by microwave and RF radiation and must be protected from excess exposure to it. The level at which a sensation of warmth is felt is around 30 mW/cm^2 which equals 300 W/m^2. Permitted exposure levels are given in Table 8.6 for occupational exposure. In general RF and microwave equipment will be designed to low levels of emissions; typically less than 1 mW/cm^2 (=10 W/m^2) should be the target. Significant engineering expertise and knowhow is required to achieve equipment designs that meet all requirements.

8.4.2.1 Meeting EMC Requirements

Industrial dielectric heating equipment radiates energy which could interfere with radio communication, control, and measuring devices and other external equipment.

Table 8.6 Reference Levels for Occupational Exposure to Time-Varying Electric and Magnetic Fields (unperturbed rms values)

Occupational Exposure Levels to Time-Varying electric and magnetic fields				
Frequency Range	E-Field Strength (V/m)	H-Field Strength (A/m)	B-Field (μT)	Equivalent Plane Wave Power Density S_{eq} (W/m²)
Up to 1 Hz	–	1.6×10^5	2×10^5	–
1–8 Hz	20,000	$1.63\times10^5/f^2$	$2\times10^5/f^2$	–
8–25 Hz	20,000	$2\times10^4/f^2$	$2.5\times10^4/f$	–
0.025–0.82 kHz	500/f	20/f	25/f	–
0.82–65 kHz	610	24.4	30.7	–
0.065–1 MHz	610	1.6/f	2.0/f	–
1–10 MHz	610/f	1.6/f	2.0/f	–
10–400 MHz	61	0.16	0.2	10
400–2000 MHz	$3f^{1/2}$	$0.008f^{1/2}$	$0.01f^{1/2}$	f/40
2–300 GHz	137	0.36	0.45	50

f is the frequency numeral as it occurs in the range given in column one, e.g., “13.56” for RF of 13.56 MHz, e.g., “900” for microwaves 900 MHz, e.g., “2.45” for microwave 2450 MHz.
International Commission on Non-Ionising Radiation Protection (ICNIRP) guidelines.

Table 8.7 Common RF and microwave ISM band frequencies

ISM bands				
Center Frequency	Bandwidth	Upper Frequency	Lower Frequency	Region
13.560 MHz	14 kHz	13.553 MHz	13.567 MHz	Worldwide
27.120 MHz	326 kHz	26.957 MHz	27.283 MHz	Worldwide
40.680 MHz	40 kHz	40.660 MHz	40.700 MHz	Worldwide
915.000 MHz	26 MHz	902.000 MHz	928.000 MHz	Region 2
2.450 GHz	100 MHz	2.400 GHz	2.500 GHz	Worldwide
5.800 GHz	150 MHz	5.725 GHz	5.875 GHz	Worldwide

Region 2: Americas, Greenland, and some of the eastern Pacific Islands.

ISM bands are allocated frequencies assigned for industrial scientific and medical uses. There are strict frequency bandwidths at these allocated frequencies; however, more lenient EMC emissions levels are allowed at these frequencies and radio communications are expected to tolerate some level of interference within these bands. The common ISM bands and frequency limits are given in Table 8.7. Outside these bands (e.g., if the operating frequency drifts) it becomes very difficult to comply with regulated emissions levels. In general EMC limits are more difficult to meet than safety ICINRP guideline limits.

8.5 Comparison of Techniques and an Overview of Opportunities

A critical factor in selecting the best volumetric heating technique is the electrical properties of the material, that is, the electrical conductivity or loss factor and the dielectric constant. Selecting the method/frequency that is possible or best suits the loss profile is a starting point. Choosing the technique that is easiest to apply and most cost-effective to achieve the desired process follows. This might lead to ohmic heating as a frequent first choice since costs are modest, however, there are many products which will not heat ohmically or where the contact with the electrodes cannot be done effectively. RF and microwave heating can always be considered where their heating mechanisms are feasible.

Ohmic heating relies upon the product to be heated having a suitable electrical conductivity, such that an electric current can pass through it generating heat in proportion to the current and the resistivity of the fluid (I^2R losses). Suitable conductivities can range from 0.01 S/m to 10 S/m, and applied voltages can range from 50 V to 5 kV, at frequencies typically from 10 Hz to 100 kHz. With ohmic heating, contact with the electrodes transferring the electricity to the product must be guaranteed and as such it is suitable for liquids and pumpable products. It is not suited for gases or liquid/gas mixes and is only rarely implemented with solids where particular measures are taken to maintain electrode contact or current conduction. The most common processes using ohmic heating are pasteurization and sterilization. Energy delivery into the product is very efficient; typically 95% (for 50 Hz systems) of the energy drawn from the electrical supply enters the product. Equipment and running costs for ohmic plant are less than those for RF or microwave heating and ohmic plant can compete with steam process costs when steam infrastructure and steam distribution and efficiency costs are taken into consideration (Pereira and Vicente, 2010). For pumpable products ohmic will often be the preferred option due to its lower cost of construction and higher operating efficiency.

RF and microwave heating rely on dipole rotation and charge movement within the product as heating mechanisms; these mechanisms occurring at high frequencies (10 MHz–10 GHz) give rise to the volumetric heating. RF covers frequencies from 10 to 300 MHz typically and conductivity and charge movement mechanisms are more prevalent. Microwave covers frequencies above 300 MHz and dipole rotation mechanisms predominate. Microwave and RF energy is transmitted to the product as an electromagnetic wave and does not require contact with the product. Microwave and RF energy can interact with products through other media, e.g., packaging or containers or surrounding fluids which are normally chosen to be microwave or RF transparent, i.e., materials that do not themselves heat up. RF and microwaves are suitable for heating of solids and liquids and mixtures or these. They do not heat gases. These heating processes are relatively expensive to implement due to the cost of the RF and microwave generation and transmission equipment, the applicator requirements and costs to safely apply these energy forms, and the electrical energy running costs. Energy delivery to the product typically ranges from 50% to 75% of the supply electricity consumed. RF and microwave techniques are very often combined to good effect with convective and radiative heating techniques, where the energy input is split and beneficial characteristics of the combined heating methods can be realized. The most common applications are cooking, baking, drying, defrosting/tempering.

Where both RF and microwave are viable options, considerations such as penetration depth and uniformity of heating and product handling will become deciding factors. These will often lead to a preference for RF heating (e.g., in defrosting operations) as the longer wavelength leads to better uniformity and better penetration in most foods. Product infeed and outfeed and emissions control is easier for RF than for microwaves due to its long wavelength. A disadvantage of RF, however, is that it can be more expensive to implement and is more tailored to a single product or product form. RF systems need to be specifically tuned/impedance matched to the process being operated for efficient power delivery. If a variety of product types are to be heated often microwave heating will be preferred. Microwave equipment is more readily available and often utilizes a more general purpose cavity albeit often with compromised heating uniformity. The physical properties and form of the product, and the required process conditions for a particular application will also feed into the choice of preferred method. Sometimes both RF and microwave processes will be equally well suited and no clear reason for one above the other will be evident. Sometimes all three technologies will be candidates.

Whichever the technique and whatever the product, volumetric heating offers the advantage of rapid uniform heating, without any hot surfaces, which with good applicator design generally leads to good quality, high throughputs, and high heating efficiency. Volumetric heating is particularly applied where conventional nonvolumetric heating techniques are found to be deficient. Environmental benefits for novel thermal processes using microwave RF and ohmic heating were evaluated by Pereira and Vicente (2010) who assessed that many applications can provide food products of improved quality with reduced environmental impact and with reduced processing costs.

Volumetric heating can produce food characteristics and types that cannot normally be produced by conventional heating techniques. Improved product quality, product and particulate identity, nutrient content, natural appearance and color, and enhanced taste and texture are often realized. The combination of dielectric and conventional heating gives the processor new tools and an ability to control food characteristics giving effects which cannot be achieved by traditional methods. This opens the door to novel processes, new products, new textures and colors or absence of color. Novel and emerging processes continue to be developed and will develop further with innovative engineering and judicious use of the principals and characteristics of the different methods. Attention is drawn to some of these areas.

Puffing or expansion of products can be achieved with microwave processing. A single mode high electric field applicator is very effective at puffing cereals and grains. The process can proceed in a tightly regulated way with fluidized grains being puffed/cooked and then emerging in an airstream when puffed due to buoyancy difference with unpuffed grains. Other high electric field applicators can also be used for filament type products or for snack food products. Heating under pressure with microwaves and depressurizing has been used in industry for expansion of starch beads to produce biodegradable packaging. This process could readily find application for novel food products.

The large penetration depth and the precise energy delivery and uniformity which can be engineered for RF and ohmic heating suggest these could be applied with advantage to dough proving processes where a large mass of dough requires very controlled and uniform heating.

Effects of electric fields in food preservation and processing is growing area of development. The pulsed electric field (PEF) process is now an established technology however novel processes are emerging utilizing moderate electric fields which utilize lower electric fields than PEF and which cross over with the fields generally employed with ohmic heating. Machado et al. (2010) demonstrated by preventing temperature rise that the electric fields of 50V/cm to 280V/cm in isolation was responsible for nonthermal death of *Escherichia coli* cells. This opens the door for synergistic processes based on thermal but also electric field effects. AC electric fields of 600V/m have shown promise in accelerating aging of young wines (Zeng et al., 2008). Some other new process enhancements using ohmically applied electric fields are reported for extraction processes and some fermentation processes. Hu et al. (2012) reports increased glucose consumption and increased ethanol production with ohmic heating applied to fermentation of Sake yeast.

Ohmic tempering and thawing has been demonstrated where measures are taken to ensure sufficient electrical conductivity in frozen blocks and where electrode contact is actively maintained. Fast tempering with reduced textural changes compared with conventional thawing and reduced drip losses are reported (Seyhun, 2014).

Ohmic heating has been shown to be very effective for fast blanching and cooking of vegetables. Icier et al. (2006) has studied peroxidase enzyme inactivation in pea puree with ohmic heating and found faster and more uniform inactivation with ohmic heating achieving an inactivation time of 54s compared with 300s with conventional water blanching. Ohmic blanching has the potential to blanch whole or large vegetables pieces reducing dicing and reducing blanching waste water arisings. Fast cooking of a range of vegetables has been demonstrated using batch ohmic equipment (see Figures 8.8 and 8.10) including cooking of pinto beans in 20min compared to 45min and cooking of whole beetroot in 10min. Ohmic heating has been identified along with IR heating and power ultrasound as a novel method for peeling of tomatoes which can eliminate lye (NaOH) from the process with environmental benefits (Rock et al., 2012). The economics are not yet compared to the conventional process.

Ohmic heating has also been successfully applied to baking bread for Japanese crumb production where a reduced crust is required. Similarly, microwave and RF-assisted baking of bread has been implemented with low temperature air/steam for bread baking to produce product which appears crustless, i.e., with no color development on the surface.

Ohmic heating of solid products is also attracting interest for cooking of regular shaped meat joints and pork loins and a viable process is a real prospect. BCH Ltd in the United Kingdom are active in this field and methods have been investigated, pilot equipment built and tested and protocols have been studied (Zell et al., 2010). Ohmic heating for tempering and thawing of block frozen foods where sufficient conductivity is present is another area of current development with potential fast cycle times and low microbial growth.

All three volumetric techniques have been applied to continuous cooking of pumped products, meat slurries, sausage, petfood, blood products, scrambled egg, all with product emerging from a pipe section as a continuous cooked form ready for packing/slicing or subsequent processing. Rapid cooking and pasteurization can be achieved in a few minutes with very close temperature uniformity across the product. Large cross sections can be engineered depending on the technique used and shaped cross sections can be engineered for novelty shaped foods. For improved economics both in terms of running costs and the capital cost of equipment ohmic heating as a first stage followed by RF or microwave heating as a subsequent stage can be an improved process.

Another novel process, engineered by C-Tech Innovation, has combined ohmic heating with microwave heating for food waste minimization for use on military ships reducing galley and dining hall waste to a sterile crumb, where small size and very high energy density was needed for the dewatering and sterilizing requirement.

Novel cooking operations can be designed based upon some of the distinct differences in product interactions and techniques for applying the electric field. An RF applicator that can cook and heat your apple pie while not melting the adjacent ice-cream is a thought provoking demonstration that gives an insight into some possibilities. Similarly, an RF defroster system can be configured to defrost a cake from deep frozen while not melting the cream on top.

Heating in pack and using susceptor materials with microwaves and RF is becoming increasingly well-developed allowing multicomponent prearranged food combinations to be heated and cooked with improved quality compared to some years ago. Arrangements for in-pack deployment of ohmic electrodes are being patented and developed, resulting in novel options for reheating, cooking, and pasteurizing foods (Sastry et al., 2013).

There will be a growing role within food processing and food manufacture for volumetric heating processes due to the unique processing characteristics that they embody and the benefits they bring which cannot be achieved by conventional processes. This is foreseen for both the established volumetric heating processes on their own but even more so for the processes which combine volumetric heating with conventional techniques and the novel processes which utilize electric field effects. Equipment costs for all the volumetric methods are falling. Low cost RF and small-scale RF generators are in development. Microwave technology is already relatively cheap but solid state technology is reducing in cost and increasing in power. Ohmic heating power supplies and control equipment costs are reducing and new ohmic electrode materials are also being actively developed. Food scientists, engineers, and marketing interests will fuel the development of new processes and new products. Food safety and food security concerns will lead to increased adoption of advanced processes using these methods.

References

Anderson, A.K., Finkelstein, R., 1919. A study of the electropure process of treating milk. Journal of Dairy Science 2, 374–406.

Anderson, D.R., 2008. Ohmic heating as an alternative food processing technology (MSc Thesis). Kansas State University.

Anon, 2006. Kingsmill Cuts Off Crusts in Bid for a Bigger Slice. The Grocer (17.06.06.).

Anon, 2007. Greater Visibility for the Invisible. The Grocer (20.01.07.).

Avasoo, M., Johansson, L., 2011. Evaluation Processing Technologies for Strawberry jam (MSc Thesis in Food Technology). Lund University, Sweden.

Awuah, G.B., Koral, T., Guan, D., 2014. Radio-Frequency baking and roasting of food products. In: Awuah, G.B., Ramaswamy, H.S., Tang, J. (Eds.), Radio-Frequency Heating in Food Processing: Principles and Applications. CRC Press, pp. 231–246.

Benabderrahmane, Y., Pain, J.P., 2000. Thermal behaviour of a solid/liquid mixture in an ohmic heating sterilizer-slip phase model. Chemical Engineering and Science 55, 1371–1384.

Castro, I., Teixeira, J.A., Salengke, S., Sastry, S.K., Vicente, A.A., 2004. Ohmic heating of strawberry products: electrical conductivity measurements and ascorbic acid degradation kinetics. Innovative Food Science and Emerging Technologies 5, 27–36.

Cho, H.Y., Sastry, S.K., Yousef, A.E., 1999. Kinetics of inactivation of *Bacillus subtilis* spores by continuous or intermittent ohmic and conventional heating. Biotechnology and Bioengineering 62 (3), 368–372.

Christian, G., Leadley, C.E., 2006. New Technologies Bulletin, 32. Campden & Chorleywood Food Research Association.

Datta, A.K., Sumnu, G., Raghavan, G.S.V., 2005. Dielectric properties of foods. In: Rao, M.A., Rizvi, S.S.H., Datta, A.K. (Eds.), Engineering Properties of Foods, third ed. CRC Press–Taylor and Francis Group.

De Halleux, D., Piette, G., Buteau, M.L., Dostie, M., 2005. Ohmic cooking of processed meats: Energy evaluation and food safety considerations. Canadian Biosystems Engineering 47, 3.41–3.47.

Dove Thermal Engineering. www.dovethermal.co.uk http://www.dovethermal.co.uk/Process_Heating_Systems.html.

Eliot-Godereaux, S.C., Zuber, F., Goullieux, A., 2001. Processing and stabilisation of cauliflower by ohmic heating technology. Innovative Food Science and Emerging Technologies 2, 279–287.

Erdogdu, S.B., Palazoglu, T.K., Gokmen, V., Senyuva, H.Z., Ekiz, H.I., 2007. Reduction of acrylamide formation in French fries by microwave pre-cooking of potato strips. Journal of the Science of Food and Agriculture 87, 133–137.

Halden, K., de Alwis, A.A.P., Fryer, P.J., 1990. Changes in the electrical conductivity of foods during ohmic heating. International Journal of Food Science Technology 25, 9–25.

Hu, H., Yonezawa, Y., Matsuda, A., Ishida, N., Noguchi, A., 2012. Influences of ultrasound and ohmic heating on growth of Sake yeast. Food Science and Technology International 18 (5), 611–616.

Icier, F., Yildiz, H., Baysal, T., 2006. Peroxidase inactivation and colour changes during ohmic blanching of pea puree. Journal of Food Engineering 74, 424–429.

Jones, F., 1897. Apparatus for Electrically Treating Liquids. US Patent 1592735.

Jones, P.L., 1981. Heat and Mass Transfer in Radio-frequency Dryer (PhD Thesis). Loughborough University, England.

Kichlu, K., Kichlu Kamini., 1988. Method for the treatment of fish and meat. - European Patent 0278592 A2.

Khraisheh, M.A.M., McMinn, W.A.M., Magee, T.R.A., 2004. Quality and structural changes in starchy foods during microwave and convective drying. Food Research International 37 (5), 497–503.

Leizerson, S., Shimoni, E., 2005a. Stability and sensory shelf life of orange juice pasteurized by continuous ohmic heating. Journal of Agricultural and Food Chemistry 53 (10), 4012–4018.

Leizerson, S., Shimoni, E., 2005b. Effect of ultrahigh-temperature continuous ohmic heating treatment on fresh orange juice. Journal of Agricultural Food Chemistry 53 (9), 3519–3524.

Lima, M., Sastry, S.K., 1999. The effects of ohmic frequency on hot-air drying rate and juice yield. Journal of Food Engineering 41, 115–119.

Lyng, J., McKenna, B.M., 2007. Ohmic pasteurisation of meat & meat products. In: Kutz, M. (Ed.), Handbook of Farm, Dairy, and Food Machinery. Elsevier.

Machado, L.F., Pereria, R.N., Martins, R.C., Teixeira, J.A., Vincente, A.A., 2010. Moderate electric fields can inactivate *Escherichia coli* at room temperature. Journal of Food Engineering 96, 520–527.

Mans, J., Swientek, B., 1993. Electrifying progress in aseptic technology. Prepared Foods 162 (9), 151–156.

Mitchell, F.R.G., de Alwis, A.A.P., 1989. Electrical conductivity meter for food samples. Journal of Physics E: Scientific Instruments 22, 554–556.

McKenna, B.M., Lyng, J., Brunton, N., Shirsat, N., 2006. Advances in Radio Frequency and ohmic heating of meats. Journal of Food Engineering 77, 215–229.

McKenna, S., 2006. Mac Foods. Meat tenderising apparatus and a method of operation therefor. WO Patent 2006/043085.

Meredith, R., 1998. Microwave interaction with dielectric materials. In: Engineer's Handbook of Industrial Microwave Heating. The Institution of Electrical Engineers London, Power Series 25.

Metaxas, A.C., Meredith, R.J., 1993. Industrial Microwave Heating. In: IEE Power Engineering Series, 4.

Ng, W.X., Tiwari, B.K., Musa, H., 2014. Industrial application of novel thermal technology on manufacturing of caramel colour. In: 7th Annual Manchester Metropolitan University Postgraduate Research Conference, Manchester, 2014.

Ozkan, N., Ho, I., Farid, M., 2004. Combined ohmic and plate heating of hamburger patties – quality of cooked patties. Journal of Food Engineering 63, 141–145.

Palaniappan, S., Sastry, S.K., 1991. Electrical conductivities of selected solid foods during ohmic heating. Journal of Food Process Engineering 14, 221–236.

Palaniappan, S., Sastry, S.K., Richter, E., 1992. Effects of electroconductive heat treatment and electrical pre-treatment on thermal death kinetics of selected microorganisms. Biotechnology and Bioengineering 39 (2), 225–232.

Parrott, D.L., 1992. Use of ohmic heating for aseptic processing of food particulates. Food Technology 36 (12), 68–72.

Pereira, R.N., Vicente, A.A., 2010. Environmental impact of novel thermal and non-thermal technologies in food processing. Food Research International 43, 1936–1943.

Perkin, R.M., 1983. The drying of porous materials with electromagnetic energy generated at radio frequency and microwave frequencies. ECRC/M1646. Electricity Council Research Centre, UK.

Perkin, R.M., Maloney, N.J., Jones, I.H., Kell, D., Ellis, D., 2012. C-Tech Innovation. Electromagnetic heating reactor and improvements. US Patent 2012/0132649.

Rahman, M.S., 1999. Handbook of Food Preservation. Dekker, New York, pp. 521–532.

Rock, C., Yang, W., Goodrich-Schneider, R., Feng, H., 2012. Conventional and alternative methods for tomato peeling. Food Engineering Reviews 4, 1–15.

Rowley, A.T., 2001. Radio frequency heating. In: Richardson, P. (Ed.), Thermal Technologies in Food Processing. Woodhead Publishing.

Ruan, R., Ye, X., Chen, P., Doona, C.J., Taub, I., 2001. ohmic heating. In: Richardson, P. (Ed.), Thermal Technologies in Food Processing. CRC Press Woodhead Publishing.

Sahin, S., Sumnu, G.S., 2008. Alternative Frying Technologies. In: Sumnu, G.S., Sahin, S. (Eds.), Advances in Deep-Fat Frying of Foods. CRC Press, pp. 289–302.

Sastry, S.K., Heskitt, B.F., Jun, S., Somavat, R., 2013. Ohio State University. Ohmic Heating Packet. US Patent 2013/062332.

Seyhun, N., 2014. ohmic heating as thawing & tempering technology. In: Ramaswamy, H.S., Marcotte, M., Sastry, S.K., Abdelrahim, K. (Eds.), ohmic Heating in Food Processing. CRC Press.

Simpson, D.P., 1983. Electricity Council. Apparatus for heating electrically conductive flowable media. US Patent 4,434,537.

Simpson, D.P., Stirling, R., 1995. EA Technology. Ohmic Heater Including electrodes arranged along a flow axis to reduce leakage current. US Pat. 5440667.

Simpson, D.P., 1994. EA Technology. Internal resistance ohmic heating apparatus for fluids, UK Patent No. GB2268671A.

Skudder, P.J., 1993. ohmic heating. In: Willhoft, E. (Ed.), Aseptic processing & packaging of particulate foods. Springer.

Smith, F.J., 1979. Microwave energy applications. Newsletter XII (6), 6–12.

Stirling, R., 1992. Electricity Council. Surface Fouling Resistant Materials. UK Patent GB2231761.

Stirling, R., 1993. EA Technology. Ohmic heating apparatus using electrodes formed of closed micro-porosity material. US Pat. 5226106.

Stirling, R., Coombes, S.A., 1990. Electricity Association Services Limited. Apparatus for heating electrically conductive flowable material flowing through a pipeline. US Pat. 4959525.

U.S. Food & Drug Administration, 2011. Safe practices for food processes. Kinetics of microbial inactivation for alternative food processing technologies – ohmic and inductive heating.

Vikram, V.B., Ramesh, M.N., Prapulla, S.G., 2005. Thermal degradation kinetics of nutrients in orange juice heated by electromagnetic and conventional methods. Journal of Food Engineering 69, 31–40.

Von Hippel, A., 1954. Dielectric Materials and Applications. Technology Press of M.I.T. and Wiley and Sons, New York.

Wang, W.C., Sastry, S.K., 2000. Effects of thermal and electrothermal pretreatments on hot air drying rate of vegetable tissue. Journal of Food Process Engineering 23 (4), 299–319.

Wharton, Y., 2014. Microwave Chemistry. In: Letcher, T., Scott, J., Patterson, D. (Eds.), Chemical Processes for a Sustainable Future. Royal Society of Chemistry.

Wiggins, C., Cauvain, S.P., 2015. Proving baking and cooling. In: Cauvain, S.P. (Ed.), Technology of Breadmaking. Springer, pp. 141–173.

Zeng, X.A., Yu, S.J., Zhang, L., Chen, X.D., 2008. The effects of AC electric field on wine maturation. Innovative Food Science and Emerging Technologies 9, 463–468.

Yang, T., Cohen, J.S., Kluter, R.A., Tempest, P., Manvell, C., Blackmore, S.J., Adams, S., 1997. Microbiological and sensory evaluation of six comically heated stew type foods. Journal of food quality 20, 303–313.

Zell, M., Lyng, J.G., Cronin, D.A., Morgan, D.J., 2010. Ohmic cooking of whole turkey meat – effect of rapid ohmic heating onselected product parameters. Food Chemistry 120, 724–729.

Zhang, L., Lyng, J.G., Brunton, N.P., 2006. Quality of radio frequency heated pork leg and shoulder ham. Journal of Food Engineering 75, 275–287.

Zhong, T., Lima, M., 2003. The effect of ohmic heating on vacuum drying rate of sweet potato tissue. Bioresource Technology 87 (3), 215–220.

Zhu, S.M., Zareifard, M.R., Chen, C.R., Marcotte, M., Grabowski, S., 2010. Electrical conductivity of particle-fluid mixtures in ohmic heating – measurement and simulation. Food Research International 43 (6), 1666–1672.

Emerging Nonthermal Food Preservation Technologies

9

Kumari Shikha Ojha[1,2], *Brijesh K. Tiwari*[1], *Colm O'Donnell*[2], *Joseph P. Kerry*[3]
[1]Food Biosciences, Teagasc Food Research Centre, Dublin, Ireland; [2]School Biosystems and Food Engineering, University College Dublin, Dublin, Ireland; [3]Food Packaging Group, School of Food and Nutritional Sciences, University College Cork, Cork, Ireland

9.1 Introduction

The main objective of food preservation technologies employed in the food industry is to control and eliminate microorganisms responsible for the safety and quality of food products. Over past several decades, the focus of the food industry has shifted toward novel nonthermal technologies to reduce the thermal impact of classical heat-based technologies. This shift is attributed to various reasons, including (1) consumer demand for safe and nutritious food products; (2) process and energy efficiency; and (3) legislative and regulatory requirements aiming to minimize the use of chemical preservatives. Research on novel technologies is ongoing around the world with numerous potential food industry applications. Food processors are increasingly investigating new and innovative food preservation technologies for potential commercial applications. Emerging food processing and preservation technologies, including high pressure processing (HPP), pulsed ultraviolet (UV) light, dense phase carbon dioxide, pulse electric field, power ultrasound, cold atmospheric plasma, and ozone have gained much attention in recent years. These technologies offer numerous advantages, including improved product safety profile, shelf life, and nutritional and sensory properties. Emerging nonthermal technologies can be employed to extend shelf life, where mild processing can provide a significant benefit for short shelf-life products but with minimal effects on quality parameters particularly for products valued for their taste and nutritional profiles. Even though these technologies have been demonstrated to have numerous advantages applications in the food industry to date has been limited due to several factors. Cost and throughput of these technologies along with regulatory approvals of these technologies act as potential barriers to market uptake. However, adoption of these emerging nonthermal technologies should result in added value to product quality or increased equipment performance at lower cost in order to be acceptable to both the food industry and market. The objective of this chapter is to summarize various commercial applications of emerging nonthermal technologies in the food industry intended for food preservation. Emerging nonthermal preservation techniques considered in this chapter include UV light, cold plasma, pulsed electric field, ultrasound, dense phase carbon dioxide, and ozone processing of various food commodities. Various technological and adoption challenges are also discussed.

Innovation and Future Trends in Food Manufacturing and Supply Chain Technologies
http://dx.doi.org/10.1016/B978-1-78242-447-5.00009-5

9.2 Emerging Nonthermal Technologies

Several novel nonthermal technologies have been widely investigated for preservation purposes. In addition, a number of commercial food processing plants are currently successfully exploiting novel technology for niche applications. The literature suggests that these technologies will assist food processors to meet both consumer demands for high-quality, nutritious, and safe products and also industry's requirement for energy efficient processes. However, some of these technologies are still not routinely employed at a commercial scale due to various factors. In the following section, some of these key emerging technologies are discussed along with their commercial applications.

9.2.1 *Power Ultrasound*

Power ultrasound is widely employed in the research and development of various industrial processes including several unit operations carried out in the food industry. The history of scientific advances and discovery of ultrasound are rooted in the study of sound, with Sir Isaac Newton first proposing his theory of sound waves in 1687 (Mason, 1998). The sound ranges employed can be divided into high-frequency, low-energy diagnostic ultrasound, and low-frequency, high-energy power ultrasound. Ultrasound can also be broadly classified as low intensity sonication (<1 W/cm^2) and high intensity sonication (10–1000 W/cm^2). The former is typically used as a nondestructive analytical technique for quality assurance and process control, especially in the case of physicochemical properties such as composition, structure, and physical state of matter, whereas high intensity sonication is employed for extraction and processing applications. Various physical and chemical phenomena, including agitation, vibration, pressure, shock waves, shear forces, microjets, compression, rarefaction, acoustic streaming, cavitation, and radical formation are responsible for ultrasonic effect (Ölmez and Kretzschmar, 2009; Teagasc, 2013). The main driving force for the preservation effects of ultrasound processing is acoustic cavitation. When ultrasound propagates through any medium, it induces a series of compression and rarefaction in the molecules of the medium. Such alternative pressure changes cause formation and ultimately collapse of bubbles in a liquid medium. This phenomenon of the creation, expansion, and implosive collapse of microbubbles in ultrasonically irradiated liquids is known as "acoustic cavitation." The collapse of cavitation bubbles is a near adiabatic process and results in the generation of high temperature and localized high pressure within the bubble (Suslick et al., 1986).

Ultrasound processing finds several commercial applications in various chemical and allied industries. Within the food industry, ultrasound processing has emerged as an alternative food processing option to conventional heat-based pasteurization and sterilization of food products (O'Donnell et al., 2010). Literature suggests that the antimicrobial efficacy of ultrasound is relatively low in some conditions, and under specific conditions ultrasound can be an effective alternative decontamination technique (Bilek and Turantaş, 2013). In most cases, ultrasound alone or in combination with heat (thermosonication), pressure (manosonication), and heat plus pressure (manothermosonication) has been shown to be effective for inactivation of microorganisms in a range of food products (Mason et al., 2005). Ultrasound processing has been identified as a potential technology to meet the United States Food and Drug Administration (USFDA) requirement of a 5 log reduction of *Escherichia coli* in liquid foods (Tiwari and Mason, 2012). For example, D'amico et al. (2006) achieved a 5 log reduction of *Listeria monocytogenes* and a 6 log reduction in *E. coli* O157:H7 in apple cider using a continuous flow ultrasound treatment combined with mild heat (57 °C) for 18 min. Ultrasound processing of liquid foods has been reported to be effective against food-borne pathogens found in a range of juices (Cheng et al., 2007; Ferrario et al., 2015; Martínez-Flores et al., 2015; Moody et al., 2014; Valero et al., 2007).

There are many reports indicating ultrasound-assisted inactivation of enzymes responsible for quality deterioration of liquid and solid foods, including peroxidase, pectin methyl esterase, lipase, peroxidase, and polyphenol oxidase (Cheng et al., 2013; Islam et al., 2014; Jang and Moon, 2011) producing undesirable changes in many foods during processing and storage periods when used alone or with temperature and pressure. However, in some cases it has been reported that the sonication enhances enzyme activity as in the case of polyphenol oxidase activities in *Panax ginseng* cells (Wu and Lin, 2002). Jang and Moon (2011) observed an increase in activity of monophenolase, diphenolase, and peroxidase with ultrasound treatment alone in an ultrasonic bath (40 kHz), while combined treatment of ascorbic acid (1%) and ultrasound showed a significant decrease in enzymatic activities in the case of fresh-cut Fuji apples.

The promising effect of sonication alone or in combination with heat or pressure has been demonstrated at laboratory scale. However, scale-up remains a significant challenge for industrial adoption. For application of power ultrasound at an industrial scale, it is essential to have energy efficient ultrasound processors to ensure the economic feasibility of the technology. Nonstandardized reporting of extrinsic and intrinsic control parameters, including ultrasound power level, amplitude, processing time, and raw material characteristics is evident in many publications. Standardized reporting in terms of energy density, probe types, and sample volumes would facilitate technoeconomic assessment prior to industrial application (O'Donnell et al., 2010). There are few reported industrial scale applications of power ultrasound for food preservation purposes. Commercially, ultrasound has been demonstrated as an effective technique for surface decontamination of fresh produce, and when used in combination with other disinfectants it can improve process efficiency while reducing the usage of chemical disinfectant. For example, Huang et al. (2006) achieved 2.5 and 4.3 log reductions in *Salmonella* and *E. coli* O157:H7 counts on apples when combined ultrasound (170 kHz) and ClO_2 (20–40 mg/L) treatments depending on the exposure time were applied. Table 9.1 shows some examples for use of high-power ultrasound alone or in combination with other chemical disinfectants for microbial reductions in various fruits and vegetables. Ultrasound as a physical disinfectant technique for fruit and vegetable has been recommended for the food industry (Bilek and Turantaş, 2013). Commercially, ultrasonic baths (Figure 9.1(a) and (b)) are employed to reduce microbial load of fresh fruits and vegetables at retail level and horn-based systems (Figure 9.1(c)) are employed for liquid processing.

Table 9.1 **Application of Ultrasound Applications Alone or in Combination with Other Techniques for Microbial Reductions for Some Fruits and Vegetables**

Product	Treatments	Ultrasound (US) Parameters	Microbial Reductions[a] (log_{10} CFU/g Sample)	References
Strawberry	US alone	350 W/L; 40 kHz; 20 °C; 10 min	TVC: 0.6YMC: 0.5	Cao et al. (2010)
Lettuce	US alone	280 W/L; 20 kHz; 53 min	*Escherichia coli* O157:H7: 4.4 in wash water	Elizaquível et al. (2012)
Strawberry	US alone	120 W; 35 kHz; 15 °C sample/water: 1/25	TVC: 0.6YMC: 1.4	Alexandre et al. (2012)
Red bell pepper	US alone	120 W; 35 kHz; 15 °C sample/water: 1/25	*Listeria innocua*: 1.98	Alexandre et al. (2013)
Iceberg lettuce	US alone	10 W/L; 32–40 kHz; 10 min sample/water:1/20	*Salmonella* Typhimurium: 1.5	Seymour et al. (2002)
		US + chlorinated water (25 ppm free chlorine)	*S.* Typhimurium: 2.7	
Shredded carrot	US alone US + chlorinated water (200 ppm free chlorine)	45 kHz; 1 min	TVC: 1.3YMC: 0.9TVC: 1.0YMC: 0.9	Alegria et al. (2009)
Cherry tomatoes	US alone US + peracetic acid (40 mg/L)	45 kHz; 10 min; 25 °C	*Salmonella enterica* Typhimurium: 0.8 *S. enterica* Typhimurium: 3.9	São José and Vanetti (2012)
Lettuce	US + ClO_2 (5 and 10 ppm)	170 kHz; 6–10 min	*Salmonella* spp.: 2.2–2.9 *E. coli* O157:H7: 1.3–2.2	Huang et al. (2006)
Apple			*Salmonella* spp.: 3.1–4.2 *E. coli* O157:H7: 2.2–3.8	

Table 9.1 Continued

Product	Treatments	Ultrasound (US) Parameters	Microbial Reductions[a] ($\log_{10}$ CFU/g Sample)	References
Spinach leaves	US + acidified sodium chloride (200 mg/L)	200 W/L; 21.2 kHz; 2 min	*E. coli* O157:H7: 4	Zhou et al. (2009)
Lettuce	US + lactic/citric/malic acid (2%)	30 W/L; 40 kHz; 5 min	*E. coli* O157:H7: 2.7 *S.* Typhimurium: 3.2 *L. monocytogenes*: 2.9	Sagong et al. (2011)
Truffles	US + ethanol (70%)	35 kHz; 4 °C; 10 min	TVC: 4 mold count: <1.7 yeast count: <0.5 *Pseudomonas* spp.: >4 *Enterobacteriaceae*: 3.6 lactic acid bacteria: 3.5	Rivera et al. (2011)
Plum fruit	US + ClO_2 (40 mg/L) US + peracetic acid (40 mg/L)	100 W; 40 kHz; 20 °C; 10 min sample/water: 1/5	TVC (mesophilic):3TVC (psychrotrophic):2.9YMC: 2	Chen and Zhu (2011)

TVC, total viable counts (mesophilic); YMC, yeast and mold count; CFU, colony-forming unit.
[a]Microbial reductions in given range.
Adapted from Bilek and Turantaş (2013).

9.2.2 Pulsed Electric Field

Pulsed electric field (PEF) is an emerging nonthermal technique that has gained increasing attention in recent decades for various unit operations. Food industry applications include food dehydration, enzyme activity modification, wastewater treatment, preservation of various food products, and improvement of juice and metabolite extraction from various plant tissues (Bajgai and Hashinaga, 2001; Mohamed et al., 2012). PEF processing involves the application of short pulses (1–10 μs) of high-intensity electric field (typically 20–80 kV/cm) to fluid foods placed between two electrodes either in batch or continuous flow system (Mosqueda-Melgar et al., 2008). PEF has shown the ability to inactivate enzymes, pathogenic and spoilage microorganisms at ambient temperatures, and thus extend the overall shelf life of different food products without significant changes in sensorial and nutritional attributes (Bendicho et al., 2002;

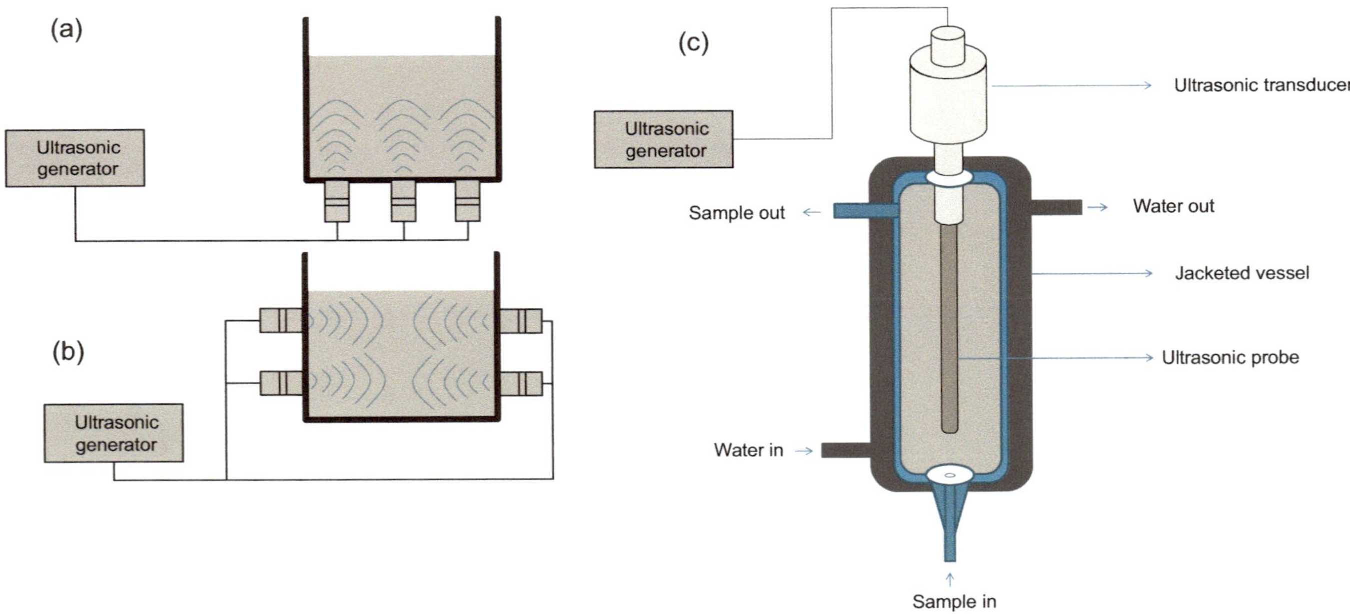

Figure 9.1 Various types of ultrasound-based systems employed for washing of fruits and vegetables (a and b) and liquid foods (c). Ultrasonic bath (a and b); ultrasonic horn-based (c).
Modified from Alexandru et al. (2013).

Table 9.2 Cost Benefits Analysis of PEF Compared with HPP and Thermal Processing

Parameters	PEF	HPP	Thermal Processing
Total production cost (¢/L)	3.7	10.7	1.5
Total CO_2 equivalent emissions (kg CO_2/year)	700,000	773,000	90,000
Total electricity consumption (kW/year)	38,200	920,000	1,020,000

Adapted from Sampedro et al. (2014).

Griffiths and Walkling-Ribeiro, 2014; Sampedro et al., 2005). PEF has potential application for pasteurization of a wide range of liquid foods including juices, milk, yogurt, soups, liquid eggs, and beer (Agcam et al., 2014; Buckow et al., 2014; Milani et al., 2015). Despite several promising applications, there are significant challenges encountered in large-scale commercial applications. One of the limiting factors is the electric field strength (20–50 kV/cm) required to achieve microbial inactivation. The electric field strength required is possible at a laboratory or pilot scale; however, commercial scale units require higher voltages or special electrode arrangements to achieve the desired conditions for microbial inactivation. Another limiting factor for PEF processing is the fact that PEF is very effective against vegetative bacteria compared with spores; hence, PEF can be exploited for fluid foods. Technological challenges persists for solid food applications. The reluctance of the food industry to adopt the PEF technique for commercial production is due to high capital investment, complexity of PEF process control parameters, and lack of consistent data in microbial reduction for a 5 log reduction of a target pathogen (Jin et al., 2015). In spite of these limitations and challenges, to date there are over 50 commercial units operated worldwide for various food applications including liquid food preservation (Toepfl, 2012). Cost benefit analyses, including environmental impact assessment of a commercial scale PEF, HPP, and thermal pasteurization for orange juice as a model were carried out by Sampedro et al. (2014) and validated for a 2-month shelf life under refrigeration conditions. This study showed that both the HPP and PEF processing techniques are costly and have higher environmental impact in terms of greenhouse gas emissions (CO_2 production) compared with conventional thermal processing as shown in Table 9.2 (Sampedro et al., 2014). However, nutritional and quality attributes may compensate higher production cost. Hence, technoeconomic and cost benefit analysis is essential prior to future wider adoption of PEF in the food industry.

9.2.3 Dense Phase Carbon Dioxide

Dense phase carbon dioxide (DPCD) is a nonthermal technique for food preservation that can ensure food safety and extended shelf life with minimal impact on food quality.

Dense phase carbon dioxide processing or supercritical CO_2 is a collective term for liquid CO_2 and supercritical CO_2 or high pressurized carbon dioxide. DPCD has been demonstrated in commercial applications for the extraction of bioactives compounds from plant tissues. The unique properties of DPCD make it an appealing medium for food preservation, which has a strong potential as an antimicrobial agent because it involves the use of nontoxic gas, which can be easily removed by simple depressurization. DPCD has significant lethal effects on microorganisms in food and inactivates spoilage enzymes with a minimal effect on end product quality. A possible mechanism for microbial inactivation is a result of the physical disruption of microbes due to rapid pressure release and expansion of CO_2 within the cell. DPCD is also effective against various food enzymes responsible for spoilage. Inactivation of enzymes by DPCD may be due to the pH lowering effect of extracellular fluids as outlined above and conformational changes in the secondary and tertiary structure of protein (Damar and Balaban, 2006). DPCD is attracting increased interest from the food industry for inactivation of microorganisms and enzymes in various solid and liquid foods, including whole eggs, shrimps, spinach leaves, fruit juices, and beer (Balaban and Ferrentino, 2012; Dagan and Balaban, 2006; Ferrentino and Spilimbergo, 2011; Garcia-Gonzalez et al., 2009; Ji et al., 2012; Zhang et al., 2011).

In terms of commercial application in the food industry, there are several patents issued outlining the use of DPCD for preservation of various liquid foods. Praxair, Inc. is the recognized industrial leader for scaling up and commercializing continuous DPCD processing equipment with a capacity of 151 L/min. The continuous DPCD processing unit was commercialised under the trade name of Better Than Fresh™ (BTF) system (Connery et al., 2005). The BTF system has been validated for fruit juices for microbial safety, enzyme inactivation, and shelf life (Omer et al., 2010). Processing of fruit juices using BTF system is similar to other DPCD systems consisting of three major stages: (1) the pressurized mixing stage; (2) the reaction stage; and (3) the depressurization stage as shown in Figure 9.2.

9.2.4 Pulsed Light

Since the mid-1990s, a number of researchers have investigated pulsed light as an emerging nonthermal technique for sterilization and/or decontamination, which aims to reduce the spoilage and pathogenic microorganisms without minimal effects on its quality (Bank et al., 1990). In scientific literature, this technique has been recognized by several synonyms, including pulsed ultraviolet light (Sharma and Demirci, 2003), high intensity broad-spectrum pulsed light (Roberts and Hope, 2003), pulsed light (Rowan et al., 1999), and pulsed white light (Marquenie et al., 2003). Pulsed light refers to intense and short time pulses of broad spectrum "white light" (ultraviolet, visible, and near infrared region) produced by xenon lamps (Charles et al., 2013). Though UV disinfection has been applied in drinking water sterilization and surface and air decontamination, pulsed UV light is relatively new and has been demonstrated as an effective technique for microbial inactivation. There has been a debate on the efficacy of continuous versus pulsed UV light against foodborne pathogens. In a study, Luo et al. (2014) demonstrated that pulse frequencies of 8 Hz exhibited better *E. coli* inactivation compared to continuous UV light.

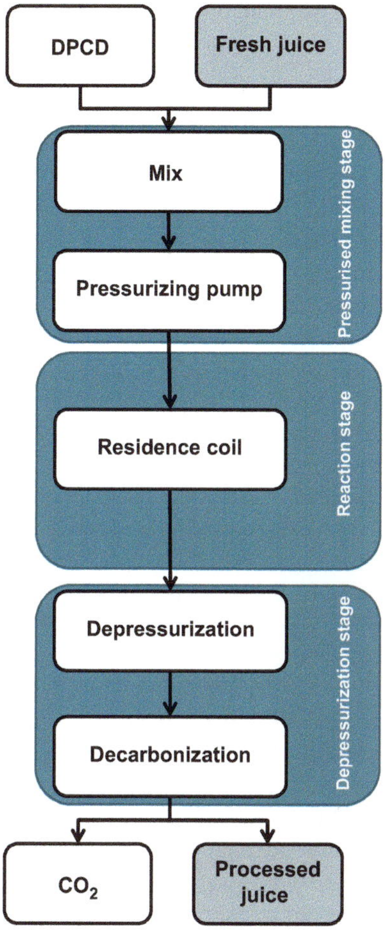

Figure 9.2 Schematic diagram of $DPCO_2$ processing for freshly squeezed orange juice.

Pulsed light has been successfully used in numerous food processing applications at laboratory scale, including pasteurization (Heinz et al., 2001), microbial decontamination (bacteria, yeasts, molds, spores, and viruses) of food, food contact surfaces, and packaging materials. Published results have demonstrated the inactivation of both pathogenic and spoilage microorganisms in a variety of food products, including fruit juices and milk (Krishnamurthy et al., 2007; Pataro et al., 2011), egg surface (Hierro et al., 2009), and ready-to-eat cooked meat products (Hierro et al., 2011). Pulsed light technology in combination with other nonthermal processing technologies has also been investigated. High-intensity light pulses in combination with pulsed electric field were found to be effective for inactivation of *E. coli* in apple juice under optimum combination without affecting the quality of product (Caminiti et al., 2011). High-intensity light pulses in combination with thermosonication were investigated for inactivation of *E. coli* in orange juice (Muñoz et al., 2011). Although pulsed light is not widely used in the food industry, it has been successfully used in large-scale commercial applications for bottle cap decontamination. A number of commercial-scale in-line food packaging decontamination systems based on pulsed light are being used throughout the world (Keener and Krishnamurthy, 2014).

9.2.5 Ozone Processing

Ozone has been used commercially as a disinfectant for the treatment of drinking water since 1906 (Hill and Rice, 1982) and is increasingly employed in the food industry for produce preservation and sanitizing of food-contact surfaces. The interest in ozone as a potential disinfectant is mainly due to its broad spectrum antimicrobial efficacy. In the fresh produce industry, chlorine is the most commonly used disinfectant, which is applied in a form of sodium hypochloride (NaOCl) because of its cost effectiveness and efficacy as an antibacterial and antimicrobial agent (Joshi et al., 2013), or less commonly as chlorine dioxide (ClO_2). The reaction of chlorine with other organic compounds leading to formation of carcinogens has raised issues centered around the chemical safety of fresh produce. Thus, new disinfectants including ozone, organic acids, peroxyacetic acid, electrolyzed oxidizing water, and hydrogen peroxide have gained widespread commercial interest (Joshi et al., 2013; Ölmez and Kretzschmar, 2009; Rico et al., 2007). Unlike chlorine, ozone is considered as a Generally Recognized As Safe chemical. Ozone can be applied both in aqueous and gaseous form for surface decontamination of various food products including fruits and vegetables via washing or storage in ozone-rich atmospheres (Cullen et al., 2009; Cullen et al., 2010). Microbial studies to date show that reductions of spoilage and pathogenic species most commonly associated with food products can be achieved.

Commercially, ozone has been successfully employed for a range of food industry applications including disinfection of fruits and vegetables. The SAMRO AG plant in Burgdorf, Switzerland is a good example of using several novel technologies for decontamination technique for fruits, vegetables, and meat products. This technique involves sequential operations of ozone in combination with UV-C radiation, electrolyzed water, ultrasound washing, and modified air packaging to produce higher quality food products with enhanced shelf life. Briefly, this technique involves washing of fruit and vegetables with ozone-containing water, followed by UV radiation in a disinfection tunnel. Sequential application of ozone gas at high relative humidity followed by UV light treatment is termed as Ventafresh Disinfection Tunnel (Steffen et al., 2010). Figure 9.3 shows a disinfection process flowchart employed at commercial establishment. Similarly, Steffen et al. (2010) demonstrated the use of ozone both in gas and aqueous phases for the preparation of complete meals in a central food processing plant using Ventafresh technology involving defrosting of frozen foods overnight in water with 8–15 mg/L ozone and finally packed in boxes previously washed in ozone water (Figure 9.4).

9.2.6 Cold Atmospheric Plasma

Cold atmospheric plasma (CAP) or ionized air systems offer potential for bacterial inactivation (Laroussi et al., 2000). CAPs are a stable, light-emitting gas

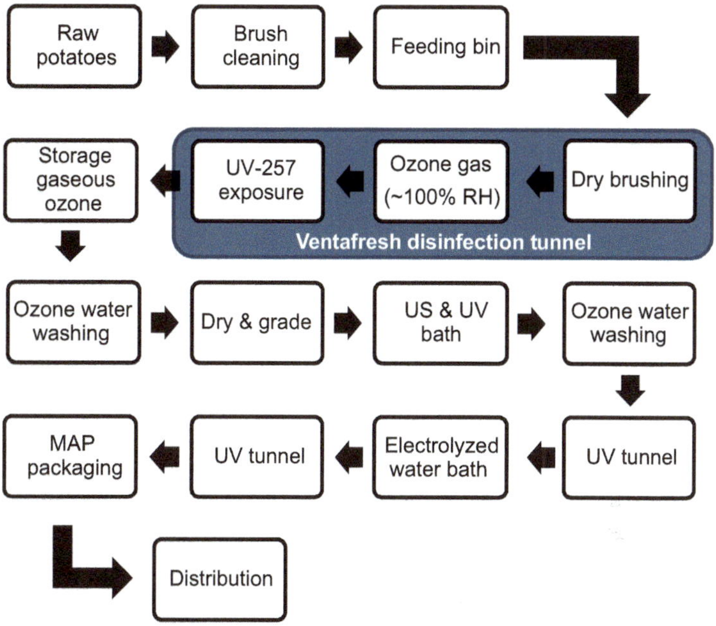

Figure 9.3 Ventafresh technology for disinfection of fresh vegetables.
Adapted from Steffen et al. (2010).

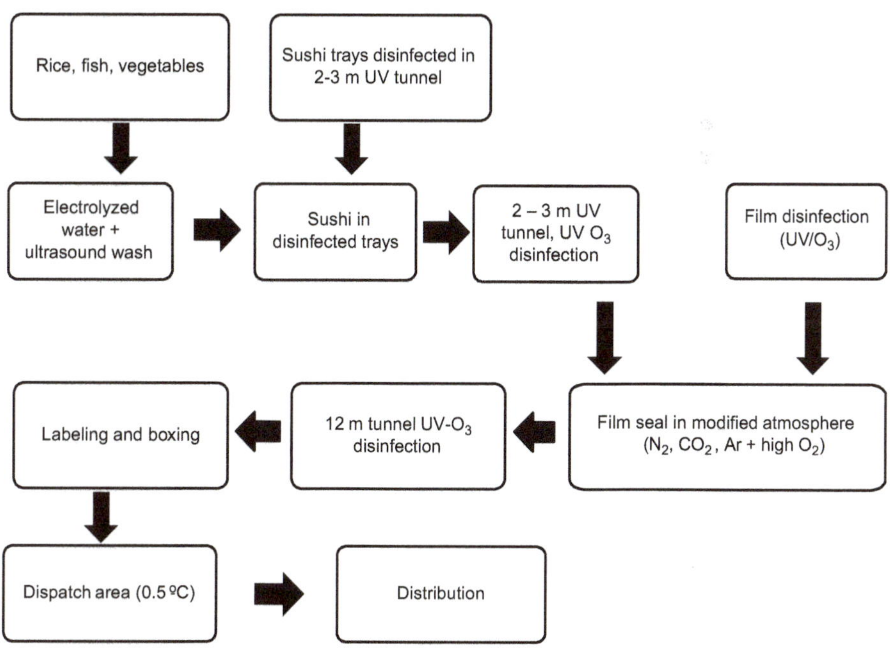

Figure 9.4 Ventafresh process for sushi for 7-day shelf life.
Adapted from Steffen et al. (2010).

that can be delivered to heat-sensitive surfaces such as plant and animal tissues. Decontamination of fresh produce surface is often achieved by washing with chemical disinfectants as discussed previously (see Section 9.2.5); however, these chemical disinfectants rely heavily on potable water usage. The additional impetus of reducing water usage, as well as chemical emissions, changes the focus of decontamination technologies toward nonwater-based or reduced water usage. CAP systems have shown promise as a decontamination tool on a variety of food surfaces (Klockow and Keener, 2009; Misra et al., 2011; Niemira, 2012). Application of nonthermal plasmas for decontamination is currently the subject of significant research with tremendous commercial potential, using either needle-electrode (Deng et al., 2006; Laroussi and Akan, 2007) or parallel-plate (Klockow and Keener, 2009; Nersisyan and Graham, 2004; Ragni et al., 2010) plasma systems. CAP systems are capable of efficiently inactivating a wide range of microorganisms, including *E. coli*, *Listeria*, and *Bacillus subtilis* spores as well as fungi and virus (Fridman et al., 2008; Kong et al., 2009). CAP has numerous potential applications in the food industry, including surface decontamination of fresh meat, poultry, fish, and fruits and vegetables (Fernandez et al., 2013; Noriega et al., 2011; Ziuzina et al., 2014). This technology has also been successfully employed for the surface decontamination of packaging material and improving functional properties of polymeric materials (Güleç et al., 2006; Lee et al., 2015; Ozdemir et al., 1999; Pankaj et al., 2014). The "in-pack decontamination of fresh-cut produce" project (SAFEBAG) which has been funded by the European Commission (EU Framework 7), aims at developing a novel continuous in-pack decontamination system for fresh produce. SAFEBAG has delivered an alternative, which is free from harmful chemicals and can significantly reduce microbial load of fresh fruits and vegetables. Figure 9.5 shows an industrial continuous plasma prototype for decontamination of fresh produce.

Figure 9.5 Industrial continuous plasma prototype for decontamination of fresh produce.

9.3 Conclusions and Future Trends

Emerging nonthermal technologies have been extensively investigated as a novel technique for food preservation. In addition, a number of commercial food processing companies are currently successfully exploiting these technologies for niche applications. However, the technologies are still not routinely employed at a commercial scale. This is mainly due to the high capital investment and relatively low throughput required to employ these technologies and integrate them into existing production lines. The key processing advantages afforded by these technologies such as reduced operational costs should encourage the future adoption of emerging nonthermal food preservation technologies in the food processing industries. These technologies will also assist food processors to meet consumer demands for safe, nutritious, and high quality products. The applications of novel technologies are likely to grow steadily in the future because of increased consumer demands for minimally processed foods. More complex considerations arise for combinations of technologies, particularly with respect to optimization of practical applications. Informed selection of appropriate systems based on detailed sequential studies is necessary.

References

Agcam, E., Akyıldız, A., Evrendilek, G.A., 2014. Comparison of phenolic compounds of orange juice processed by pulsed electric fields (PEF) and conventional thermal pasteurisation. Food Chemistry 143, 354–361.

Alegria, C., Pinheiro, J., Gonçalves, E.M., Fernandes, I., Moldão, M., Abreu, M., 2009. Quality attributes of shredded carrot (*Daucus carota* L. cv. Nantes) as affected by alternative decontamination processes to chlorine. Innovative Food Science & Emerging Technologies 10 (1), 61–69.

Alexandre, E.M., Brandão, T.R., Silva, C.L., 2012. Efficacy of non-thermal technologies and sanitizer solutions on microbial load reduction and quality retention of strawberries. Journal of Food Engineering 108 (3), 417–426.

Alexandre, E.M., Brandão, T.R., Silva, C.L., 2013. Impact of non-thermal technologies and sanitizer solutions on microbial load reduction and quality factor retention of frozen red bell peppers. Innovative Food Science & Emerging Technologies 17, 99–105.

Alexandru, L., Cravotto, G., Giordana, L., Binello, A., Chemat, F., 2013. Ultrasound-assisted extraction of clove buds using batch- and flow-reactors: a comparative study on a pilot scale. Innovative Food Science & Emerging Technologies 20, 167–172.

Bajgai, T.R., Hashinaga, F., 2001. High electric field drying of Japanese radish. Drying Technology 19 (9), 2291–2302.

Balaban, M.O., Ferrentino, G., 2012. Dense Phase Carbon Dioxide: Food and Pharmaceutical Applications. John Wiley & Sons.

Bank, H., John, J., Schmehl, M., Dratch, R., 1990. Bactericidal effectiveness of modulated UV light. Applied and Environmental Microbiology 56 (12), 3888–3889.

Bendicho, S., Barbosa-Cánovas, G.V., Martín, O., 2002. Milk processing by high intensity pulsed electric fields. Trends in Food Science & Technology 13 (6–7), 195–204.

Bilek, S.E., Turantaş, F., 2013. Decontamination efficiency of high power ultrasound in the fruit and vegetable industry, a review. International Journal of Food Microbiology 166 (1), 155–162.

Buckow, R., Chandry, P.S., Ng, S.Y., McAuley, C.M., Swanson, B.G., 2014. Opportunities and challenges in pulsed electric field processing of dairy products. International Dairy Journal 34 (2), 199–212.

Caminiti, I.M., Palgan, I., Noci, F., Muñoz, A., Whyte, P., Cronin, D.A., Morgan, D.J., Lyng, J.G., 2011. The effect of pulsed electric fields (PEF) in combination with high intensity light pulses (HILP) on *Escherichia coli* inactivation and quality attributes in apple juice. Innovative Food Science & Emerging Technologies 12 (2), 118–123.

Cao, S., Hu, Z., Pang, B., Wang, H., Xie, H., Wu, F., 2010. Effect of ultrasound treatment on fruit decay and quality maintenance in strawberry after harvest. Food Control 21 (4), 529–532.

Charles, F., Vidal, V., Olive, F., Filgueiras, H., Sallanon, H., 2013. Pulsed light treatment as new method to maintain physical and nutritional quality of fresh-cut mangoes. Innovative Food Science & Emerging Technologies 18, 190–195.

Chen, Z., Zhu, C., 2011. Combined effects of aqueous chlorine dioxide and ultrasonic treatments on postharvest storage quality of plum fruit (*Prunus salicina* L.). Postharvest Biology and Technology 61 (2), 117–123.

Cheng, L., Soh, C., Liew, S., Teh, F., 2007. Effects of sonication and carbonation on guava juice quality. Food Chemistry 104 (4), 1396–1401.

Cheng, X-f., Zhang, M., Adhikari, B., 2013. The inactivation kinetics of polyphenol oxidase in mushroom (*Agaricus bisporus*) during thermal and thermosonic treatments. Ultrasonics Sonochemistry 20 (2), 674–679.

Connery, K., Shah, P., Coleman, L., Hunek, B., 2005. Commercialization of Better Than Fresh™ Dense Phase Carbon Dioxide Processing for Liquid Food. ISSF, Orlando.

Cullen, P.J., Tiwari, B.K., O'Donnell, C.P., Muthukumarappan, K., 2009. Modelling approaches to ozone processing of liquid foods. Trends in Food Science and Technology 20 (3–4), 125–136.

Cullen, P.J., Valdramidis, V.P., Tiwari, B.K., Patil, S., Bourke, P., O'Donnell, C.P., 2010. Ozone processing for food preservation: an overview on fruit juice treatments. Ozone: Science and Engineering 32 (3), 166–179.

D'amico, D.J., Silk, T.M., Wu, J., Guo, M., 2006. Inactivation of microorganisms in milk and apple cider treated with ultrasound. Journal of Food Protection 69 (3), 556–563.

Dagan, G.F., Balaban, M.O., 2006. Pasteurization of beer by a continuous dense–phase CO_2 System. Journal of Food Science 71 (3), E164–E169.

Damar, S., Balaban, M.O., 2006. Review of dense phase CO_2 technology: microbial and enzyme inactivation, and effects on food quality. Journal of Food Science 71 (1), R1–R11.

Deng, X., Shi, J., Kong, M.G., 2006. Physical mechanisms of inactivation of Bacillus subtilis spores using cold atmospheric plasmas. Plasma Science, IEEE Transactions on 34 (4), 1310–1316.

Elizaquível, P., Sánchez, G., Selma, M., Aznar, R., 2012. Application of propidium monoazide-qPCR to evaluate the ultrasonic inactivation of *Escherichia coli* O157:H7 in fresh-cut vegetable wash water. Food Microbiology 30 (1), 316–320.

Fernandez, A., Noriega, E., Thompson, A., 2013. Inactivation of *Salmonella enterica* serovar Typhimurium on fresh produce by cold atmospheric gas plasma technology. Food Microbiology 33 (1), 24–29.

Ferrario, M., Alzamora, S.M., Guerrero, S., 2015. Study of the inactivation of spoilage microorganisms in apple juice by pulsed light and ultrasound. Food Microbiology 46, 635–642.

Ferrentino, G., Spilimbergo, S., 2011. High pressure carbon dioxide pasteurization of solid foods: current knowledge and future outlooks. Trends in Food Science & Technology 22 (8), 427–441.

Fridman, G., Friedman, G., Gutsol, A., Shekhter, A.B., Vasilets, V.N., Fridman, A., 2008. Applied plasma medicine. Plasma Processes and Polymers 5 (6), 503–533.
Garcia-Gonzalez, L., Geeraerd, A.H., Elst, K., Van Ginneken, L., Van Impe, J.F., Devlieghere, F., 2009. Inactivation of naturally occurring microorganisms in liquid whole egg using high pressure carbon dioxide processing as an alternative to heat pasteurization. The Journal of Supercritical Fluids 51 (1), 74–82.
Griffiths, M.W., Walkling-Ribeiro, M., 2014. Chapter 7-Pulsed electric field processing of liquid foods and beverages. In: Sun, D.-W. (Ed.), Emerging Technologies for Food Processing, second ed. Academic Press, San Diego, pp. 115–145.
Güleç, H.A., Sarloglu, K., Mutlu, M., 2006. Modification of food contacting surfaces by plasma polymerisation technique. Part I: determination of hydrophilicity, hydrophobicity and surface free energy by contact angle method. Journal of Food Engineering 75 (2), 187–195.
Heinz, V., Alvarez, I., Angersbach, A., Knorr, D., 2001. Preservation of liquid foods by high intensity pulsed electric fields—basic concepts for process design. Trends in Food Science & Technology 12 (3), 103–111.
Hierro, E., Barroso, E., De la Hoz, L., Ordóñez, J.A., Manzano, S., Fernández, M., 2011. Efficacy of pulsed light for shelf-life extension and inactivation of *Listeria monocytogenes* on ready-to-eat cooked meat products. Innovative Food Science & Emerging Technologies 12 (3), 275–281.
Hierro, E., Manzano, S., Ordóñez, J.A., de la Hoz, L., Fernández, M., 2009. Inactivation of *Salmonella enterica* serovar Enteritidis on shell eggs by pulsed light technology. International Journal of Food Microbiology 135 (2), 125–130.
Hill, A., Rice, R., 1982. Historical background, properties and applications. In: Handbook of Ozone Technology and Applications, vol. 1.
Huang, T.-S., Xu, C., Walker, K., West, P., Zhang, S., Weese, J., 2006. Decontamination efficacy of combined chlorine dioxide with ultrasonication on apples and lettuce. Journal of Food Science 71 (4), M134–M139.
Islam, M.N., Zhang, M., Adhikari, B., 2014. The inactivation of enzymes by ultrasound—a review of potential mechanisms. Food Reviews International 30 (1), 1–21.
Jang, J.-H., Moon, K.-D., 2011. Inhibition of polyphenol oxidase and peroxidase activities on fresh-cut apple by simultaneous treatment of ultrasound and ascorbic acid. Food Chemistry 124 (2), 444–449.
Ji, H., Zhang, L., Liu, S., Qu, X., Zhang, C., Gao, J., 2012. Optimization of microbial inactivation of shrimp by dense phase carbon dioxide. International Journal of Food Microbiology 156 (1), 44–49.
Jin, T.Z., Guo, M., Zhang, H.Q., 2015. Upscaling from benchtop processing to industrial scale production: more factors to be considered for pulsed electric field food processing. Journal of Food Engineering 146, 72–80.
Joshi, K., Mahendran, R., Alagusundaram, K., Norton, T., Tiwari, B., 2013. Novel disinfectants for fresh produce. Trends in Food Science & Technology 34 (1), 54–61.
Keener, L., Krishnamurthy, K., 2014. Shedding light on food safety: applications of pulsed light processing. Food Safety Magazine 20 (3), 28–33.
Klockow, P.A., Keener, K.M., 2009. Safety and quality assessment of packaged spinach treated with a novel ozone-generation system. LWT-Food Science and Technology 42 (6), 1047–1053.
Kong, M.G., Kroesen, G., Morfill, G., Nosenko, T., Shimizu, T., Van Dijk, J., Zimmermann, J., 2009. Plasma medicine: an introductory review. New Journal of Physics 11 (11), 115012.
Krishnamurthy, K., Demirci, A., Irudayaraj, J., 2007. Inactivation of *Staphylococcus aureus* in milk using flow–through pulsed UV–light treatment system. Journal of Food Science 72 (7), M233–M239.

Laroussi, M., Akan, T., 2007. Arc–free atmospheric pressure cold plasma jets: a review. Plasma Processes and Polymers 4 (9), 777–788.
Laroussi, M., Alexeff, I., Kang, W.L., 2000. Biological decontamination by nonthermal plasmas. Plasma Science, IEEE Transactions on 28 (1), 184–188.
Lee, T., Puligundla, P., Mok, C., 2015. Inactivation of foodborne pathogens on the surfaces of different packaging materials using low-pressure air plasma. Food Control 51, 149–155.
Luo, W., Chen, A., Chen, M., Dong, W., Hou, X., 2014. Comparison of sterilization efficiency of pulsed and continuous UV light using tunable frequency UV system. Innovative Food Science & Emerging Technologies 26 (0), 220–225.
Marquenie, D., Geeraerd, A., Lammertyn, J., Soontjens, C., Van Impe, J., Michiels, C., Nicolaı, B., 2003. Combinations of pulsed white light and UV-C or mild heat treatment to inactivate conidia of *Botrytis cinerea* and *Monilia fructigena*. International Journal of Food Microbiology 85 (1), 185–196.
Martínez-Flores, H.E., Garnica-Romo, M.G., Bermúdez-Aguirre, D., Pokhrel, P.R., Barbosa-Cánovas, G.V., 2015. Physico-chemical parameters, bioactive compounds and microbial quality of thermo-sonicated carrot juice during storage. Food Chemistry 172, 650–656.
Mason, T., Riera, E., Vercet, A., Lopez-Bueza, P., 2005. Application of ultrasound. Emerging Technologies for Food Processing 32, 3–351.
Mason, T.J., 1998. Power ultrasound in food processing – the way forward. In: Povey, M.J.W., Mason, T.J. (Eds.), Ultrasound in Food Processing. Thomson Science, London, pp. 105–126.
Milani, E.A., Alkhafaji, S., Silva, F.V., 2015. Pulsed electric field continuous pasteurization of different types of beers. Food Control 50, 223–229.
Misra, N., Tiwari, B., Raghavarao, K., Cullen, P.J., 2011. Nonthermal plasma inactivation of food-borne pathogens. Food Engineering Reviews 3 (3–4), 159–170.
Mohamed, M.E., Ayman, H., Eissa, A., 2012. Pulsed Electric Fields for Food Processing Technology. INTECH Open Access Publisher.
Moody, A., Marx, G., Swanson, B.G., Bermúdez-Aguirre, D., 2014. A comprehensive study on the inactivation of *Escherichia coli* under nonthermal technologies: high hydrostatic pressure, pulsed electric fields and ultrasound. Food Control 37, 305–314.
Mosqueda-Melgar, J., Elez-Martínez, P., Raybaudi-Massilia, R.M., Martín-Belloso, O., 2008. Effects of pulsed electric fields on pathogenic microorganisms of major concern in fluid foods: a review. Critical Reviews in Food Science and Nutrition 48 (8), 747–759.
Muñoz, A., Palgan, I., Noci, F., Morgan, D., Cronin, D., Whyte, P., Lyng, J., 2011. Combinations of high intensity light pulses and thermosonication for the inactivation of *Escherichia coli* in orange juice. Food Microbiology 28 (6), 1200–1204.
Nersisyan, G., Graham, W., 2004. Characterization of a dielectric barrier discharge operating in an open reactor with flowing helium. Plasma Sources Science and Technology 13 (4), 582.
Niemira, B.A., 2012. Cold plasma decontamination of foods. Annual Review of Food Science and Technology 3, 125–142.
Noriega, E., Shama, G., Laca, A., Diaz, M., Kong, M.G., 2011. Cold atmospheric gas plasma disinfection of chicken meat and chicken skin contaminated with *Listeria innocua*. Food Microbiology 28 (7), 1293–1300.
O'Donnell, C.P., Tiwari, B.K., Bourke, P., Cullen, P.J., 2010. Effect of ultrasonic processing on food enzymes of industrial importance. Trends in Food Science and Technology 21 (7), 358–367.
Ölmez, H., Kretzschmar, U., 2009. Potential alternative disinfection methods for organic fresh-cut industry for minimizing water consumption and environmental impact. LWT-Food Science and Technology 42 (3), 686–693.
Omer, M.K., Alvseike, O., Holck, A., Axelsson, L., Prieto, M., Skjerve, E., Heir, E., 2010. Application of high pressure processing to reduce verotoxigenic *E. coli* in two types of dry-fermented sausage. Meat Science 86 (4), 1005–1009.

Ozdemir, M., Yurteri, C.U., Sadikoglu, H., 1999. Physical polymer surface modification methods and applications in food packaging polymers. Critical Reviews in Food Science and Nutrition 39 (5), 457–477.

Pankaj, S.K., Bueno-Ferrer, C., Misra, N., Milosavljević, V., O'Donnell, C., Bourke, P., Keener, K., Cullen, P., 2014. Applications of cold plasma technology in food packaging. Trends in Food Science & Technology 35 (1), 5–17.

Pataro, G., Muñoz, A., Palgan, I., Noci, F., Ferrari, G., Lyng, J., 2011. Bacterial inactivation in fruit juices using a continuous flow Pulsed Light (PL) system. Food Research International 44 (6), 1642–1648.

Ragni, L., Berardinelli, A., Vannini, L., Montanari, C., Sirri, F., Guerzoni, M.E., Guarnieri, A., 2010. Non-thermal atmospheric gas plasma device for surface decontamination of shell eggs. Journal of Food Engineering 100 (1), 125–132.

Rico, D., Martin-Diana, A.B., Barat, J., Barry-Ryan, C., 2007. Extending and measuring the quality of fresh-cut fruit and vegetables: a review. Trends in Food Science & Technology 18 (7), 373–386.

Rivera, C.S., Venturini, M.E., Oria, R., Blanco, D., 2011. Selection of a decontamination treatment for fresh *Tuber aestivum* and *Tuber melanosporum* truffles packaged in modified atmospheres. Food Control 22 (3), 626–632.

Roberts, P., Hope, A., 2003. Virus inactivation by high intensity broad spectrum pulsed light. Journal of Virological Methods 110 (1), 61–65.

Rowan, N., MacGregor, S., Anderson, J., Fouracre, R., McIlvaney, L., Farish, O., 1999. Pulsed-light inactivation of food-related microorganisms. Applied and Environmental Microbiology 65 (3), 1312–1315.

Sagong, H.-G., Lee, S.-Y., Chang, P.-S., Heu, S., Ryu, S., Choi, Y.-J., Kang, D.-H., 2011. Combined effect of ultrasound and organic acids to reduce *Escherichia coli* O157: H7, *Salmonella* Typhimurium, and *Listeria monocytogenes* on organic fresh lettuce. International Journal of Food Microbiology 145 (1), 287–292.

Sampedro, F., McAloon, A., Yee, W., Fan, X., Geveke, D., 2014. Cost analysis and environmental impact of pulsed electric fields and high pressure processing in comparison with thermal pasteurization. Food and Bioprocess Technology 7 (7), 1928–1937.

Sampedro, F., Rodrigo, M., Martinez, A., Rodrigo, D., Barbosa-Cánovas, G., 2005. Quality and safety aspects of PEF application in milk and milk products. Critical Reviews in Food Science and Nutrition 45 (1), 25–47.

São José, J.F.B., Vanetti, M.C.D., 2012. Effect of ultrasound and commercial sanitizers in removing natural contaminants and *Salmonella enterica* Typhimurium on cherry tomatoes. Food Control 24 (1), 95–99.

Seymour, I., Burfoot, D., Smith, R., Cox, L., Lockwood, A., 2002. Ultrasound decontamination of minimally processed fruits and vegetables. International Journal of Food Science & Technology 37 (5), 547–557.

Sharma, R., Demirci, A., 2003. Inactivation of *Escherichia coli* O157:H7 on inoculated alfalfa seeds with pulsed ultraviolet light and response surface modeling. Journal of Food Science 68 (4), 1448–1453.

Steffen, H., Duerst, M., Rice, R.G., 2010. User experiences with ozone, electrolytic water (active water) and UV-C light (Ventafresh technology) in production processes and for hygiene maintenance in a Swiss sushi factory. Ozone: Science & Engineering 32 (1), 71–78.

Suslick, K.S., Hammerton, D.A., Cline, R.E., 1986. Sonochemical hot spot. Journal of the American Chemical Society 108 (18), 5641–5642.

Teagasc, 2013. In: Irish Consumer and Industry Acceptance of Novel Food Technologies: Research Highlights, Implications and Recommendations, vol. 1. Teagasc Food Research Centre, Dublin, Ireland. pp. 68.

Tiwari, B.K., Mason, T.J., 2012. Chapter 6-Ultrasound processing of fluid foods. In: Cullen, P.J., Brijesh, K.T., Valdramidis, Vasilis (Eds.), Novel Thermal and Non-Thermal Technologies for Fluid Foods. Academic Press, San Diego, pp. 135–165.

Toepfl, S., 2012. Pulsed electric field food processing–industrial equipment design and commercial applications. Stewart Postharvest Review 8 (2), 1–7.

Valero, M., Recrosio, N., Saura, D., Munoz, N., Martí, N., Lizama, V., 2007. Effects of ultrasonic treatments in orange juice processing. Journal of Food Engineering 80 (2), 509–516.

Wu, J., Lin, L., 2002. Ultrasound–induced stress responses of *Panax ginseng* Cells: enzymatic browning and phenolics production. Biotechnology Progress 18 (4), 862–866.

Zhang, L., Liu, S., Ji, H., Zhang, C., Deng, C., Cao, W., Mao, W., Gao, J., 2011. Inactivation of polyphenol oxidase from Pacific white shrimp by dense phase carbon dioxide. Innovative Food Science & Emerging Technologies 12 (4), 635–641.

Zhou, B., Feng, H., Luo, Y., 2009. Ultrasound enhanced sanitizer efficacy in reduction of *Escherichia coli* O157:H7 population on spinach leaves. Journal of Food Science 74 (6), M308–M313.

Ziuzina, D., Patil, S., Cullen, P.J., Keener, K., Bourke, P., 2014. Atmospheric cold plasma inactivation of *Escherichia coli, Salmonella enterica* serovar Typhimurium and *Listeria monocytogenes* inoculated on fresh produce. Food Microbiology 42, 109–116.

Index

Note: Page numbers followed by "f" and "t" indicate figures and tables respectively.

A

Aerosolization, 154
Air handling systems, 108–109, 109t
Air Radio Frequency Assisted (ARFA), 229–231, 231f
Alternating (AC) current, 204
ANN. *See* Artificial neural network (ANN)
Antimicrobial surfaces, 160
ARFA. *See* Air Radio Frequency Assisted (ARFA)
Artificial neural network (ANN), 143
Assembly-line analogues, 25
Automatic storage and retrieval systems (AS/RS), 70

B

Batch ohmic heaters, 215, 217f
Benzalkonium chloride (BAC), 164
Better Than Fresh™ (BTF) system, 264
BIM. *See* Building information modeling (BIM)
Bird coil (spiral), 97–98, 97f
Bird deterrent wire, 97–98, 97f
Bird nets, 98, 98f
Bird slides, 98, 99f
Bird spikes, 97
Bird wire mesh, 98, 98f
Brick floors, 115
Bricks-and-mortar construction, 42
BTF system. *See* Better Than Fresh™ (BTF) system
Building information modeling (BIM), 72–73

C

CAP. *See* Cold atmospheric plasma (CAP)
Capacity recovery, 138
CFD. *See* Computational fluid dynamics (CFD)
Cloud-based technology, 23–25
Cogeneration, 68
Cold atmospheric plasma (CAP), 266–268, 268f
Cold plasma, 156
Collinear design, 214
Combustion-isotope ratio mass spectrometry (GC-C-IRMS), 14–15
Commercial software tools, 147
Communication, 25–26
Comprehensive Assessment System for Built Environment Efficiency (CASBEE), 64
Computational fluid dynamics (CFD), 73
Concrete flooring, 113–114
Continuous flow ohmic heating, 212–215, 214f–216f
Convective surface heating, 239
Conveyor systems, 30
Cooperative communication, 23
Cutting-edge communication technologies, 23

D

Data-mining system processes, 22
Decontamination process, 28
  nonthermal methods
    chemical methods, 152–156
    physical methods, 156–157
  thermal methods, 157, 158t–159t
Defrosting, 240–242
Dense phase carbon dioxide (DPCD), 263–264, 265f
Desorption electrospray ionization (DESI), 13–14
Dielectric drying, 238–239, 239f–240f
Dielectric heating
  benefits and applications of
    cooking and baking, 242–243, 243f–244f
    dielectric drying, 238–239, 239f–240f
    fats, melting of, 243, 244f

Dielectric heating (*Continued*)
frozen products, defrosting and tempering of, 240–242, 241f, 242t, 243f
novel processes. *See* Novel Processes
postbaking, 239–240, 241f
Dielectric process component, 237
Direct analysis in real time (DART), 13–14
Doors
exterior doors, 89–90
hygienic design, need for, 89
interior doors, 90
protection, 91–92
types, 90–91, 91f
DPCD. *See* Dense phase carbon dioxide (DPCD)
Drains
channels, 117–118, 118f
gullies, 116–117
use and location, 116

E

Electrical conductivity, 210–212, 211f, 211t–212t
Electric bird deterrent systems, 99, 99f
Electromagnetic compatibility (EMC), 227–228
Electropure process, 206
Embed data quality management, 142
Emerging bird control techniques, 95–99, 96f
bird coil (spiral) and bird deterrent wire, 97–98, 97f
bird nets and bird wire mesh, 98, 98f
bird slides, 98, 99f
bird spikes, 97
electric bird deterrent systems, 99, 99f
laser bird repelling devices and power strobe lights, 97
visual bird deterrents, 96
Emerging nonthermal technologies
cold atmospheric plasma (CAP), 266–268, 268f
defined, 258–268
dense phase carbon dioxide (DPCD), 263–264, 265f
ozone processing, 266, 267f
power ultrasound, 258–259, 260t–261t, 262f
pulsed electric field (PEF), 261–263, 263t
pulsed light, 264–266
Energy intensity, 138
Equipment reengineering, 145
Exterior doors, 89–90

F

Fiberglass batting, 94, 94f
Floor, 113–116, 114f
brick floors, 115
concrete flooring, 113–114
metal sheeting, 116
polymethylmethacrylate floors, 115
polyurethane floors, 114–115
resin flooring, 114
tile floor, 115
vinyl sheeting, 115
wood floors, 116
Fogging technology, 154
Food and Drugs Administration (FDA), 26
Food contact surfaces
decontamination, increased focus on, 163
methods and principles
decontamination. *See* Decontamination
other measures, 160, 161t
prevention, 160–163
natural alternatives, 165
potential toxicity, traditional compounds due to, 164
specific challenges, 165
Food factory, design and build for
adaptability, 60–61, 60f
anticipate climate change, 70–71
automation and robotics, increasing trend to, 56–57
building new food factory, 50–53
cleaning and maintenance, 63
computational fluid dynamics (CFD), 73
Food Factory of the Future programs, 53–54
food terrorism acts, protect against, 71
future expansion, 58–59
future regulatory and standard expectations, 54–56
food factory design and construction, standards on, 55–56
food safety regulations, 54
new environmental/health/safety and building regulations, 56
new factory, construction of, 75–76, 76f
past/present and future, food safety management systems of, 54–55

increased worker safety and welfare, 62–63
in/on and at-line sensors, 57–58
new technologies, 56–58
performing simulations, 71
present and future segregation requirements, 59–60
rationales to new building, 50–53
legacy facilities, 50–53
reduce present and future costs, 54
renovation
drivers in favor, 41–42
need to, 42–45
planning and execution of, 45–49
sustainable renovation, 49
sustainability, 63–70
artificial lighting, energy savings on, 66, 66f
electrical power and heat, on-site generation of, 68
energy best practices, application of, 65
energy recovery, 68
HVAC system, energy savings on, 67
polluting substances, reducing emission of, 69–70
reducing geographical footprint, 70
refrigeration systems, energy savings on, 67–68
water, savings on, 68–69
three-dimensional visualization programs, 72–73, 72f
versatility, 61–62
visual communication and remote diagnostics, 57
Food fraud
analytical verification, 10
defined, 3
emerging issues and future trends, 10–16
fraud analytical measurements. *See* Fraud analytical measurements
fraud vulnerability assessments, 11–12, 11f
occurrence and usage of databases, 6–8, 7t
quality management systems, 9
traceability, 9–10
types of, 3–6, 4f
composition, 3–4
cultivar and variety identification, 4–5
geographical origin, 5
processing, 6
production system, 5–6
products of animal origin, species identification, 5
vulnerability, fraud triangle, 8–9, 9f
Food plant building concepts
pests entrance and harboring prevent, 81–99
building foundations, 83–84, 85f
doors. *See* Doors
emerging bird control techniques. *See* Emerging bird control techniques
external lighting, 94
external walls, 84–86, 85f–86f
factory site selection and orientation, 82–83, 82f
floors and drains, 94–95
insulated piping, 94, 94f
loading and unloading docks, 92–93, 92f–93f
roofs, 86–87, 87f–88f
vegetation, 83, 83f–84f
waste handling facilities, 95
windows, 87–89
Food preservation
ohmic heating, 204
advantages and disadvantages, 218–220
applications, 220–222, 221f
control, 215–218, 218f
design. *See* Ohmic heater design
food quality, 206–208, 208f
microbial death kinetics, 208–209
overview, 205–206, 205f
opportunities, techniques and overview of, 250–253
product oriented technologies
antifreeze proteins, 186–188, 187f
dehydrofreezing, 188–189, 188f
hydrofluidization, 189
ice nucleation proteins, 189
magnetic field freezing/cells alive system (CAS), 189–190
perfusion, 190
pressure shift freezing, 190–192, 191f
superchilling, 192–193
supercooling, 193–194, 194t
ultrasound assisted freezing, 194, 195f
radio frequency and microwave heating, 204, 222–247

Food preservation (*Continued*)
dielectric heating, benefits and applications of, 236–247
electrical breakdown and avoidance, 226–227
generation and application of, 227–236
loss factor and loss tangent, 223–224, 224f
penetration depth, 226
physical principles, overview of, 222–223, 223f
power dissipation, 224–225
refrigeration based technologies
acoustic refrigeration, 175–177, 176f
air cycle, 177–179, 178f
barocaloric refrigeration, 179–180
electrocaloric cooling, 180–181
hydraulic refrigeration, 181
magnetic refrigeration, 179
optical cooling, 182
stirling cycle variations, 182–183, 183f
thermionic refrigeration, 183–184
thermoelectric devices, 184–185, 184f
vapor compression, water in, 186
vortex tube cooling, 185, 185f
safety
meeting EMC requirements, RF and microwave, 248–250, 249t
ohmic heating, 247–248
traditional thermal processing techniques, 203
volumetric heating, 203
Food Safety Modernization Act (FSMA), 81–82
FoodSHIELD database, 8
Fourier transform (FT), 12
Fraud analytical measurements, 12–16
cell phone-based detection and crowd analysis, 15–16, 16f
lab-on-a-chip, 15
mass spectrometry, 13–15
NMR and MRI, 12–13, 14f
spectroscopic equipment, 12
FSMA. *See* Food Safety Modernization Act (FSMA)

G

Gas chromatography combustion-isotope ratio mass spectrometry (GC-C-IRMS), 14–15
GC-C-IRMS. *See* Gas chromatography combustion-isotope ratio mass spectrometry (GC-C-IRMS)
GFSI. *See* Global Food Safety Initiative (GFSI)
Global Food Safety Initiative (GFSI), 9, 11f
Good manufacturing practice (GMP), 50, 128–129
Gullies, 116–117

H

HACCP. *See* Hazard Analysis and Critical Control Points (HACCP)
Hazard Analysis and Critical Control Points (HACCP), 9
Hybrid heating technologies, 237
Hygiene concepts
air handling systems, 108–109, 109t
food plant building concepts, pests entrance and harboring prevent, 81–99
building foundations, 83–84, 85f
doors. *See* Doors
emerging bird control techniques. *See* Emerging bird control techniques
external lighting, 94
external walls, 84–86, 85f–86f
factory site selection and orientation, 82–83, 82f
floors and drains, 94–95
insulated piping, 94, 94f
loading and unloading docks, 92–93, 92f–93f
roofs, 86–87, 87f–88f
vegetation, 83, 83f–84f
waste handling facilities, 95
windows, 87–89
interior building layout
barrier and zoning concept, 100, 100f–101f
microbial contamination control, zoning for, 100–105
movable process equipment cleaning, auxiliary area for, 106–108
nonmicrobial contamination control, zoning for, 105–106, 105f
production area, size reduction of, 106, 107f
interior construction elements
basic/medium hygiene areas, beams and columns construction for, 110, 110f–112f

drains. *See* Drains
floor, 113–116, 114f
floor slab, 112–113
food factories, walls in, 112, 113f
high hygiene areas, beams and columns construction for, 111–112, 112f
zone H process areas, relocation process piping and utilities
arguments, 123–124, 124f
walls and ceilings uncluttered, methods to, 125–132
zone M process areas, process piping and utilities within
cable routing, 119–122, 120f–123f
installation, 118–119, 119f
Hygienic engineering/design, experts/ contractors and department specialized in
need for, 74
new factory, construction of, 75–76, 76f
project execution, 74–75

I

Increased tolerance, 164
Information Technology (IT), 148
communication technology, 23, 24f
components, 22–23, 22f, 24f
computing technology, 22–23
example, 34–35
food processing and packaging, 28–32, 29f, 31f
nanotechnology, 32–34, 33f
supply chain management, 23–26, 25f
traceability, 26–27, 27f
Inspection machines, 28
Instrumentation/Systems and Analysis society (ISA), 148
Insulated piping, 94, 94f
Interior building layout
barrier and zoning concept, 100, 100f–101f
microbial contamination control, zoning for, 100–105
segregation dry/wet processing, 103–105, 104f
segregation raw/decontaminated RTE products, 103
zone B/M and H areas, 100–101, 102t–103t
movable process equipment cleaning, auxiliary area for, 106–108
nonmicrobial contamination control, zoning for, 105–106, 105f
production area, size reduction of, 106, 107f
Interior construction elements
basic/medium hygiene areas, beams and columns construction for, 110, 110f–112f
drains. *See* Drains
floor, 113–116, 114f
floor slab, 112–113
food factories, walls in, 112, 113f
high hygiene areas, beams and columns construction for, 111–112, 112f
Interior doors, 90
Isolator technology, 106

L

Leadership in Energy and Environmental Design (LEED), 64
Lifecycle Assessment (LCA), 64
Listeria monocytogenes, 165, 258–259
Logistics management, 23, 25
Lossy dielectrics, 204

M

Magnetic resonance imaging (MRI), 12–13
Material specification, changes in, 145
Maximum residue level (MRL), 164
Mean conductivity, 210
Metal sheeting, 116
Microwave heating, 204, 222–247
dielectric heating, benefits and applications of, 236–247
electrical breakdown and avoidance, 226–227
generation and application
heating, cavities for, 233–236, 234t, 235f–236f
power generation, 231–232, 233f
generation and application of, 227–236
loss factor and loss tangent, 223–224, 224f
penetration depth, 226
physical principles, overview of, 222–223, 223f
power dissipation, 224–225
Mid infrared spectrometers (MIR), 12
MRI. *See* Magnetic resonance imaging (MRI)

Multilevel traceability, 26
Multimode microwave cavities, 235, 235f
Multitapped transformer, 217–218
Multivariate analytical approaches
common plant systems, embedding predictive capability within, 146–147, 146f
data sources and data sets establishing, 139–142, 140t
preparing data sets, challenges in, 140–142, 141f
suitably robust data collection infrastructure, attributes of, 142
developing trends, 147–148
historical performance, reliable process models from, 143–145, 144f–145f
process execution and equipment, 137
single-variable focused analysis, 136
typical producer, opportunity for, 137–139
complex products/processes, 138
significant mechanical complexity, packaging processes with, 139

N

National Center for Food Protection and Defense (NCFPD), 6, 7t, 8
NCFPD. *See* National Center for Food Protection and Defense (NCFPD)
Near infrared spectrometers (NIR), 12
Neutral electrolyzed water (NEW), 155
NMR. *See* Nuclear magnetic resonance (NMR)
Nonthermal methods
chemical methods, 152–156
physical methods, 156–157
Novel processes
advanced meat cooking processes, 246, 246f
alternative cooking methods, 245–246
controlled surface color, baking with, 245
dough products, proving for, 246–247
puffing, 245
Nuclear magnetic resonance (NMR), 12–13

O

Ohmic heater design, 209–215
batch ohmic heaters, 215, 217f
continuous flow ohmic heating, 212–215, 214f–216f
electrical conductivity, 210–212, 211f, 211t–212t
OMAC. *See* Organization for Machine Automation and Control (OMAC)
Organization for Machine Automation and Control (OMAC), 148
Ozone processing, 266, 267f

P

Parallel plate arrangement, 214
Parallel rod design, 214
Pectin esterase (PE) activity, 207
PEF process. *See* Pulsed electric field (PEF) process
Peroxy compounds, 152–154
Polymethylmethacrylate floors, 115
Polyurethane floors, 114–115
Postbaking, 239–240, 241f
Power ultrasound, 258–259, 260t–261t, 262f
Prerequisite programs (PRPs), 54–55
Product oriented technologies
antifreeze proteins, 186–188, 187f
dehydrofreezing, 188–189, 188f
hydrofluidization, 189
ice nucleation proteins, 189
magnetic field freezing/cells alive system (CAS), 189–190
perfusion, 190
pressure shift freezing, 190–192, 191f
superchilling, 192–193
supercooling, 193–194, 194t
ultrasound assisted freezing, 194, 195f
Proton transfer reaction-mass spectrometry (PTR-MS), 5, 14–15
PTR-MS. *See* Proton transfer reaction-mass spectrometry (PTR-MS)
Puffing, 245
Pulsed electric field (PEF), 252, 261–263, 263t
Pulsed light, 264–266

R

Radio frequency (RF), 204, 222–247
dielectric heating, benefits and applications of, 236–247
electrical breakdown and avoidance, 226–227
generation and application

applicators and control, 229, 230f
power generators, 227–228, 227f–228f
strayfield and staggered throughfield, 231, 232f
throughfield, 229–231, 230f–231f
generation and application of, 227–236
loss factor and loss tangent, 223–224, 224f
penetration depth, 226
physical principles, overview of, 222–223, 223f
power dissipation, 224–225
Radio frequency identification (RFID), 30–34, 31f
Rapid Alert System for Food and Feed (RASFF), 8
RASFF. *See* Rapid Alert System for Food and Feed (RASFF)
Ready-to-eat (RTE) products, 81, 101, 103
Recipe adjustments, 144
Reduced susceptibility, 164
Refrigeration based technologies
acoustic refrigeration, 175–177, 176f
air cycle, 177–179, 178f
barocaloric refrigeration, 179–180
electrocaloric cooling, 180–181
hydraulic refrigeration, 181
magnetic refrigeration, 179
optical cooling, 182
stirling cycle variations, 182–183, 183f
thermionic refrigeration, 183–184
thermoelectric devices, 184–185, 184f
vapor compression, water in, 186
vortex tube cooling, 185, 185f
Renewable energy production resources, 68
Renovation, food factory
drivers in favor, 41–42
need to, 42–45
planning and execution of, 45–49
hygiene measures before onset of, 46–47
hygiene measures, practical implementation of, 47–49, 47f
renovation activities, inconveniences during, 46
retrofit, specific inconveniences during, 45
sustainable renovation, 49
Resin flooring, 114
RFID. *See* Radio frequency identification (RFID)
Robust signal design, 23

S

Salmonella, 165
Scalability, 25
Staggered through-field electrode system, 231
Supervisory Control and Data Acquisition (SCADA) systems, 147
Sustainability, 63–70
artificial lighting, energy savings on, 66, 66f
electrical power and heat, on-site generation of, 68
energy best practices, application of, 65
energy recovery, 68
HVAC system, energy savings on, 67
polluting substances, reducing emission of, 69–70
reducing geographical footprint, 70
refrigeration systems, energy savings on, 67–68
water, savings on, 68–69

T

Temperature-sensitive materials, 239
Tempering, 240–242
Thermal methods, 157, 158t–159t
Tile floor, 115
Trace-back, 26, 31–32
Trace forward, 31–32

U

Uniform heating, 237

V

Ventafresh Disinfection Tunnel, 266
Vinyl sheeting, 115
Visual bird deterrents, 96

W

Walls/ceilings uncluttered, methods to
adjacent areas, position equipment in, 130, 131f
applying multilevel concepts, gravity flow by, 132

Walls/ceilings uncluttered, methods to (*Continued*)
building structure, integral part services of, 125, 125f–126f
ceiling mounted utility columns, 130, 131f
false ceilings and mezzanines, piping and utilities to, 125–127, 126f–128f
pipe trains, use of, 127, 129f
technical room/corridor, piping and utilities relocation in, 128–129, 129f–130f
wireless measurement/communication and control techniques, cables reducing by using, 130
Wireless-based networking systems, 21
Wood floors, 116

Z

Zinc oxide (ZnO), 156
Zone H process areas
relocation process piping and utilities
arguments, 123–124, 124f
walls and ceilings uncluttered, methods to, 125–132
Zone M process areas
process piping and utilities within
cable routing, 119–122, 120f–123f
installation, 118–119, 119f

Kolibri 360 GmbH
Ferdinand-Jühlke-Straße 7
99095 Erfurt, Deutschland
produktsicherheit@kolibri360.de